Science and Technology of Semiconductor-On-Insulator Structures and Devices Operating in a Harsh Environment

NATO Science Series

A Series presenting the results of scientific meetings supported under the NATO Science Programme.

The Series is published by IOS Press, Amsterdam, and Kluwer Academic Publishers in conjunction with the NATO Scientific Affairs Division

Sub-Series

I. Life and Behavioural Sciences	IOS Press
II. Mathematics, Physics and Chemistry	Kluwer Academic Publishers
III. Computer and Systems Science	IOS Press
IV. Earth and Environmental Sciences	Kluwer Academic Publishers
V. Science and Technology Policy	IOS Press

The NATO Science Series continues the series of books published formerly as the NATO ASI Series.

The NATO Science Programme offers support for collaboration in civil science between scientists of countries of the Euro-Atlantic Partnership Council. The types of scientific meeting generally supported are "Advanced Study Institutes" and "Advanced Research Workshops", although other types of meeting are supported from time to time. The NATO Science Series collects together the results of these meetings. The meetings are co-organized bij scientists from NATO countries and scientists from NATO's Partner countries – countries of the CIS and Central and Eastern Europe.

Advanced Study Institutes are high-level tutorial courses offering in-depth study of latest advances in a field.
Advanced Research Workshops are expert meetings aimed at critical assessment of a field, and identification of directions for future action.

As a consequence of the restructuring of the NATO Science Programme in 1999, the NATO Science Series has been re-organised and there are currently Five Sub-series as noted above. Please consult the following web sites for information on previous volumes published in the Series, as well as details of earlier Sub-series.

http://www.nato.int/science
http://www.wkap.nl
http://www.iospress.nl
http://www.wtv-books.de/nato-pco.htm

Series II: Mathematics, Physics and Chemistry – Vol. 185

Science and Technology of Semiconductor-On-Insulator Structures and Devices Operating in a Harsh Environment

edited by

Denis Flandre

Université Catholique de Louvain,
Microelectronics Laboratory,
Louvain-la-Neuve, Belgium

Alexei N. Nazarov

National Academy of Science of Ukraine,
Institute of Semiconductor Physics,
Kyiv, Ukraine

and

Peter L.F. Hemment

University of Surrey,
School of Electronics and Physical Sciences,
Guildford, U.K.

Kluwer Academic Publishers

Dordrecht / Boston / London

Published in cooperation with NATO Scientific Affairs Division

Proceedings of the NATO Advanced Research Workshop on
Science and Technology of Semiconductor-On-Insulator Structures and Devices
Operating in a Harsh Environment
Kiev, Ukraine
26–30 April 2004

A C.I.P. Catalogue record for this book is available from the Library of Congress.

ISBN 1-4020-3012-6 (PB)
ISBN 1-4020-3011-8 (HB)
ISBN 1-4020-3013-4 (e-book)

Published by Kluwer Academic Publishers,
P.O. Box 17, 3300 AA Dordrecht, The Netherlands.

Sold and distributed in North, Central and South America
by Kluwer Academic Publishers,
101 Philip Drive, Norwell, MA 02061, U.S.A.

In all other countries, sold and distributed
by Kluwer Academic Publishers,
P.O. Box 322, 3300 AH Dordrecht, The Netherlands.

Printed on acid-free paper

TABLE OF CONTENTS

vi

RELIABILITY AND OPERATION OF SOI DEVICES IN HARSH ENVIRONMENT

RADIATION EFFECTS

CHARACTERIZATION AND SIMULATION OF SOI DEVICES OPERATING UNDER HARSH ENVIRONMENT

NOVEL SOI DEVICES AND SENSORS OPERATING AT HARSH CONDITIONS

viii

PREFACE

This proceedings volume archives the contributions of the speakers who attended the NATO Advanced Research Workshop on "Science and Technology of Semiconductor-On-Insulator Structures and Devices Operating in a Harsh Environment" held at the Sanatorium Puscha Ozerna, Kyiv, Ukraine, from 25th to 29th April 2004.

The semiconductor industry has maintained a very rapid growth during the last three decades through impressive technological achievements which have resulted in products with higher performance and lower cost per function. After many years of development semiconductor-on-insulator materials have entered volume production and will increasingly be used by the manufacturing industry. The wider use of semiconductor (especially silicon) on insulator materials will not only enable the benefits of these materials to be further demonstrated but, also, will drive down the cost of substrates which, in turn, will stimulate the development of other novel devices and applications. In itself this trend will encourage the promotion of the skills and ideas generated by researchers in the Former Soviet Union and Eastern Europe and their incorporation in future collaborations.

This volume contains the extended abstracts of both oral and poster papers presented during the four-day meeting, under the headings of:
- Technology and Economics
- Semiconductor-On-Insulator Material Technologies
- Reliability and Operation of SOI Devices in a Harsh Environment
- Radiation Effects
- Characterization and Simulation of SOI Devices Operating in a Harsh Environment
- Novel SOI Devices and Sensors Operating under Harsh Conditions

These high-quality papers were presented by researchers from Japan, USA, European Union and the Eastern European countries of the Former Soviet Union, thereby fulfilling a further underlying objective of the Workshop which was to cement existing links established during the three previous NATO Silicon on Insulator Workshops held in Ukraine, in 1994, 1998, and 2000 and to develop new world-wide contacts between researchers in the attendees countries.

The meeting thus successfully achieved its scientific and networking goals and the attendees wish to express their gratitude to the NATO International Scientific Exchange Programme, whose financial support made the meeting possible, and to the National Academy of Science of Ukraine and the Science and Technology Centre of Ukraine who provided local support. The organizers offer their sincere thanks to Prof. V. Lysenko who worked unstintingly to guarantee the success of the Workshop. We would like to thank the agency "Optima" whose Director Mariya Miletska professionally helped us to organize this workshop. Our deep acknowledgements also go to Yu. Houk, Ya. Vovk, V. Stepanov, A. Rusavsky, Dr. A. Stronsky, Dr. G. Rudko, Dr. T. Rudenko, V. Torbin, V. Smirnaya, Dr. I. Tyagulsky, Dr. I. Osiyuk and Dr. A. Vasin for their clerical and technical assistance, which ensured the conference and social arrangements ran smoothly. A final special thanks to Dr. Valeria Kilchytska for her dedication in compiling this book and for very many other practical contributions.

Denis Flandre
Louvain-la-Neuve, Belgium

Alexei Nazarov
Kyiv, Ukraine

Peter Hemment
Guildford, UK

HIGH TEMPERATURE ELECTRONICS - CLUSTER EFFECTS

Colin Johnston and Alison Crossley
Department of Materials, University of Oxford, Oxford University Begbroke Science Park, Sandy Lane, Yarnton, Oxford OX5 1PF UK

Abstract: High temperature electronics is emerging as a strategic technology for many countries, particularly those with highly developed oil & gas and aerospace sectors. However, problems remain particularly in establishing a reliable and secure commercial source of component technology. Even with reliable devices the problem of packaging and system testing remains to be fully solved. One mechanism which is being used to achieve critical mass in specific areas of research is clustering. This paper describes the work of two clusters (one in Europe and one in the US) in developing high temperature electronics solutions.

Key words: High Temperature Electronics; Automotive; Aerospace; Oil & Gas

1.   INTRODUCTION

It is clear that high temperature electronics is emerging as a strategic technology for many countries, particularly those with highly developed oil & gas and aerospace sectors. However, problems remain particularly in establishing a reliable and secure commercial source of component technology. Even with reliable devices the problem of packaging and system testing remains to be fully solved. One mechanism which is being used to achieve critical mass in specific areas of research is clustering. This paper will describe the work of various clusters in developing high temperature electronics solutions.

An example of such clustering is illustrated by the "European Cluster for Electronic Control Units and Devices for High Temperature Applications". This cluster is to a large extent focused on the automotive market sector

1

D. Flandre et al. (eds.), Science and Technology of Semiconductor-On-Insulator Structures and Devices Operating in a Harsh Environment, 1-10.

although other industrial sectors will benefit from the availability of these generic demonstrators.

The next generation of automotive electronic control units (ECU) is expected to be mounted close to, or directly onto, the actuator – this means for example directly on the engine, into the transmission, or near the brake disk. Possible HT electronic applications in the car are motor control, electromagnetic valve control, direct injection systems, integrated starter-alternators, auxiliary drives, exhaust monitoring, power steering, and electronically actuated brakes[1]. For example, the Transmission Control Unit (TCU) represents an advanced mechatronic system for automatic gearbox control, directly mounted inside the gearbox, able to operate in harsh environment as described in[2].

ECU mounting locations close to the actuator mean in any case extreme environmental conditions for the built-in electronic components. These components have to bear high ambient temperatures and rapid temperature cycles between -40°C and the maximum operating temperature. At the same time they have to withstand heavy mechanical shocks and high vibration levels[3]. Further demands arise from the wish to bring power electronics to harsh environments as in different versions of hybrid vehicles. The automotive industry needs a system solution in order to realize these new features. This requires availability of semiconductor devices and circuits, passive devices together with suited packaging and assembly processes, and a dedicated method to predict reliability.

Innovative basic technologies (i.e. advanced semiconductor technologies, circuit techniques, packaging and assembly techniques, etc.) have to be employed for successful development of cost-effective solutions in this area. Three projects working on high temperature electronics for automotive applications have been initiated on European level (Figure 1). These projects involve expertise from other industrial sectors such as oil drilling and railway applications and from a cluster by exchanging information and following the same approach on reliability methodology, adapted to the specific topics.

A similar approach has also been adopted in the US where Honeywell are leading a Joint Industrial Partnership (JIP), part funded by the US governments DeepTrek programme to develop SOI based high temperature electronic components.

DeepTrek[4] is an initiative of the U.S. Department of Energy (DOE), the National Energy Technology Laboratory (NETL) and Strategic Centre For Natural Gas (SCNG). The DeepTrek program objective is to develop technology to dramatically reduce the cost/risk of drilling to depths of 20,000 feet or more.

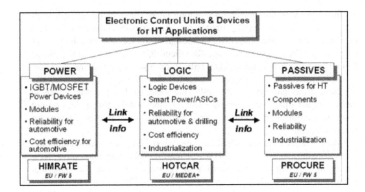

Figure 1. The European Cluster for Automotive High Temperature Electronics

The objective of the DeepTrek high temperature electronics programme is to provide a solution for the gap in the down hole industry for a functional suite of high temperature electronic components, which can be used for high temperature instrumentation in the gas and petroleum deep well domain and other smart well applications. One of the reasons for this gap is the reluctance of component manufacturers to invest in new product lines absent a reasonable assurance of a viable market. Conversely, the gas and petroleum industry needs reasonable assurance that a full suite of compatible components will be commercially available and usable in industry deep well applications. This technology also has spin-off potential in other high temperature markets (e.g., industrial and/or aerospace control systems) and enable electronic controls anywhere that temperature is a roadblock.

2. THE EUROPEAN CLUSTER FOR AUTOMOTIVE HIGH TEMPERATURE ELECTRONICS

The European cluster for automotive high temperature electronics is focused on three inter-related projects to develop generic demonstrator power, logic and passive components for automotive applications. This cluster project also addresses the critical issues of packaging, testing and reliability.

2.1   Semiconductor devices

In many cases, the temperature limiting factor for semiconductor devices is not the silicon itself. Improving the chip metallisation system (electromigration) and optimizing the design of devices and circuits enables extension of the temperature range for silicon ICs to ~200°C, which can be further extended by using SOI (silicon-on-insulator) material, see for example reference [5]. Circuits specified for 225°C are commercially available from Honeywell. Recently, the successful application of these devices in aerospace has been reported[6]. Unfortunately, the devices do not meet the required performance and restrictive cost figures of the car manufacturers.

Most semiconductor devices are designed for consumer and communication markets with less severe requirements than automotive. New technology generations and shrinks of feature sizes bring the risk that circuits will not work at temperatures above 125°C. High temperature devices cannot be developed in general by simply uprating products designed and developed for other use. Therefore, it is necessary that semiconductor manufacturers commit to high-temperature products. At present the biggest need is seen in (flash) memories, high-performance microcontrollers, and smart power drivers. These devices are tackled in the MEDEA+ project HOTCAR (High Operating Temperature System on Chip, Assembly and Reliability).

2.2   Passive devices

Nearly all types of passive components are necessary for the realization of hot spot ECUs, and further developments are required for capacitors, inductors, oscillators, and resonators[7,8]. New device generations in the 25V range withstanding 150°C for 1000 hours (power applications) or even 175°C (control units) are highly desirable. Especially Ta capacitors suffer from reduced voltage stability and wet aluminum capacitors have to fight dry-out effects. Here further work is needed. In the EU-funded project PROCURE (PROgram for the development of passive Components Used in Rough Environments), solutions for a generic set of passive devices operating at 175°C are being elaborated.

2.3   Packaging and assembly

The packaging for electronics operating at higher temperatures is a great challenge, because current technologies were optimised for a temperature region between –40 and +125°C over several decades. The main issues, HOTCAR is working on are:

1. Adaptation of material combinations with respect to their coefficient of thermal expansion (CTE). This is necessary to reduce mechanical stress caused by thermal mismatch.
2. Thermal stability of the used materials. This means for metals and their alloys not only the resistance against any kind of oxidation, but also against the formation of critical intermetallic phases. Using monometallic systems for chip metallisation and interconnects would help, but also new materials for bonding wires (e.g. Pd) are a possible option.

On the ECU level, cost issues force to consider plastic encapsulated devices instead of hermetic packages. The plastic materials must not be degraded by cracking their chemical composition or by changing their structural arrangement. The new materials must comply with legal restrictions ("green packages"). Also, an organic board material (PCB) for HT applications is being development.

In addition, solder materials as well as soldering techniques have to be provided which fit well into existing manufacturing processes. It is most likely that these solder materials are lead-free solders.

Power electronic systems need special attention in order to achieve high reliability and long lifetimes. Optimisation of assembly and cooling are mandatory. These aspects are covered by HIMRATE (High-Temperature IGBT and MOSFET Modules for Railway Traction and Automotive Electronics)[9].

2.4 Relability aspects

The required lifetimes for electronic components in automotive applications range from typically 6000 hours for passenger cars to 20,000 hours for commercial vehicles.

In order to assure quality and reliability, defined stress conditions for the qualification testing of devices for automotive applications have to be found. These accelerated tests comprise electrical, environmental and mechanical procedures like latch-up, electro-thermal induced gate leakage, temperature cycles, autoclave, temperature-humidity bias, power temperature cycling, solderability, bond shear, vibration, mechanical shocks etc. In the HT projects, detailed analyses are performed to quantify the increased stress of electronic assemblies when subjected to harsh environments in a car in order to derive suited test conditions.

Developing related test standards requires the knowledge of the underlying failure mechanisms. Generally, these mechanisms are investigated by accelerated testing and subsequently acceleration factors are calculated.

But entering the high temperature field, acceleration by simply increasing the temperature becomes a questionable approach. All three projects are working on the question, whether such methodology still can be applied without changing the failure mechanisms and provide enough acceleration to end up with satisfying test duration.

3.   DEVELOPMENT OF SOI CMOS ELECTRONICS UNDER THE DEEPTREK PROGRAM

Honeywell lead a strategic US funded programme to develop SOI components that will support not only the energy sector (oil&gas and geothermal) but also the aerospace and defence community. Honeywell are well suited to this role as they have a unparalleled position as the world's leading supplier of high temperature electronics based on their existing SOI product range as shown in Table 1. A 1.2 micron IC process with analog features has been used to develop primarily linear products, while a 0.8 micron digital IC process has been used to develop primarily digital products (Table 1). These products feature 225°C performance specifications and are designed for continuous operation for up to 5 years at 225°C[10].

Table 1. SOI Processes and Features Applied to High Temperatures

	"10V Linear" Process	5V SOI4 Digital Process
Gate Oxide	350 angstroms	150 angstroms
Max. Gate Ox. Voltage	10V	5v
Target Vtn/Vtp	1.2V / -1.2V	1.2V / -1.3V
Min transistor length	1.2 microns	0.8 microns
# of metal layers	2	3 or 4
Top Si Thickness	0.3 microns	0.3 microns
Buried Oxide	1.0 micron	0.4 microns
Partially/Fully depleted	Partially depleted	Partially depleted
Lithography	1X	5X
DMOS option	Yes: >30V VDS	No
CrSiN resistors	Yes	No
Linear Cap Implant	Yes	No
Laser trim fuse links	Yes	No
Lateral PNP VREF	Yes	No

One issue which will be addressed by the DeepTrek HTE project will be the divergence of the existing products into two different flows, one for analog and a second for digital. The analog flow in particular is problematic as it uses sub-processes and equipment that are unique to that flow. Honeywell and the US Government have since made additional investments to develop mixed-signal ASIC capability using 0.8micron SOI technology (i.e., analog features and library cells have been added to a digital IC process flow). This presents the opportunity for extending this capability from radiation-hardened aerospace applications to commercial high temperature applications. The main steps to accomplishing this are :

1. Remove process steps included strictly to render the process radiation hard (a requirement to avoid U.S. export restrictions),
2. Re-targeting the threshold voltage implants to address sub-threshold leakage and render the technology functionally useful for up to 300°C;
3. Modelling and library characterization for extension to high temperature.

Libraries and toolkits developed to support rad-hard aerospace products (such as Cadence PDK and/or mixed-signal ASIC design flows) may be extended to high temperature.

Table 2. New SOI Processes and Features for High Temperature Applications

	High Temp 5V SOI4 Digital Process (existing flow)	Radhard 5V SOI4 Mixed-Signal Process (existing flow)	High Temp 5V SOI4 Digital Process(to be developed)
Gate Oxide	150 angstroms	150 angstroms	150 angstroms
Max. Gate Ox. Voltage	5v	5v	5v
Target Vtn/Vtp	1.2V / -1.3V	0.8 / -0.75V	1.2V / -1.3V
Min transistor length	0.8 microns	0.8 microns	0.8 microns
# of metal layers	3 or 4	3 or 4	3 or 4
Top Si Thickness	0.3 microns	0.3 microns	0.3 microns
Buried Oxide	0.4 microns	0.4 microns	1.0 microns
Partially/Fully depleted	Partially depleted	Partially depleted	Partially depleted
Lithography	5X	5X	5X
DMOS option	No	Yes: >30V VDS	Yes: >30V VDS
CrSiN resistors	No	Yes	Yes
Linear Cap Implant	No	Yes	Yes
Laser trim fuse links	No	Yes	Yes
Lateral PNP VREF	No	Yes	Yes
Total Dose Hardened	No	Yes (1MRad, Si)	No

The mixed-signal IC process flow that is proposed for high-temperature is similar to the existing high temperature digital process except for supporting additional process features and devices to enable analog design (see Table 2). These include extended-drain DMOS devices (for higher source-drain voltage capability), linear capacitors, CrSiN thin-film resistors, laser-trimmable links in the top-metal layer, and lateral PNP devices for voltage reference circuits. These features have already been incorporated into the radiation hardened mixed-signal process flow. The only significant differences between the high temperature mixed-signal flow and the radhard mixed-signal flow would be in the Vt implant dose and the buried-oxide thickness.

The resulting overall approach to create designs suitable for high temperature is a combination of IC process approach and IC design techniques. (See Table 3).

Objectives of the DeepTrek High Temperature Electronics project include:

- Operation to at least 225°C long-term, with a goal for transient excursions to 300°C
- Sustainable source for reliable down-hole micro-electronics (Silicon-on-Insulator CMOS)
- Develop key building blocks for down-hole systems
- Establish foundation for future HT development : Design platforms, toolkits and libraries
- Reduced Cost/Schedule/Risk for HT down-hole electronics development.

Table 3. Summary of SOI high temperature design issues and primary mitigation strategies

Issue	Primary Mitigation Strategies
Junction Leakage	SOI process
Sub-threshold leakage	Vt adjustment
Electro-migration	Design rules to lower max. current density
Reduced mobility	Design adjustments
	Temperature compensated biasing
	Larger digital devices or de-rate clock frequency
Bias voltage drift with temperature	Design techniques (e.g., ZTC biasing)
Self-heating	Design for lower power density
	Layout floorplanning to match thermal profiles
	Metal-interconnect heat-spreading
Floating Body Effects	Partially depleted SOI with body-tie layout rules for max. distance to body-tie contacts
Back-gate transistors	Increased back-oxide thickness

The program is proposed as a three-year effort. The 5V, 0.8 micron mixed-signal process flow will be extended to high temperature suitable for supporting foundry design activity by the end of the first year. Wafer lots will be fabricated to exercise the IC process flow and provide material for BSIMSOIPD SPICE model development applicable to 225°C. During this time the digital libraries and toolkits associated with the radiation-hardened process will also be adapted and extended to 225°C, along with tookit features that support analog/mixed-siganl core cell insertion (such as providing for separate analog and mixed-signal power domains and supporting analog I/O sites). As a validation exercise a high-temperature mixed-signal ASIC "borecleaner" design will be completed to provide a vehicle for demonstrating single-chip implementation of analog and digital sub-systems on a single mixed-signal ASIC.

Several high-temperature standard products will also be developed. The exact set of products will be subject to change as the project gets underway, but initial plans call for the development of a precision (offset-compensated) op-amp, an A/D converter with 16-bit resolution, a 16-bit microcontroller, and a serial communications interface IC. All of these will be implemented in the 0.8 micron 5-volt SOI process. The program plan provides for two complete design passes for these product designs followed by manufacturing development to create the necessary specifications and documentation, test programs, burn-in boards, screening flows, etc. consistent with full-production capability for these products as high temperature COTS components that will be introduced sometime shortly after this DeepTrek project is completed.

It has also been determined that there is a high demand for a high-temperature FPGA product. Honeywell has previously developed and negotiated licensing for a radiation hardened FPGA that is a functionally equivalent to ATMEL 62010 FPGA. This has been implemented in a radiation hardened 0.35 micron technology. The exact path for creating a high-temperature FPGA has yet to be finalized, but the baseline approach is to develop a high-temperature 0.35 micron, 3.3V IC process flow and to adapt the rad-hard design to this process flow. Note that this FPGA is a volatile design that is configured on power-up by downloading from an external memory device. While this is less than ideal, there is not sufficient funding scope in the DeepTrek program to develop and qualify a high-temperature non-volatile memory process. The program will instead need do work on identifying an acceptable means for configuring the FPGA, perhaps through development or qualification of this capability by a third party. As in the case of the 0.8 micron standard products, the program plan calls for development of full production flow and COTS product introduction after the project is completed.

4.   ACKNOWLEDGEMENTS

The authors would like to acknowledge contributions to this paper from Wolfgang Wondrak, DaimlerChrysler AG, Stuttgart, Germany for assistance with the European Automotive Cluster and Bruce Ohme, Honeywell Defense and Space Electronic Systems, Plymouth, Minnesota, USA for assistance with the Deep Trek HTE programme.

REFERENCES

1. R. Thompson, J. Freytag, W. Senske, and W. Wondrak, An Automotive Industries Perspective on High Temperature Electronics, SMTA (Surface Mount Technology Association) Conference, Chicago, Il, September 2002
2. T. Riepl, G. Lugert, R. Ingenbleek, W. Runge, and L. Berchtold, Integration of Micromechanic Sensors, Actuators and Miniaturized High Temperature Electronics in Advanced Transmission Systems, Microtec 2000, September 25-27, Hannover: Proceedings pp. 599-604
3. R. W. Johnson, J. L. Evans, P. Jacobsen, R. Thompson, and M. Christopher, The changing automotive environment: high temperature electronics, Proceedings of the HITEC 2002, pp 92-97
4. www.netl.doe.gov
5. R. Constapel, J. Freytag, P. Hille, V. Lauer and W. Wondrak, "High Temperature Electronics for Automotive Applications", Int. Conf. On Integrated Power Systems (CIPS 2000), 20.-21.6.2000, Bremen VDI-ETG Fachbericht 81, pp. 46 – 52
6. P. Shrimpling, "Advances in high temperature electronics applied to aerospace saftey critical control applications", Proceedings of the HITEC 2002
7. R. Grzybowski, CARTS 1998 Proceedings, pp. 49 – 56
8. W. Wondrak, A. Boos and R. Constapel, "Design for Reliability in Automotive Electronics: Part I. Semiconductor Devices", Microtec 2000, Hannover, 25. –27 .9. 2000, pp 299
9. www.himrate.com
10. www.ssec.honeywell.com/hightemp

ON THE EVOLUTION OF SOI MATERIALS AND DEVICES

Jean-Pierre Colinge
Dept. of Electrical and Computer Engineering, Univ. of California, Davis, CA 95616, USA

Abstract: This paper retraces the evolution of SOI materials and devices over a period of
roughly twenty years. Initial SOI materials were designed as a replacement to
SOS and to fabricate 3D integrated circuits. It was soon realized that, unlike
early SOS material, SOI films could be used to make fully depleted devices.
SOI devices enjoy unprecedented flexibility because of the full dielectric
isolation. This has enabled the design and fabrication novel types of devices,
such as double-, triple- and quadruple-gate MOSFETs. These devices make it
possible to push device scaling to new horizons.

Key words: Silicon-on-Insulator, SOI materials, SOI MOSFET, multiple-gate MOSFETs

1.   INTRODUCTION

With the exception of silicon-on-sapphire (SOS), which was first
introduced in 1964, one can situate the birth date of SOI technology around
1980. At first SOI technology was only considered as a possible replacement
for SOS in some niche applications. Some SOI-forming techniques were,
however, developed for a bolder purpose: three-dimensional integration.
Some remarkable 3D integrated circuits were fabricated by Japanese
companies, but the poor yield of early SOI fabrication techniques never
permitted commercialization of such circuits. For 15 years SOI applications
were limited to some niche markets such as aerospace, military and high-
temperature electronics. A big change occurred end of the 20th century, as
high-quality SOI wafers suddenly became available in large quantities. From
then on, it took only a few years to witness the use of SOI microprocessors
in personal computers and SOI audio amplifiers in car stereo systems.

*D. Flandre et al. (eds.), Science and Technology of Semiconductor-On-Insulator Structures
and Devices Operating in a Harsh Environment, 11-26.*
© *2005 Kluwer Academic Publishers. Printed in the Netherlands.*

Jean-Pierre Colinge

2.   SOI MATERIALS

Many techniques have been developed for producing a film of single-crystal silicon on an insulator. Some of them are based on the epitaxial growth of silicon on either a silicon wafer covered with an insulator (homoepitaxial techniques) or on a crystalline insulator (heteroepitaxial techniques). Other techniques are based on the crystallization of a thin polysilicon layer by melt and regrowth (laser recrystallization and zone-melting recrystallization). Silicon-on-insulator material can also be produced from a bulk silicon wafer by isolating a thin silicon layer from the substrate through the formation and oxidation of porous silicon (FIPOS) or through the ion beam synthesis of a buried insulator layer (SIMOX). SOI material can also be obtained by thinning a silicon wafer bonded to an insulator and a mechanical substrate (BESOI). More recently, layer transfer techniques such as the Smart-Cut® and Eltran® processes have made it possible to peel-off a thin silicon layer from a wafer and its transfer onto an oxidized wafer. These layer transfer techniques are applicable to other materials than silicon and have enabled the advent of compliant substrate technologies. Figure 1 shows a timetable representing the evolution of the most important SOI materials.

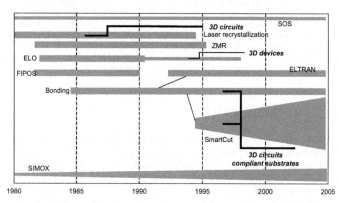

Figure 1. Evolution of SOI materials.

Silicon-on-Sapphire (SOS) has been used since the 1960's for niche applications such as rad-hard electronics. The initial properties of the materials were rather poor. The electron mobility was low and junction leakage currents were high because of low carrier lifetime. However, improvements such as the SPEAR, DSPE, and later on, the UTSI process have transformed SOS into a competitive material that can be used for thin-film SOI applications.[1,2,3,4,5] In particular, the full dielectric nature of the substrate is very well adapted to RF applications.[6]

Three-dimensional circuits were extensively studied during the 1980's in Japan, and several remarkable circuits were successfully fabricated.[7,8] These include a moving object detector [9,10] and a character recognition system [11,12,13]. These circuits were made using laser or e-beam recrystallization, which allows for the formation of SOI layers on top of processed devices. More recently, however, other techniques, such as epitaxial lateral overgrowth [14], wafer bonding [15,16], and bonding/debonding combined with Smart-Cut® [17] were proposed for the fabrication of 3D circuits.

The first commercially successful SOI (non-SOS) material was, of course, SIMOX. First developed by K. Izumi of NTT in 1978 the SIMOX process rapidly became a popular research topic and many research teams contributed to the improvement of the material. Twenty years of improving implant and annealing conditions, and the development of high-current oxygen ion implanters by NTT/Eaton, Ibis and Hitachi were necessary to obtain a product that can be used for mass production of integrated circuits. Even though the quality of the MLD-class (multiple low dose) SIMOX is quite excellent, the SIMOX technique suffers from low throughput and cannot sustain the large volume production required for SOI to become a mainstream technology.

Large volume production was finally achieved around the year 2000 by a new technique invented by M. Bruel of LETI: the Smart-Cut® process. This process is based on splitting a thin silicon layer from a donor SOI wafer and bonding it to a handle substrate. The process is often described as an "atomic scalpel" that can remove the top surface of a wafer and attach it to another substrate. This process is very flexible because it can be used with semiconductors other than silicon, and the handle substrate can be an oxidized silicon wafer, a glass/quartz substrate, or even a flexible polymer film. Furthermore, the process can be combined with various bonding/debonding techniques [18] such that processed films containing active devices can be transferred to alternative substrates.[19] Table 1 shows the different SOI fabrication techniques and several acronyms found in the literature.

The Smart-Cut® process has a much higher throughput than the SIMOX process. As a result, high-quality SOI wafers are now available (Figure 2) and companies such as AMD, IBM and Motorola are using SOI for mass production of 64-bit microprocessor chips.

14   *Jean-Pierre Colinge*

Table 1. SOI Materials and associated fabrication techniques [20]

Generic Process	Fabrication Technique (acronym)	Process Variations
Silicon heteroepitaxy	Silicon-on-Sapphire (SOS)	SPEAR DSPE UTSi
	Graphoepitaxy	
	Silicon on Cubic Zirconia (SOZ)	
	Silicon on CaF$_2$	
Thick polysilicon deposition	Dielectric Isolation (DI)	
Polysilicon melting and recrystallization	Laser Recrystallization	Selective Annealing
	Electron Beam Recrystallization	
	Zone-Melting Recrystallization (ZMR)	LEGO
Silicon homoepitaxy	Epitaxial Lateral Overgrowth (ELO)	CLSEG PACE
	Lateral Solid-Phase Epitaxy (LSPE)	MILC
Formation of porous silicon	Full Isolation by Porous Silicon (FIPOS)	
Ion beam synthesis of a buried insulator	Separation by Implanted Oxygen (SIMOX)	ITOX SIMOX MLD
	Synthesis of silicon carbide (SiCOI)	
Wafer bonding	Wafer Bonding and Etch Back (BESOI)	PACE
Layer transfer	H$^+$-Induced Splitting (Smart-Cut®)	UNIBOND® NanoCleave®
	Porous Silicon Splitting	Eltran®
SiGe epitaxy	Strained silicon-on-insulator (SSOI)	
	Strained SiGe-on-insulator (SiGeOI)	
Diamond layer formation	Silicon on Diamond (SOD)	
Preferential etching	Silicon on Nothing (SON)	

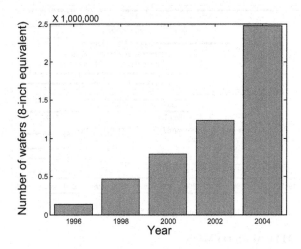

Figure 2. Estimated SOI worldwide production capability.

3. SOI MOSFETS

The first SOI transistors date back to 1964. Once the first SOI substrates (the insulator is now silicon dioxide) were available for experimental MOS device fabrication, partially depleted technology the natural choice derived from SOS experience.

3.1 "Classical" SOI MOSFETs

Partially depleted CMOS continues to be used nowadays and several commercial IC manufacturers have SOI products. Variations on the partially depleted SOI MOSFET theme include devices where the gate is connected to the floating body, which have ideal subthreshold characteristics, reduced body effect, improved current drive, and superior HF characteristics [21]. The first fully depleted SOI MOSFET date back to the early 1980's where it was quickly established that these devices exhibited superior transconductance, current drive and subthreshold swing [22,23]. Fully depleted SOI technology is being used in commercial products as well.

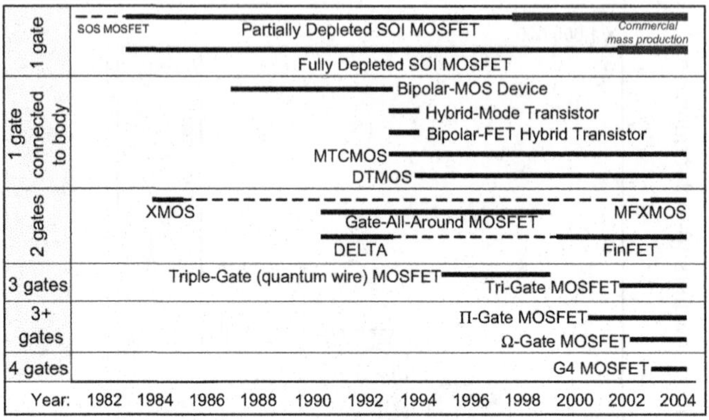

1 gate	SOS MOSFET		Partially Depleted SOI MOSFET							*Commercial mass production*		
			Fully Depleted SOI MOSFET									
1 gate connected to body			Bipolar-MOS Device									
					Hybrid-Mode Transistor							
					Bipolar-FET Hybrid Transistor							
			MTCMOS									
			DTMOS									
2 gates	XMOS				Gate-All-Around MOSFET				MFXMOS			
					DELTA				FinFET			
3 gates		Triple-Gate (quantum wire) MOSFET				Tri-Gate MOSFET						
3+ gates							Π-Gate MOSFET					
							Ω-Gate MOSFET					
4 gates									G4 MOSFET			
Year:	1982	1984	1986	1988	1990	1992	1994	1996	1998	2000	2002	2004

Figure 3. Evolution of SOI MOSFETs

3.2 MTCMOS/DTMOS

Interesting improvements of the electrical characteristics of an SOI MOSFET are obtained when a connection is made between the gate and the body (Figure 6-25). The simulation, fabrication and characterization of SOI transistors with gate-to-body connection were first reported in 1987, and the device was called the "voltage-controlled bipolar-MOS device (VCBM)".[24] Other research teams reproduced the device and named it the "hybrid bipolar-MOS device" [25,26], the "hybrid-mode SOI MOSFET" [27] or the "gate-controlled lateral BJT".[28] These early publications placed an emphasis on the high current drive of the device due to combined presence of both MOS and BJT currents. Later on emphasis was put on the dependence of the threshold voltage on body potential, and thus on the gate bias, through the classical body effect. The device was renamed by several teams to either the "multi-threshold CMOS (MTCMOS)" [29], the "dynamic threshold MOS (DTMOS)" [30], or the "varied-threshold MOS (VTMOS)".[31] These devices have ideal subthreshold characteristics, reduced body effect, improved current drive, and superior HF characteristics.[32] They are mostly used for very low-voltage (0.5 V) applications.[33,34] Because of the variation of threshold voltage with gate bias the subthreshold slope, S, of the MTCMOS/DTMOS device is very close to the ideal value of 60 mV/decade. Figure 4 shows the subthreshold drain current in a MTCMOS/DTMOS device and in the same device when the body is grounded are presented. Several analytical models developed for the MTCMOS/DTMOS device can be found in the literature.[35,36]

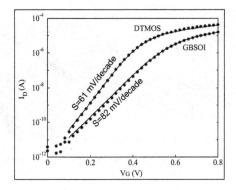

Figure 4. Threshold voltage *vs.* gate voltage in grounded-body SOI (GBSOI) and DTMOS devices. Solid lines represent equation 6.2.2 and the (*) symbols represent measured data; t_{ox}=8nm, N_A=1.65x10^{17}cm^{-3}.[37]

3.3   Double-gate MOSFETs

One of the first publication on the double-gate MOS (DGMOS) transistor concept dates back to 1984.[38] It shows that one can obtain significant reduction of short-channel effects in a device, called XMOS, where excellent control of the potential in the silicon film is achieved by using a top-and-bottom gate. The name of the device comes from its resemblance with the Greek letter Ξ. Using this configuration, a better control of the channel depletion region is obtained than in a "regular" SOI MOSFET, and, in particular, the influence of the source and drain depletion regions are kept minimal. Short-channel effects are reduced by preventing the source and drain field lines from reaching the channel region.[39] The first fabricated double-gate SOI MOSFET was the "fully DEpleted Lean-channel TrAnsistor (DELTA, 1989)", where the silicon film stands vertical on its side (Figure 5).[40] Later implementations of vertical-channel, double-gate SOI MOSFETs include the FinFET [41], the MFXMOS [42], the triangular-wire SOI MOSFET [43] and the Δ-channel SOI MOSFET [44]. Volume inversion was discovered in 1987 [45], and the superior transconductance brought about by this phenomenon was experimentally observed in 1990 in the first practical implementation of a planar double-gate MOSFET called the "gate-all-around" (GAA) device (Figure 5).[46]

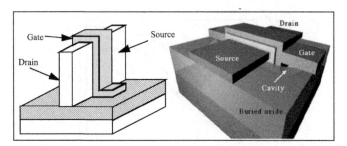

Figure 5 DELTA/FinFET double-gate MOS structure (left) and gate-all-around MOSFET (right).

3.4   Triple-gate MOSFETs

The triple-gate MOSFET is a thin-film, narrow silicon island with a gate on three of its sides. Implementations include the quantum-wire SOI MOSFET (Figure 6) [47] and the tri-gate MOSFET [48, 49]. Improved versions feature either a field-induced, pseudo-fourth gate such as the Π-gate device [50], the Ω-gate device [51] and the strained-channel multi-gate device.[52] Such devices have electrical properties between triple- and quadruple-gate devices and are sometimes called triple-plus (3⁺) gate devices or multiple-gate FETs (MuGFETs).

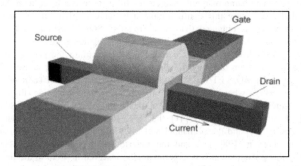

Figure 6. Triple-gate SOI MOSFET.

Quadruple-gate (surrounding-gate) devices are, theoretically, the best SOI MOSFETs in terms of current drive and suppression of short-channel effects, but their fabrication is unpractical. It is possible to design and fabricate quasi-surrounding-gate MOSFETs using a process similar to that used to fabricate triple-gate SOI MOSFETs. Such devices are called either Π-gate [53,54] or Ω-gate [55] MOSFETs (Figure 7). These devices are basically triple-gate devices with an extension of the gate electrode below the active silicon island, which increases current drive and improves short-channel effects. The gate extension can readily be formed by slightly overetching the buried oxide (BOX) during the silicon island patterning step. The gate extension forms a virtual, field-induced gate electrode underneath the device that can block drain electric field lines from encroaching on the channel region at the bottom of the active silicon. Instead the electric field lines terminate on the gate extensions. This gate structure is very effective at reducing short-channel effects. Such devices can be called 3[+] (triple-plus)-gate devices because their characteristics lie between those of triple- and quadruple-gate devices.

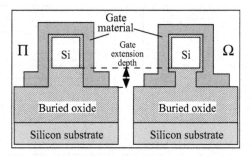

Figure 7. Π-gate (Pi-gate) and Ω-gate (Omega-gate) MOSFET cross-sections.

3.4.1   Short-channel effects

Based on Poisson's equation it is possible to calculate a "natural length", λ, that represents the extension of the electric field lines from the drain in the device body.[56,57] A device will be free of short-channel effects if λ is 5 to 10 times smaller than the effective gate length. The value of the natural length is:

$$\lambda_1 = \sqrt{\frac{\varepsilon_{si}}{\varepsilon_{ox}} t_{ox} t_{si}} \quad \text{in a single-gate SOI MOSFET,}$$

$$\lambda_2 = \sqrt{\frac{\varepsilon_{si}}{2\varepsilon_{ox}} t_{ox} t_{si}} \quad \text{in a double-gate SOI MOSFET, and}$$

$$\lambda_3 = \sqrt{\frac{\varepsilon_{si}}{4\varepsilon_{ox}} t_{ox} t_{si}} \quad \text{in a 4-gate (surrounding gate) SOI MOSFET.}$$

Based on these relationships one can estimate for each gate configuration the maximum silicon film thickness and device width that can be used while avoiding short-channel effects (Figure 8). Triple-gate devices with square channel cross section are expected to have an intermediate behaviour between those of 2 and 4 gate devices.

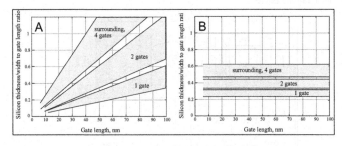

Figure 8. Maximum allowed silicon film thickness and device width *vs.* gate length for short-channel-free operation $(6\lambda > L > 8\lambda)$. A: Gate oxide thickness is held constant $(t_{ox}=1.5\ nm)$, and B: $t_{ox}=t_{si}/10$.[58]

3.4.2   Current drive

The current drive of multiple-gate SOI MOSFETs is essentially proportional to the total gate width. For instance, the current drive of a double-gate device is double that of a single-gate transistor with identical gate length and width. In triple-gate and vertical double-gate structures all individual devices need to have the same thickness and width. As a result the current drive is fixed to a single, discrete value, for a given gate length. To drive larger currents multi-fingered devices need to be used. The current drive of a multi-fingered MOSFET is then equal to the current of an individual device multiplied by the number of fins (also sometimes referred to as "fingers" or "legs") (Figure 9). Considering a pitch P for the fingers, the current per unit device width is given by:

$$I_D = I_{Do} \frac{\theta \mu_o W + 2\mu_1 t_{si}}{\mu_o P}$$

where $\theta = 1$ in a triple-gate device and $\theta = 0$ in a FinFET with no channel at the top interface; μ_o is the mobility at the (100)-oriented top interface and μ_1 is the sidewall mobility (with (110) orientation, typically [59]). W is the device width, t_{si} is the silicon film thickness, and P is the pitch.

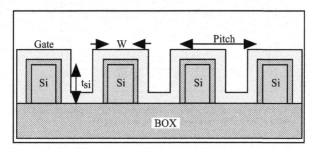

Figure 9. Cross section of a multi-fin triple-gate MOSFET.

The current drive of triple-gate and FinFET devices is shown in Figures 10 and 11 as a function of the fin pitch. If the cross section of the device is square ($t_{si}=W$) the drain current is independent on the pitch, and we have $I_D=I_{Do}$ for a triple-gate device and $I_D=I_{Do}/2$ for a FinFET with (110) sidewalls. If the sidewalls are (100) the currents are equal to $1.5 \times I_{Do}$ and I_{Do}, respectively.

Figures 10 and 11 illustrate an important point: it is possible to draw more current from multiple-gate MOSFETs than from standard, single-gate devices, provided the lithography used to pattern the fins is fine enough. They also outline the importance of the device orientation on a wafer, which determines the orientation of the sidewall channels. Comparing Figures 10 and 11, one realizes that triple-gate devices offer better performance than FinFETs. For instance, using "standard" device orientation ((100) sidewalls), and a silicon film thickness of 100 nm, a pitch smaller than 100 nm is required in order for FinFETs to outperform the current drive of a classical MOSFET, while a pitch of only 200 nm will achieve the same goal in a triple-gate device.

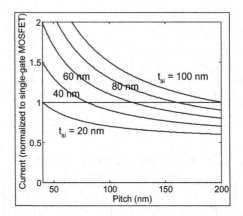

Figure 10. Normalized current drive of a triple-gate MOSFET *vs.* pitch. *W*=pitch/2; top interface mobility is 300 cm^2/Vs and sidewall mobility is 150 cm^2/Vs.

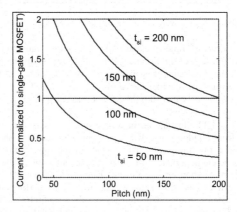

Figure 11. Normalized current drive of a FinFET *vs.* pitch. *W*=pitch/2; top interface mobility is 300 cm^2/Vs and sidewall mobility is 150 cm^2/Vs.

3.4.3 Radiation effects

The excellent radiation hardness of double-gate (gate-all around, GAA) transistors has been reported in the past.[60] In triple-gate MOSFETS the effects of ionizing radiations are still to be investigated experimentally. These effects, however, can be predicted by simulation. We know that the bottom of the silicon film is in contact with the buried oxide. Figure 11 shows the simulated subthreshold characteristics of triple-gate and Π-gate n-channel MOSFETs exposed to various radiation doses that have created various densities of positive oxide charges in the oxides. The formation of a back channel can be observed. The leakage current is similar to that observed at the back interface of a single-gate SOI MOSFETs.

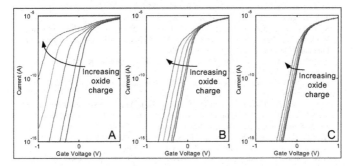

Figure 12. Subthreshold characteristics of SOI MOSFETs for different oxide charge densities (0, 2.5, 5, 7.5 and 10 x 10^{17} x q (C/cm^{-3})). A: triple gate. B: Π gate with 10 nm extension. C: Π gate with 20 nm extension. t_{ox}=2nm, t_{BOX}=60nm, t_{si}=W=30nm.

4. CONCLUSION

This paper retraces the evolution of SOI materials and devices aver the last 25 years. SOI wafers are now available in industrial quantities and the Smart-Cut® process is contemplated for the fabrication of 3D integrated circuits. SOI MOSFETs have evolved from the classical single-gate structure into multiple-gate devices that offer superior current drive and short-channel immunity.

24 *Jean-Pierre Colinge*

REFERENCES

[1] S.S. Lau, S. Matteson, J.W. Mayer, P. Revesz, J. Gyulai, J. Roth, T.W. Sigmon, T. Cass, Applied Physics Letters Vol. 34, p. 76, 1979

[2] J. Amano, K.A. Carey, Applied Physics Letters Vol. 39, p. 163, 1981

[3] P.K. Vasudev, D.C. Mayer, Materials Research Society Symposia Proceedings, Vol. 33, p. 35, 1984

[4] M. Megahed, M. Burgener, J. Cable, D. Staab, R. Reedy, Topical Meeting on Silicon Monolithic Integrated Circuits in RF Systems, Digest of Papers, p. 94, 1988

[5] R. Reedy, J. Cable, D. Kelly, M. Stuber, F. Wright, G. Wu, Analog Integrated Circuits and Signal Processing Vol. 25, Kluwer academic Publishers, p. 171, 2000

[6] R. Reedy, J. Cable, D. Kelly, M. Stuber, F. Wright, G. Wu, Analog Integrated Circuits and Signal Processing, Vol. 25, p. 171, 2000

[7] M. Nakano, Oyo Buturi, Vol. 54, no. 7, p. 652, 1985

[8] T. Nishimura, Y. Akasaka, Vol. 54, no. 12, p. 1274, 1985

[9] K. Yamazaki, Y. Itoh, A. Wada, Y. Tomita, Extended Abstracts of the 8th International Workshop on Future Electron Devices – Three Dimensional ICs and Nanometer Functional Devices, Kochi, Japan, p. 105, 1990

[10] Y. Itoh, A. Wada, K. Morimoto, Y. Tomita, K. Yamazaki K, Microelectronic Engineering, Vol .15, no. 1-4, p. 187, 1991

[11] K. Kioi, S. Toyayama, M. Koba, Technical Digest of the International Electron Devices Meeting, p. 66, 1988

[12] S. Toyoyama, K. Kioi, K. Shirakawa, T. Shinozaki, K. Ohtake, Extended Abstracts of the 8th International Workshop on Future Electron Devices – Three Dimensional ICs and Nanometer Functional Devices, Kochi, Japan, p. 109, 1990

[13] K. Kioi, T. Shinozaki, S. Toyoyama, K. Shirakawa, K. Ohtake, S. Tsuchimoto, IEEE Journal of Solid-State Circuits, Vol. 27, no. 8, p. 1130, 1992

[14] G.W. Neudeck, Electrochemical Society Proceedings, Vol. 99-3, p. 25, 1999

[15] J. Burns, L. McIlrath, J. Hopwood, C. Keast, D.P. Vu, K. Warner, P. Wyatt, Proceedings of the IEEE Onternational SOI Conference, p. 20, 2000

[16] K. Warner, J. Jurns, C. Keast, R. Kunz, D. Lennon, A. Loomis, W. Mowers, D. Yost, Proceedings of the IEEE Onternational SOI Conference, p. 123, 2002

[17] H. Moriceau, O. Rayssac, B. Aspar, B. Ghyselen, Electrochemical Society Proceedings, Semiconductor Wafer Bonding: Science, Technology, and Applications VII, ed. By S. Bengtsson, 2003

[18] C. Colinge, B. Roberds, B. Doyle, Journal of Electronic Materials, Vol. 30, p. 841, 2001

[19] C. Mazuré, Electrochemical Society Proceedings, Vol. 2003-05, p.13, 2003

[20] P.L.F. Hemment, "The SOI odyssey", Electrochemical Society Proceedings, Vol. 2003-05, p. 1, 2003

[21] V. Ferlet-Cavrois, A. Bracale, N. Fel, O. Musseau, C. Raynaud, O. Faynot and J.L. Pelloie, Proceedings IEEE Intl. SOI Conference, 24, (1999)

[22] J.C. Sturm, K.Tokunaga and J.P. Colinge, IEEE-Electron Dev. Lett. **9**, 460 (1988)

[23] J.P. Colinge, IEEE Electron Device Letters 7, 244 (1986)

[24] J.P. Colinge, IEEE Transactions on Electron Devices, Vol. 34, p. 845, 1987

[25] S.A. Parke, C. Hu, and P.K. Ko, IEEE Electron Device Letters, Vol. 14, p. 234, 1993

[26] S.S. Rofail, and Y.K. Seng, IEEE Transactions on Electron Devices, Vol. 44, p. 1473, 1997

[27] M. Matloubian, IEEE International SOI Conference Proceedings, p. 106, 1993

[28] Z. Yan, M.J. Deen, and D.S. Malhi, IEEE Transactions on Electron Devices, Vol. 44, p. 118, 1997

[29] T. Douseki, S. Shigematsu, J. Yamada, M. Harada, H. Inokawa, and T. Tsuchiya, IEEE Journal of Solid-State Circuits, Vol. 32, p. 1604, 1997

[30] F. Assaderaghi, D. Sinitsky, S. A. Parke, J. Bokor, P.K. Ko, and C. Hu, Tech. Digest of IEDM, p. 809, 1994

[31] Z. Xia, Y. Ge, Y. Zhao, Proceedings 22nd International Conference on Microelectronics (MIEL), p. 159, 2000

[32] V. Ferlet-Cavrois, A. Bracale, N. Fel, O. Musseau, C. Raynaud, O. Faynot and J.L. Pelloie, Proceedings of the IEEE Intl. SOI Conference, p. 24, 1999

[33] T. Douseki, F. Morisawa, S. Nakata and Y. Ohtomo, Extended Abstracts of the International Conference on Solid-State Devices and Materials (SSDM), p. 264, 2001

[34] A. Yagishita, T. Saito, S. Inumiya, K. Matsuo, Y. Tsunashima, K. Suguro, IEEE Transactions on Electron Devices, Vol. 49, no. 3, p. 422, 2002

[35] S.S. Rofail, Y.K. Seng, IEEE Transactions on Electron Devices, Vol. 44, no. 9, p.1473, 1997

[36] R. Huang, Y. Y. Wang, and R. Han, Solid-State Electronics, Vol. 39, no. 12, p. 1816, 1996

[37] J.P. Colinge and J.T. Park, Journal of Semiconductor Technology and Science, Vol. 3, no. 4, p. 223, 2003

[38] T. Sekigawa and Y. Hayashi, Solid-State Electronics, Vol. 27, p. 827, 1984

[39] B. Agrawal, V.K. De, and J.D. Meindl, Proceedings of 23rd ESSDERC, Ed. by J. Borel, P. Gentil, J.P. Noblanc, A. Nouhaillat, and M. Verdone, Editions Frontières, p. 919, 1993

[40] D. Hisamoto, T. Kaga, Y. Kawamoto and E. Takeda, Technical Digest of IEDM, p. 833, 1989

[41] X. Huang, W.C. Lee, C. Kuo, D. Hisamoto, L. Chang, J. Kedzierski, E. Anderson, H. Takeuchi, Y.K. Choi, K. Asano, V. Subramanian, T.J. King, J. Bokor, C. Hu, Technical digest o fIEDM, p. 67, 1999

[42] Y.K. Liu, K. Ishii, T. Tsutsumi, M. Masahara, H. Takamisha and E. Suzuki, Electrochemical Society Proceedings 2003-05, p. 255, 2003 and Y. Liu, K. Ishii, T. Tsutsumi, M. Masahara, and E. Suzuki, IEEE Electron Device Letters, Vol. 24, no. 7, p. 484, 2003

[43] T. Hiramoto, IEEE International SOI Conference Proceedings, p. 8, 2001, and T. Saito, T. Saraya, T. Inukai, H. Majima, T. Nagumo, and T. Hiramoto, IEICE Transactions on Electronics, Vol. E85-C, no. 5, p. 1073, 2002

[44] Z. Jiao and A.T. Salama, Electrochem. Society Proceedings 2001-3, 403, 2001

[45] F. Balestra, S. Cristoloveanu, M. Benachir and T. Elewa, IEEE Electron Device Letters, Vol. 8, p. 410, 1987

[46] J.P. Colinge, M.H. Gao, A. Romano, H. Maes, and C. Claeys, Technical Digest of IEDM, p. 595, 1990

[47] X. Baie, J.P. Colinge, V. Bayot and E. Grivei, Proceedings of the IEEE International SOI Conference, p. 66, 1995

[48] R. Chau, B. Doyle, J. Kavalieros, D. Barlage, A. Murthy, M. Dozky, R. Arghavani and S. Datta, Extended Abstracts of the International Conference on Solid State Devices and Materials, SSDM, p. 68, 2002

[49] B.S. Doyle, S. Datta, M. Doczy, B. Jin, J. Kavalieros, T. Linton, A. Murthy, R. Rios, R. Chau, IEEE Electron Device Letters, Vol. 24, no. 4, p. 263, 2003

[50] J.T. Park, J.P. Colinge and C. H. Diaz, IEEE Electron Device Letters, 22, 405, 2001

[51] F.L. Yang, H.Y. Chen, F.C. Cheng, C.C. Huang, C.Y. Chang, H.K. Chiu, C.C. Lee, C.C. Chen H.T. Huang, C.J. Chen, H.J. Tao, Y.C. Yeo, M.S. Liang, and C. Hu, Technical Digest of IEDM, p. 255, 2002

[52] Z. Krivokapic, C. Tabery, W. Maszara, Q. Xiang, M.R. Lin, Extended Abstracts of the International Conference on Solid State Devices and Materials, SSDM, p. 760, 2003

[53] J.T. Park and J.P. Colinge, IEEE Transactions on Electron Devices, Vol. 49, p. 2222, 2002

[54] US patent 6,359,311

[55] F.L. Yang, H.Y. Chen, F.C. Cheng, C.C. Huang, C.Y. Chang, H.K. Chiu, C.C. Lee, C.C. Chen H.T. Huang, C.J. Chen, H.J. Tao, Y.C. Yeo, M.S. Liang, and C. Hu, Technical Digest of IEDM, p. 255, 2002

[56] K.K. Young, IEEE Transactions on Electron Devices, Vol. 37, p. 504, 1989

[57] R.H. Yan, A. Ourmazd, and K.F. Lee, IEEE Transactions on Electron Devices, Vol. 39, p. 1704, 1992

[58] J.P. Colinge, Solid-State Electronics, Vol. 48, p. 987, 2004

[59] F. Daugé, J. Pretet, S. Cristoloveanu, A. Vandooren, L. Mathew, J. Jomaah and B.Y. Nguyen, Solid-State Electronics, Vol. 48, p. 535, 2004

[60] J.P. Colinge, Progress in SOI Structures and Devices Operating at Extreme Conditions, Kluwer academic Publishers, pp. 167-188, 2002

SOI TECHNOLOGY AS A BASIS FOR MICROPHOTONIC-MICROELECTRONIC INTEGRATED DEVICES

M.Yu.Barabanenkov, V.V.Aristov, V.N.Mordkovich
Institute of Microelectronics Technology and high-purity Materials, Russian Academy of Science, 142432 Chernogolovka, Moscow Region, Russia

Abstract: A survey of crucial issues relating to the origin of Silicon On Insulator based microphotonics is presented. The main problem in microphotonics is the large size of today's photonic circuits in comparison with ULSI electronic ones. To increase the level of integration in photonic components, compact building blocks are required, performing elementary optical functions. Photonic crystals or, generally, photonic band gap materials offer a way to this reduction in size. Special attention is given to optical properties of two dimensional photonic crystal slabs which are well adapted to the domain of integrated optics.

Key words: Silicon On Insulator; microphotonics; photonic crystal; photonic band gap; opaque band; transfer relations; Mie resonance; Bragg-like multiple scattering

1.   INTRODUCTION

Silicon On Insulator (SOI) and more complicated stacked systems are believed to be the platform that combines an increasingly common material for high-speed electronics with a natural waveguide structure. By reducing parasitic capacitance and leakage currents, an insulating layer provided by buried $Si_xO_yN_z$ (in particular, SiO_2) increases the circuit speed and lowers power consumption. At the same time, the optical waveguide properties of SOI were recognized and exploited to the point where SOI emerged as an attractive platform for planar lightwave circuits.

Various optical components can be bound or connected to SOI substrates which serves as an optical bench[1]. It is known that photons can do many things better than electrons. For instance, light beams have very broad

D. Flandre et al. (eds.), Science and Technology of Semiconductor-On-Insulator Structures and Devices Operating in a Harsh Environment, 27-37.

bandwidth (practically unlimited information capacity), low transmission losses, do not dissipate heat, and are immune to cross-talk and harsh environment. However, there is another more challenging problem in optical devices. This arises from the fact that integration, even on SOI-based structures, can limit the operating speed of microelectronic devices because of the inadequacies of present day interconnects. The high density (nowadays a chip may contain 5 km/cm^2 of interconnection tracks) of metal lines may introduce a signal delay as a result of wiring due to capacitive and inductive coupling and signal cross-talk. The use of optical interconnects is a major research topic[2] the objective being to achieve so-called Photonic Band Gap (PBG) materials for integration of waveguides with Photonic Crystals (PC).

The advent of PBG materials has opened up many opportunities for microphotonics due to the unique properties of PCs. Developments in the physics of PCs focus on the following main overlapping directions : (i) the physics of PBG materials and systems (optical properties, bandgap origination and formation mechanisms, fabrication processes); (ii) the follow-on of existing optics topics now carried out in PCs; (iii) novel physical concepts and experiments which might be carried out in PBG materials (i.e. modification of atom-field interaction, light localization); (iv) novel physics in order to meet new demands including general applications (i.e. advances towards the limit of communications).

The first item is of prime interest in this paper.

2.   PHOTONIC CRYSTALS

PCs are artificial electromagnetic (EM) materials possessing some form of spatial periodicity. The best known materials[3] are probably conservative (non-absorbing) dielectrics with a periodically varying dielectric permittivity ε. Like the spectrum of a Schrödinger operator with a periodic potential, the periodic Maxwell operator $\nabla \times (1/\varepsilon)\nabla \times$ has a gap in the spectrum, usually referred to as PBG. The spectrum can be classified according to the Bloch momentum $\bar{k}$. The energy (frequency) levels ω_n are continuous functions of $\bar{k}$ in the first Brillouin zone. A PBG is a full gap between the nth and (n+1)th levels when an inequality $\omega_{n+1}(\bar{k}) > \omega_n(\bar{k}')$ holds for all $\bar{k}$ and $\bar{k}'$. A direct gap corresponds to the case when $\bar{k} = \bar{k}'$. If the inequality holds only for a certain $\bar{k}$, PC exhibits a gap in the spectrum for a fixed direction of the incident waves, the so-called stop gap. Ordinary propagation of EM waves is forbidden in all directions in full band gaps. PBG materials were predicted theoretically[4,5] as a means to describe two fundamentally new optical phenomena: namely, the localization of light and the inhibition of spontaneous emission over a broad frequency range.

The usefulness of PCs derives to a large extent from two facts: (i) suitably engineered PCs exhibit PBG and (ii) defect structures embedded in PCs lead to the defect peak appearance in the gap.

The appearance of a photonic gap depends on the dielectric contrast ratio, lattice structure, filling factor *f*, and topology. Three dimensional (3D) PCs exhibit full PBG only if several requirements are satisfied[6,3]. These are light absorption should be absent, the refractive index contrast must be high, the volume fraction of the high-index material should be small, the crystal should preferably have many unit cells in all directions, and the disorder of the crystal should be negligible. In practice PCs obviously suffer from fluctuations in the position and size of building blocks. These fluctuations affect the gap width. As a full gap opens in spectra of 3D PCs between high (for example, eighth and ninth) energy bands[7], it is expected that this gap will be far more sensitive to the site and size randomness than a gap between lower-laying bands (2D PCs). Nonuniformities as small as under 2% of the lattice constant destroy the PBG completely[8]. Note that the PBG is more sensitive to the variation in the radii of spheres (artificial opal PCs) than in the variations of their displacements.

The above mentioned combined requirements to the PBG existence proved to be tough, especially in the optical domain. Consequently, substantial effort in the fabrication of 3D PCs with a full PBG could not but trigger an increased interest into 2D PC (more tolerable to the disorder[9-11]) slabs and their future expansion into the third dimension by the woodpile technique[12-14].

2.1 Two dimensional photonic crystals

2D PCs are fabricated within planar waveguides. 2D PBGs control the propagation of light within a corresponding plane of propagation. Index guiding confines light in the third dimension. For such structures, advanced planar micro-structuring techniques borrowed from semiconductor technology can greatly simplify the fabrication process.

PC slabs (i.e. rods of a finite height) have a band gap but not of a traditional sort. A "band gap" in this case is a range of frequencies in which no guided modes exist. It is not a true band because there are still radiation modes (that are extended infinitely in the region outside the slab) at those frequencies. The slab thickness plays an important role in determining whether a PC slab has a band gap in its guided modes. It was postulated[15] and proved[16,6] that the optimal thickness is of the order of half 2D gap-bottom wavelength. PC slabs with optimal thickness retain or approximate many of the properties of true 2D PCs.

2D PCs are periodic arrangements of parallel dielectric or air rods. For in-plane propagation, two types of EM modes can be defined according to whether the electric (E polarization) or magnetic (H polarization) fields is parallel to the rod axis.

For a given type of lattice symmetry (quadratic, hexagonal, graphite or honeycomb), the PBG depends on the form of the basis ("atom") and on the dielectric contrast. A full PBG was first demonstrated for a triangular lattice of air rods in a dielectric material[17,18] and for a square lattice of air columns[18]. The triangular lattice of air cylinders (not for dielectric circular rods in air with $\varepsilon = 16$ and $f = 0.35$) was found[18] to possess a full PBG for a large volume fraction of air in dielectric ($f = 0.65 - 0.8$; the rods are just touching when their filling fraction is 0.91). The full PBG is maximum for a circular cross section of air rods. Any deviation from the circular shape of cylinders strongly reduces the width of the gaps[19]. A large full PBG corresponds to a cylinder diameter close to the lattice parameter[18,20]. Such structures require etching of cylinders separated by very thin layers which is difficult to realize when the aspect ratio is high.

To avoid the fabrication of thin layers, a 2D graphite structure of dielectric rods in air was designed[21] with the cylinder-to-lattice ratio far from the close packed condition. In the graphite structure, cylinders are centered on the vertices of a plane lattice of hexagons (each third cylinder is removed from the triangular lattice; the Bravais lattice remains hexagonal). Unlike the triangular structure, the graphite one exhibits a full PBG for dielectric ($GaAs$, $\varepsilon = 13.6, f = 0.3$) rods in air.

As mentioned above, a square lattice exhibits a full PBG only in the case of air rods whether the rods have a circular or a square cross section[22]. In the latter case it requires a higher dielectric contrast ($\varepsilon = 12.3$ versus 7.3). Unlike a traditional square lattice of air rods with a square cross section, the so-called chessboard[23] lattice exhibits a full PBG even for $\varepsilon = 8.9$ and preserves the symmetry and simplicity of the square lattice. The case of $f = 0.5$ identifies the close packing condition for both dielectric and air rods (square columns rotated by 45° with respect to the square axes of the lattice), and makes the chessboard lattice invariant with respect to dielectric and air interchange.

2.2   Mechanisms of opaque band formation

Theoretical investigation of PCs which play a key role in the characterization of existing and designing novel structures, are generally based on methods of electronic band structure theory (for example, the plane-wave and augmented-plane-wave, or Korringa-Kohn-Rostoker methods[24]). These studies have shown[7,18] that, unlike 3D lattices, band diagrams for orthogonal polarized waves in 2D lattices are very different. This means that it is not sufficient for the E and H gaps to open, but need to be sufficiently large to overlap. It was difficult to give simple arguments predicting or explaining the existence of a full PBG because the overlap of E and H bands often involves the gaps between the first and second H bands with the gap between higher E bands[18,23]. For example, two full PBGs occur in the case of a square[23] symmetry as a result of the overlap of the second E_2 and the first H_1 gaps and the E_3 and the H_2 gaps, respectively. The overlap of higher gaps (E_6 and H_3, E_6 and H_5) results in two full PBGs of a graphite structure[21].

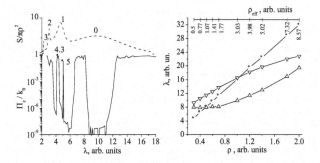

Figure 1. The calculated[29] relative power of EM radiation transmitted through N=18 layers of cylinders arranged in a square lattice with a period Λ (solid line) and the total scattering cross section S on a unit length of an infinitely long cylinder (dashed line) versus the wavelength of the EM waves, assuming wave normal incidence α=0, ε=8.41 (TiO_2), and arbitrary units for the wavelength λ, ρ = 0.6, Λ = 4. Two black dots with figures mark two specific wavelengths.

Figure 2. The edges (triangles) and the peak (numbered by zero in Fig.1) of the monopole resonance scattering on a single cylinder (circles) as a function of the radius of cylinders (18 layers of cylinders with a fixed square lattice period Λ = 4). The effective radius of cylinders is defined by the following relationship $\rho_{eff} = \rho[\, S(\lambda)/\pi\rho^2\,]^{1/2}$.

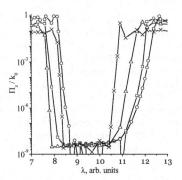

Figure 3. The calculated relative power of EM radiation transmitted through a 2D PC (the parameters see in Fig.1) at different incidence angles: $\alpha = 0$ (o), $20°$ (□), $45°$ (△), $70°$ (×).

Transmission measurements on PCs and the observed opaque frequency regions were usually compared with the dispersion relations, and hence, with the state densities obtained by the band calculations[25]. However, the transmission spectrum is not at all a replica of the state density. This is because[26] (i) the strength of the coupling between the incident/transmitted radiation and the internal field at the surface of the structure substantially modifies the spectrum; (ii) band calculations have no information on the Bragg reflection spectra which correspond to incident wavelengths shorter than the lattice period parallel to the surface; (iii) band structure calculations are not applicable to interpret optical properties of thin PCs possessing a few layers of rods[20]. So, the physics of formation and frequency position of opaque bands in the transmission spectra is beyond the scope of the band structure concept of solid-state physics. At the same time, the understanding of contributions of macroscopic nonresonant Bragg-like multiple scattering and microscopic single-scattering Mie resonances to the formation of spectral gaps in PCs has remained a challenge since the tight-binding parametrization[27] of 2D PCs. The Mie resonances was suggested[11] as a dominant mechanism in 2D dielectric crystals and the Bragg-like scattering for metallic ones. Note that already at the early stage of PC theory, a strong depletion (pseudogap) in state densities for the photon modes was proposed to be treated as a remnant of the single-sphere Mie resonances[24]. The so-called transfer relations[28] got more insight into the origin of opacity of dielectric 2D PCs. They gave the invariant imbedding relations as a Riccati equation for the matrix wave reflection coefficient from a 2D PC slab and an associated differential equation for the matrix wave transmission coefficient through a slab, and a differential equation for a local field inside a slab. A plane EM wave was considered in the computations[29], with the wavevector $\vec{k}_o$ which incidents from a background dielectric medium, $\varepsilon^{bac} = 1$, at an angle α on a slab of 2D PC consisting of N layers (perpendicular to the z axis of the rectangular coordinate system x, y, z) of rods (infinitely extended along the y axis) with the "radius" ρ. The EM wave field energy flux, Π_z, transmitted through a PC slab was evaluated[29] in the case of non-absorptive rods, $\Im\varepsilon = 0$, by solving the Riccati equation with the aid of the Poynting theorem.

The solid line in Fig.1 represents a transmission spectrum with two opaque bands separated by a transparency window. The first band ($8.2 < \lambda < 12.4$) was denoted[29] as the main gap because its wavelength position follows the wavelength of the lowest (main) Mie resonance in a single cylinder (see the peak marked with zero). Besides, only four layers of the cylinders in which the Bragg-like multiple scattering can be fully developed, give the onset of this gap formation. So the main gap is intrinsic to the bulk crystal, that is, to a set of layers (i.e. gratings), while the additional gap ($3 < \lambda < 6.5$) is inherent in a grating. As the wavelength value increases from that of the zero Mie resonance, the total scattering cross section S smoothly decreases and, correspondingly, full transparency of the photonic crystal slab can be observed. On the contrary, the other slope of the dashed curve has several peaks. It is seen that the width of the dipole resonance (labelled 1) correlates with the width of the additional spectral gap in which there are two narrow

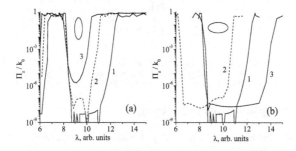

Figure 4(a,b). The calculated relative power of EM radiation transmitted through a 2D square lattice (the parameters see in Fig.1) of elliptical cylinders with the major axis of the cylinder cross section parallel (a) and perpendicular (b) to the direction of wave normal incidence. The eccentricity is equal to zero (curves 1), 0.75 (dashed curve 2 (a)), 0.87 (curve 3 (a)); 0.66 (dashed curve 2 (b)), and 0.8 (curve 3 (b))..

transmission peaks at λ = 4.3, 5. Notice that a smoothed valley between the monopole (0) and the dipole (1) peaks sets the smoothness of the long wavelength edge of the additional gap. At the same time, the octupole (2) and quadrupole (3) resonances cause a splitting of the other edge.

Fig.2 shows that the wavelength of the monopole resonance single scattering (numbered by 0 in Fig.1) falls into the main gap and it still remains there while the filling fraction of volume occupied by cylinders (per unit length) ranges from 3 to 20% (the cylinder radii vary from 0.4 to 1) for the fixed lattice period. In the scale of the upper axis in Fig.2 it is seen that the main gap becomes most pronounced when the effective radius of the total scattering cross section at the monopole Mie resonance is approximately equal to the half period of the square lattice, i.e. when "effective" cylinders are closely packed. The main gap tends to vanish as soon as the effective radius exceeds the lattice constant.

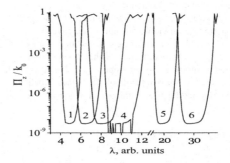

Figure 5. The calculated relative power of EM radiation transmitted through a 2D square lattice of circular cylinders (ε = 8.41) with fixed cylinder-to-lattice ratio ($\zeta = \rho/\Lambda$) and filling factor f = 7%. ζ = 0.3/2 (curve 1), 0.4/2.6 (2), 0.5/3.3 (3), 0.6/4 (4), 1.2/8 (5), and 1.8/12 (6).

2.3   Opaque band engineering

It is worth discussing the possibility of opaque band engineering. First of all, the spectrum (Fig.1) of a perfect PC exhibits transmission peaks within the opaque band. Such peaks are traditionally ascribed to PC imperfection. However, the peaks within the additional gap of the spectrum were explained[29] in terms of the high-order Mie resonances in a single cylinder. The other computations[29] demonstrated that (i) the larger the refractive index of the widely used materials, the broader and deeper is an opaque band in the transmission spectra of 2D PCs composed of cylinders; (ii) the opaque band is slightly deformed if the incidence angle of EM waves is grossly changed (Fig.3); (iii) the opaque band exists if we strongly compress or extend the cylinders (Fig.4) (it is true for elliptical dielectric cylinders with a sufficient magnitude of eccentricity but it is not the case for air cylinders); (iv) it is possible to shift the opaque band for any fixed magnitudes of dielectric contrast and filling factor; all we need is to keep the cylinder-to-lattice ratio constant (Maxwell's equations enjoy scale invariance) (Fig.5). Last, the rectangular lattice of rods whose two successive layers are oriented perpendicular to each other is actually a 3D PC which can be scaled down in size to operate in the optical region. Layer-by-layer frequency spectra of such woodpile structures[12,13] can be easily calculated by the abovementioned Riccati equation[29] (the case of *H*-polarized incident wave).

3.   CONCLUSION

SOI structures were first introduced for CMOS applications to enhance radiation tolerance and to reduce the parasitic capacitance to the silicon substrate. However, they soon proved equally suitable for the purpose of guided optics. Nowadays a number of classical optical components (waveguides, mirrors, filters, bends, splitters, interferometers) have their on-chip (SOI) analogs fabricated with the aid of 2D PCs. Besides, the SOI platform offers resonant grating coupling between an external wave and a local wave to achieve narrowband filtering for such applications as wavelength division multiplexing using grating couplers, Bragg-reflectors, and integrated couplers/Bragg reflectors. Nevertheless, we emphasize the specific nature of SOI structures as an optical bench due to their high vertical dielectric contrast. First, high contrast is useful for large structures in which the periodicity is seldom broken (long straight waveguides). For waveguides with many bends and other components, i.e. when the periodicity is broken, the losses can be lower for low dielectric contrast layers (like *InP* or *GaAs-AlGaAs*-based heterostructures). Second, ideal for on-chip transmission SOI-based PC waveguides have coupling problems with silica fiber due to both large size difference and different optical impedance of these two systems[2]. Coupling efficiency can be improved by tapering edge couplers[30] or by butt-coupling an in-plane waveguide (out-of-plane scattering losses of light via air rods of 2D PC) to a vertical fiber.

ACKNOWLEDGMENTS

One of the authors (M.Yu.B.) gratefully acknowledges partial financial support from the Russian Science Support Foundation.

REFERENCES

1. T.Bestwick, ASOC Silicon Integrated Optics Technology, in: *Proc. SPIE "Optoelectronic Integrated Circuits and Packaging III"*, edited by M.R.Feldman, J.G.Grote and M.K.Hibbs-Brenner, (San Diego, USA, 1999) **3631**, pp.182-190
2. L.Pavesi, Will Silicon be the Photonic Material of the Third Millenium?, *J.Phys.:Condens.Matter,*. **15**, pp.R1169-R1196 (2003)
3. A.Tip, A.Moroz, and JMCombes, Band Structure of Absorptive Photonic Crystals, *J. Phys. A: Math. Gen.* **33**, pp.6223–6252 (2000)
4. E.Yablonovitch, Inhibited Spontaneous Emission in Solid-State Physics and Electronics, *Phys. Rev. Lett.*, **58(20)** pp.2059-2062 (1987)
5. S.John, Strong Localization of Photons in Certain Disordered Dielectric SuperLattices, *Phys. Rev. Lett.*, **58(23)** pp.2486-2489 (1987)
6. M.Wubs and A.Lagendijk, Local Optical Density of States in Finite Crystals of Plane Scatterers, *Phys. Rev. E* **65**, p.046612 (2002)
7. A.Moroz, Ch.Sommers, Photonic Band Gaps of Three-Dimensional Face-Centered Cubic Lattices, *J.Phys.:Condens.Matter,*. **11**, pp.997-1008 (1999)
8. Zh.-Y.Li, Zh.-Q.Zhang, Fragility of Photonic Band Gaps in Inverse-Opal Photonic Crystals, *Phys. Rev. B*, **62(3)**, pp.1516-1519 (2000)
9. H.-Y. Ryu, J.-Ki Hwang, and Y.-H. Lee, Effect of Size Nonuniformities on the Band Gap of Two-Dimensional Photonic Crystals, *Phys. Rev. B* **59**(8), pp.5463-5469 (1999)
10. E.Lidorikis, M.M.Sigalas, E.N.Economou, and C.M.Soukoulis, Gap Deformation and Classical Wave Localization in Disordered Two-Dimensional Photonic-Band-Gap Materials, *Phys. Rev. B*, **61**(20) pp.13458-13464 (2000)
11. M.Bayindir, E.Cubukcu, I.Bulu, T.Tut, E. Ozbay, and C.M.Soukoulis, Photonic Band Gaps, Defect Characteristics, and Waveguiding in Two-Dimensional Disordered Dielectric and Metallic Photonic Crystals, *Phys. Rev. B*, **64**, p.195113 (2001)
12. E.Ozbay, A.Abeyta, G.Tuttle, M.Tringides, R.Biswas, C.T.Chan, C.M.Soukoulis, K.M.Ho, Measurement of a Three-Dimensional Photonic Band Gap in a Crystal Structure Made of Dielectric Rods, *Phys. Rev. B*, **50**(3), pp.1945-1948 (1994)
13. S.Noda, K.Tomoda, N.Yamamoto, Full Three-Dimensional Photonic Bandgap Crystals at Near-Infrared Wavelengths, *Science*, **289**(5479), pp.604-606 (2000)
14. B.Gralak, M.de Dood, G.Tayeb, S.Enoch, and D.Maystre, Theoretical Study of Photonic Band Gaps in Woodpile Crystals, , *Phys. Rev. E*, **67**, p.066601 (2003)
15. S.G. Johnson, S.Fan, P.R.Villeneuve, J.D.Joannopoulos, and L.A.Kolodziejski, Guided Modes in Photonic Crystal Slabs, *Phys. Rev. B*, **60**(8), pp.5751-5758 (1999)
16. N.Kawai, K.Inoue, N.Carlsson, N.Ikeda, Y.Sugimoto, K.Asakawa, and T.Takemori, Confined Band Gap in an Air-Bridge Type of Two-Dimensional AlGaAs Photonic Crystal, *Phys. Rev. Lett*, **86**(11), pp.2289-2292 (2001)
17. R.D.Meade, K.D.Brommer, A.M.Rappe, and J.D.Joannopoulos, Existence of a Photonic Band Gap in Two Dimensions, *Appl. Phys. Lett.*, **61**(4), pp.495-497 (1992)
18. P.R.Villeneuve, M.Piche, Photonic Band Gaps in Two-Dimensional Square and Hexagonal Lattices, *Phys. Rev. B*, **46**(8), pp.4969-4972 (1992)

19. R.Padjen, J.M.Gerard, J.Y.Marzin, Analysis of the Filling Pattern Dependence of the Photonic Bandgap for Two-Dimensional Systems, *J. Mod. Opt.*, **41**(2), pp.295-310 (1994)
20. J.Schilling, A.Birner, F.Muller, R.B.Wehrspohn, R.Hillebrand, U.Gosele, K.Busch, S.John, S.W.Leonard and H.M. van Driel, Optical Characterization of 2D Macroporous Silicon Photonic Crystals with Bandgaps Around 3.5 and 1.3 μm, *Opt. Mat.*, **17**, pp.7-10 (2001)
21. D.Cassagne, C.Jouanin, D.Bertho, Photonic Band Gaps in a Two-Dimensional Graphite Structure, *Phys. Rev. B*, **52**(4), pp.R2217-R2220 (1995)
22. P.R.Villeneuve, M.Piche, Photonic Band Gaps in Two-Dimensional Square Lattices: Square and Circular rods, *Phys. Rev. B*, **46**(8), pp.4973-4975 (1992)
23. M.Agio, L.C.Andreani, Complete Photonic Band Gap in a Two-Dimensional chessboard Lattice, *Phys. Rev. B*, **61**(23), pp.15519-15522 (2000)
24. Z.Zhang, S.Satpathy, Electromagnetic Wave Propagation in Periodic Structures: Bloch wave Solution of Maxwell's Equations, *Phys. Rev. Lett*, **65**(21), pp.2650-2653 (1990)
25. E.Yablonovitch, T.J.Gmitter, K.M.Leung, Photonic Band Structure: the Face-Centered-Cubic Case Employing Nonspherical Atoms, *Phys. Rev. Lett.*, **65**(17) pp.2295-2298 (1991)
26. K.Sakoda, Transmittance and Bragg Reflectivity of Two-Dimensional Photonic Lattices, *Phys. Rev. B*, **52**(12), pp.8992-9002 (1995)
27. E.Lidorikis, M.M.Sigalas, E.N.Economou, C.M.Soukoulis, Tight-Binding Parametrization for Photonic Band Gap Materials, *Phys. Rev. Lett*, **81**(7), pp.1405-1408 (1998)
28. Yu.N.Barabanenkov, V.L.Kouznetsov, M.Yu.Barabanenkov, Transfer Relations for Electromagnetic Wave Scattering From Periodic Dielectric One-Dimensional Interface, in *Progress in Electromagnetic Research*, edited by J.A.Kong (EMW Publishing, Cambridge, Massachusetts USA, 1999) **24**, pp.39-75
29. Yu.N.Barabanenkov, M.Yu.Barabanenkov, Method of Transfer Relations in Theory of Multiple Resonant Scattering of Waves as Applied to Diffraction Gratings and Photonic Crystals, *J. Exper. Theor.Phys.*, *96(4)*, pp.674-683 (2003)
30. P.Sanchis, J.Martí, J.Blasco, A.Martínez, and A.García, Mode matching technique for highly efficient coupling between dielectric waveguides and planar photonic crystal circuits, *Opt. Express*, **10**(24), pp.1391-1397 (2002)

SMART CUT TECHNOLOGY: THE PATH FOR ADVANCED SOI SUBSTRATES

H. Moriceau[1], C. Lagahe-Blanchard[1], F. Fournel[1], S. Pocas[1], E. Jalaguier[1], P. Perreau[1], C. Deguet[1], T. Ernst[1], A. Beaumont[1], N. Kernevez[1], J.M. Hartman[1], B. Ghyselen[2], C. Aulnette[2], F. Letertre[2], O. Rayssac[2], B. Faure[2], C. Richtarch[2] and I. Cayrefourq[2]

1 CEA-DRT - LETI/DIHS - CEA/GRE 17, rue des Martyrs, 38054 Grenoble Cedex 9 France
2 SOITEC, Parc technologique des Fontaines 38190 Bernin, France

Abstract: In microelectronics, photonics, opto-electronics, high frequency or high power device applications, the needs for specific substrate solutions are more and more required. Smart Cut™ technology appears as the technological answer that enables the industrial to provide engineered substrate solutions tailored to the applications. For instance a large spectrum of SOI type structures are today in volume manufacturing. At present the industrial is focused on composite substrates. This paper focuses on the realization of advanced SOI, strained SOI, SOQ substrates and many other examples of engineered substrates. Highlights are given on the most recent developments.

Key words: Smart Cut™ technology, SOI, SOQ, SOIM, sSOI, SGOI, GeOI, strained silicon

1.   INTRODUCTION

The Smart Cut™ technology slices the wafer by ion implantation, lifting off a thin layer from the donor substrate and transferring it onto a new substrate. Initiated by Bruel[1], this technology has been developed to manufacture SOI and has achieved a high degree of maturity with the UNIBOND™ mass production of SOI wafers. As a generic layer transfer technology, the Smart Cut technology expands the range of applications from single crystal thin film transfer to conventional hetero-epitaxy onto any type of substrates. The flexibility of the Smart Cut technology extends beyond the microelectronics world to applications in photonics, opto-

D. Flandre et al. (eds.), Science and Technology of Semiconductor-On-Insulator Structures and Devices Operating in a Harsh Environment, 39-52.

electronics, high frequency and high power devices, thanks to the development of composite substrates. The Smart Cut technology is today the subject of many studies and developments world wide. Many materials and crystal associations have been investigated to create a broad range of original structures. Different types of bonding interfaces (including dielectrics, semiconductors, metal) have been studied. A selection of results will be shown.

2.   THE SMART CUT TECHNOLOGY DESCRIPTION

First developed for Si layer transfers to manufacture SOI, the Smart Cut technology is used to transfer ultra-thin single crystal layers of starting substrate material onto another wafer (or more generally to a stiffener).

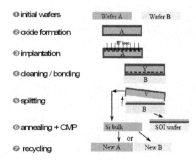

Figure 1. Main steps of the Smart Cut process

This process is based on ion implantation and wafer bonding, differing from traditional layer transfer techniques, which are mainly based on wafer bonding and etch-back or epitaxial lift-off. A schematic process flow is shown for UNIBOND SOI wafer fabrication (fig.1). As a first step, ion implantation (hydrogen, helium…) induces, in the "donor" wafer, the formation of an in-depth weakened layer, parallel to the surface and located at the mean ion penetration depth. At the second step, the implanted wafer and a second wafer are bonded together. Then the detachment (third step) is induced in the weakened layer leading to transfer a thin film. A final step consists in a soft touch polishing to remove any roughness at the bonded structure surface. Both the top Si and buried oxide layer thickness are widely flexible and the process is scalable from 100mm to 300mm wafer diameters.

More over it uses standard IC process equipment and is well suited to high volume manufacture. Besides the remainder of the donor wafer can be recycled.

3.　　PURE UNIBOND SOI WAFERS

The thickness of the top Si layer as well as the thickness of the buried insulating layer can vary to cover the full range of applications. Developed to address the most advanced silicon devices, thin-film and ultra-thin film UNIBOND SOI wafers are adapted for partially and for fully depleted CMOS applications.

Figure 2. Trend to thinner SOL and box layers. TEM observation of 75nm top silicon and 20nm box SOI structures.

Wafers are available with top-layer thickness ranging from several microns down to 20nm. At present there is increased interest in thinner buried oxide layers and silicon over-layers. In figure 2, a 75nm Si top layer and a 20nm buried oxide layer are observed by TEM on a cross section highlighting the high quality of such bondings performed via an hydrophilic way.

The thin- and ultra-thin SOI wafers are well suited to advanced, high-speed, low voltage, low-power consumption products such as microprocessors and complex wireless chips.

4.　　SILICON ONTO INSULATING SUBSTRATES

For some applications, such as displays, micro-wave devices…, it may be required to get thin silicon films onto insulating substrates, such as fused silica, sapphire, high resistivity silicon wafers… Because of different thermal coefficients between films and substrates, specific bonding conditions have to be tuned to succeed. Silicon-on-Quartz (SOQ) wafers are manufactured to create a thin, mono-crystalline silicon film onto quartz (fused silica) handle wafers. In such a case, surface micro-roughness has to be smoothed enough to enable an efficient bonding at low temperature. In fact, thermal treatments have to be optimized to avoid high stresses at the bonding interface because of the very different thermal expansion coefficients ($\sim$0.5 10^{-6} K^{-1} for fused silica and $\sim$2.4 10^{-6} K^{-1} for Si). For instance, transfers by the Smart Cut technique of thin Si films have been demonstrated onto fused silica wafers (SOQ), as illustrated in figure 3. Such process is easily extendable to Silicon-on-Glass structures. They are used for applications requiring high resistivity or transparent substrates as displays, optronic applications and high-frequency devices.

Figure 3. 200mm SOQ. 2000A of Si onto a fused silica wafer (quartz).

As more generally claimed, the Smart Cut technique is a good mean to transfer active silicon films onto low cost wafers. In some cases, insulating substrates like quartz, glass, sapphire...are not appropriate: handling and wafer detection issues due to transparency, risks of metallic contamination, maximum temperature of use or thermal mismatch with silicon limiting the maximum temperature use...

In this case, it is still possible to use for many RF applications 100% Si handle substrates as long as their resistivity is high enough. Simulations and experimental results have shown that resistivities beyond 1 kohm-cm offer significant gains for substrate loss in the 1-30 GHz range[2].

5.   SILICON ONTO SEVERAL BURIED LAYERS: FROM THE SOI TO THE SOIM TECHNOLOGY

Thanks to improvements in bonding techniques, it is now possible to make single crystalline silicon over-layers on top of a large variety of insulating layers (e.g. SiO_2 and Si_3N_4 multi-layers) and large size substrates. For instance, 200nm thick Si films have been successfully transferred onto 400nm thick insulating SiO_2 and Si_3N_4 layers by the Smart Cut technique referenced as Silicon On Insulating Multi-layers (SOIM) structures[3].

Specifically some of the SOIM structures have been made by bonding oxidized silicon wafers to Si wafers coated by CVD deposited Si_3N_4 films. The thickness of each insulating film is in the [0.05µm-0.4µm] thickness range, which allows to carry out SOIM bonded structures without any need of chemical mechanical polishing for Si_3N_4 films whatever their thickness up to ~0.5µm. A wet chemical cleaning step is efficient enough to obtain hydrophilic surfaces before bonding, whatever the Si_3N_4 film thickness in the [0nm-400nm] range. After cleaning, Si_3N_4 surface micro-roughness values were measured by AFM at about 0.35nm RMS using 5µm*5µm scanning investigations, for both 200nm and 400nm thick Si_3N_4 films. Various structures have been bonded by contacting Si_3N_4 surfaces either to bare Si wafers or to SiO_2 or again to Si_3N_4 film surfaces.

In order to monitor the chemical bond evolutions at the interface versus the annealing temperature, infrared multiple internal transmission (FTIR-MIT) measurements have been successfully performed using prism coupling as previously described by Maleville et al.[4].

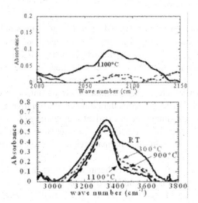

Figure 4. FTIR-MIT spectra evolution versus annealing temperature for a (SiO_2/Si_3N_4) SOIM structure.

FTIR-MIT results have shown that the $(Si_3N_4$ onto $SiO_2)$ structures present also a bonding mechanism evolution in three different temperature ranges[3]. Below 300°C, a strong decrease of the Si-OH group density (3400-3600)cm^{-1} can be observed. In the [300°C - 900°C] temperature range, a slight water desorption is still occurring. Above 900°C, a slight increase of the peaks linked to Si-H (2075cm^{-1}, figure 4-top)[5] and NH groups (3340cm^{-1}, figure 4-bottom)[6] is also evidenced. These behaviors have been attributed to water decomposition which induces a slight increase of Si-H or N-H bonds. In this temperature range, Si-N$_x$-O$_y$-Si (x, y = 0 or 1) covalent groups are mainly involved, leading to strong bonding energies. After splitting, SOIM structures were annealed at 1100°C to strengthen the bonding. Bonding of a high quality is pointed out by TEM observations as shown in figure 5 for a (Si$_3$N$_4$ to SiO$_2$) bonding interface.

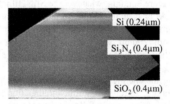

Figure 5. Bonding of a high quality is pointed out by cross section TEM observations for a (Si$_3$N$_4$ to SiO$_2$) bonding interface.

Varying the ratio between SiO$_2$ and Si$_3$N$_4$ film thickness induces a wafer curvature modification. Built with only 400nm thick box and without any silicon nitride layer, the SOI structure is very slightly convex. By introducing a 200nm thick Si$_3$N$_4$ layer and keeping the SiO$_2$ (400nm) thickness constant, the structure becomes slightly concave[3]. Use of a silicon nitride layer appears therefore as a simple route to control bows of bonded structures. In addition to the considerations of the relative SiO$_2$ to Si$_3$N$_4$ thickness ratio, it is a simple mean to adapt the bending of such SiO$_2$-Si$_3$N$_4$ stacked structures. Moreover the control of such a bending can be done through the temperature of the thermal treatment used to strengthen the bonding.

SOIM structures are attractive because of the high thermal power dissipation of Si$_3$N$_4$ films, as confirmed by its high thermal conductivities (30 W m^{-1}K^{-1} instead of ~1 W m^{-1}K^{-1} for SiO$_2$). SOIM structures could be used in micro-electronic and micro-technology applications soon, where thermal dissipations are to be taken into account. This is the case of devices submitted to a self heating effect. Such approach has been mainly developed to build structures with various insulating buried layers (e.g. SiO$_2$ and Si$_3$N$_4$ layers). For instance, to achieve a high electronic quality at the buried interface of the top Si layer and because thermal SiO$_2$ layers are well known for their good interface quality, it is worth using a very thin thermal SiO$_2$ layer as a first buried layer as illustrated in figure 6 by a SEM cross section observation of a Si- SiO$_2$- Si$_3$N$_4$- SiO$_2$- Si$_{bulk}$ structure.

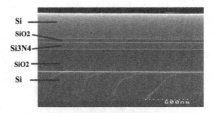

Figure 6. SEM cross section of a Si- SiO_2- Si_3N_4- SiO_2- Si_{bulk} structure : it is worth using a very thin thermal SiO_2 layer as a first buried layer to obtain a good electrical interface quality.

6.   MULTIPLE SOI LAYERS STACKED BY MULTIPLE USE OF THE SMART CUT TECHNOLOGY

Beyond these simple SOI wafers, structures made of multi-layers are more and more emergent. Wafer bonding, based on molecular or hydrogen bonds, requires pretty good substrate surface quality, especially in terms of roughness and surface defect density. Both the roles of surface micro-roughness and surface cleaning before bonding have been investigated. It highlights that SOI surfaces, as-achieved by the Smart Cut technique, are well suited for direct bonding. Standard SOI wafer exhibits the same surface roughness than silicon wafers.

As a matter of fact, wafer bonding can be applied similarly between two SOI wafers. The flexibility of both the bonding and the Smart Cut techniques can be used to stack more complex structures. For instance several (Si-SiO_2) bilayers can be subsequently transferred onto the same structure as illustrated in figure 7. It can be done either by transferring a single layer several times onto a same wafer, or by transferring stacks of SOI layers already formed onto either a single substrate or a SOI wafer. So, multiple SOI layers stacks can be easily made using SOI wafers as top and base wafers thanks to the Smart Cut technique, as shown in figure 7, where [Si/SiO_2] bilayers were stacked up, four times, onto 150mm silicon substrates[7]. The SEM view of the final structure reveals sharp interfaces and perfect intra-layer insulation. Thanks to the high versatility of such technique, thickness of each layer can be easily adapted, depending only on the targeted application. In the same manner, materials of each layer can be selected.

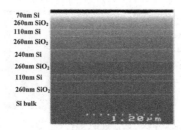

70nm Si	
260nm SiO₂	
110nm Si	
260nm SiO₂	
240nm Si	
260nm SiO₂	
110nm Si	
260nm SiO₂	
Si bulk	

Figure 7. Four (Si-SiO$_2$) bilayers subsequently transferred onto the same structure.

The Smart Cut technique can then be considered as a basic process step for obtaining multilayer SOI wafers, allowing different crystalline and / or amorphous layers to be stacked. Microroughness of SOI wafers is suitable for subsequent bonding, allowing stacked SOI layers made by multiple use of the Smart Cut technology. It opens a way to the 3D-integration applications. Moreover, this capability enables to enlarge the field of applications from 3D process integration to optical filters, micro machining, sensors...

7.   STRAINED-SILICON ON SIGE-ON-INSULATOR (SGOI) AND STRAINED SOI (sSOI)

Wafer level strained silicon is a very attractive option to further boost the performance of CMOS ICs. The production of strained silicon substrates requires the formation of a relaxed silicon germanium template on top of which a thin layer of strain silicon will be grown. The Ge content of the SiGe template defines the lattice parameter of the SiGe and thus, of the subsequent strain silicon layer. Buffer layers, in the [2-3 μm] thickness range and, for instance, formed of graded $Si_{1-w}Ge_w$ ($0 < w < x$) layers, have been used to get gradual relaxations of the lattice mismatches between pure silicon wafer and $Si_{1-x}Ge_x$ film lattice parameters. This should ensure the localization of the misfit dislocations inside these buffer layers, and the growth of single crystal $Si_{1-x}Ge_x$ fully relaxed templates with very few threading dislocations. This approach still suffers of the high level of defectivity contained in the buffer layer as well as the undesirable propagation of threading dislocations into the relaxed SiGe upper layer. Hereafter advantages of combining strained silicon epilayers and SOI structures are highlighted. They allow producing performing structures for the next microelectronic applications. Two technical approaches can be proposed thanks to the Smart Cut technique as schematically shown in figure 8[8].

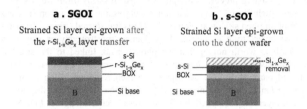

Figure 8. Two different structures for the realization of strained SOI.

7.1 The SGOI (Strained-Si on SiGe On Insulator) approach

A first process flow consists in transferring thin layers of the $Si_{1-x}Ge_x$ relaxed templates from the starting specific substrates onto final wafers: a SOI-like structure is achieved (figure 8a). Strained Si films are then epi-grown onto such transferred $Si_{1-x}Ge_x$ relaxed templates as illustrated in figure 9[9].

Figure 9. Strained Si films as observed by TEM cross sections when epi-grown onto transferred relaxed-$Si_{1-x}Ge_x$ films.

The strain intensity of the thin top silicon layer has been shown to be a function of the germanium percentage, inducing an enhancement of both the electron mobilities and also hole mobilities, and thus of the MOSFET performance, if Ge concentrations beyond 20% are used. SGOI substrates are well suited for partially depleted SOI type architectures.

7.2   The sSOI (Strained-Silicon-On-Insulator) approach

In an alternative process flow, thin strained Si (s-Si) films are first epi-grown onto such $Si_{1-x}Ge_x$ relaxed templates before the implantation step and a bilayer of strained silicon and relaxed SiGe is transferred (figure 8b). Thin silicon top layer strain is measured by Raman spectroscopy as shown in figure 10. It is worth noting that the phonon peak linked to the strained silicon is shifted by ~ $6cm^{-1}$, in a tensile mode (corresponding to ~1.5GPa)[10]. Transferred parts of $Si_{1-x}Ge_x$ templates are then removed, using chemical etch for instance, and strained Si films sit alone directly on the buried insulating layers as observed by TEM cross sections (figure 11). Such structures are referred as sSOI wafers.

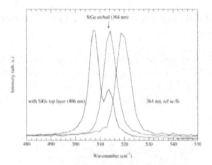

Figure 10. Raman spectra showing that the strain in the strained Si layer is maintained through the Smart Cut transfer ("with SiGe top layer" spectrum) and after SiGe overlayer removal ("SiGe etched" spectrum).

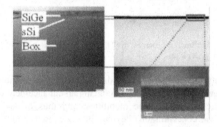

Figure 11. Strained Si films as observed by TEM cross sections before (left hand side) and after (right hand side) removal of relaxed-Si1-xGex films.

Raman characterization have been performed to check that the Si over layers in the sSOI structures preserve the same strain than they had before transfer. Phonon peaks linked to the strained Si layer are shifted from the relaxed silicon reference peak by the same wave number before and after the removal of the $Si_{1-x}Ge_x$ relaxed template (figure 10).

More, it has been shown that the stress is not affected by high temperature annealing. For instance, investigations have been carried on with sSOI structures, which have been annealed in the [700°C-1000°C] temperature range. Some of the Raman spectroscopy results after thermal treatment are plotted in figure 12. Whatever is the annealing temperature, phonon peaks linked to strained Si layers are all shifted from the relaxed silicon reference peak by the same wave numbers leading to conclude that the strain is preserved.

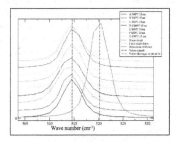

Figure 12. Raman spectra showing that the strain in the strained Si layer is maintained trough subsequent thermal treatment in the [700°C-1000°C] range.

Ultra-thin strained silicon layers are typically 20nm thick, which makes sSOI structures the perfect substrate solution for fully depleted SOI architectures.

8. GeOI (GERMANIUM-ON-INSULATOR)

Germanium on a silicon substrate is a very promising development as Ge offers a higher mobility than Si and should be better suited for the use of high-k gate oxides. In addition to microelectronics, GeOI is of interest for the manufacturing of solar cells (for instance when combined with a GaAs epitaxial step). Ge can be considered as a good template for GaAs and related compounds, because of its low lattice mismatch and its chemical stability under epitaxial conditions.

We present first results on formation of thin film GeOI 4" structures made by the Smart Cut technology. Compared to SOI manufacturing, the development of GeOI requires adaptation of processes to Ge material.

For the development of GeOI two routes can be considered. The first one concerns germanium bulk wafers. One of the main advantages is that such wafers are free of dislocations. But at the same time, they suffer of being brittle.

The second route proposes the use of germanium epilayers. This solution is attractive because it seems to be more suitable for large diameter wafers (300mm). But till now, it suffers of the quality of the epilayers which needs to be improved.

At present, encouraging results have been obtained for these two approaches[11]. For instance, Ge 100mm / Si 200 mm heterostructures (figure 13) have been realized by starting from a bulk Ge donor (from Umicore) and Ge200mm / Si 200 mm heterostructures (figure 14) have been realized by starting from Ge epilayered donor (from Léti-CEA). At this time, first results and structural characterizations reveal quite good GeOI qualities. It can be assumed that the route to the 200 mm GeOI is now opened.

Figure 13. Ge 100mm / Si 200 mm heterostructures realized by starting from a bulk Ge donor (from Umicore)

Figure 14. Ge 200mm / Si 200 mm hetero-structures realized by starting from Ge epilayered donor (from Léti-CEA)

9. OTHER APPLICATIONS

Moreover, SOI-like patterned layers are heterogeneous structures that may find a broad set of applications, ranging from double gate transistors to System-on-Chip architectures. They may combine different technologies, such as fully depleted SOI, partially depleted SOI and bulk silicon, depending on the design requirements of a given chip. It is currently at the research stage.

10. CONCLUSION

A high degree of understanding has been achieved on basic physical and chemical phenomena. Many applications of advanced SOI structures, based onto various stacked hetero-structures, have been demonstrated and some of them are now major industrial realities. As a layer transfer technology, Smart Cut has significantly enlarged the field of engineered composite substrates and opens new exciting application perspectives in the field of microelectronics, photonics, opto-electronics, high frequency and high power devices.

ACKNOWLEDGEMENTS

Acknowledgements are due to V. Paillard for Raman characterization. Work on strained Si has been partly supported by the French Ministry of Industry in the framework of the Micro-Nano-Technology program "Smart Strain".

REFERENCES

1. M. Bruel, Electron. Lett., 31 (14), 1201, (1995)
2. D. Lederer, C. Desrumeaux, François Brunier and J.-P. Raskin, 2003 IEEE Intern. SOI Conference (2003)
3. O. Rayssac, H. Moriceau, M. Olivier, I. Stoemenos, A. M. Cartier, B. Aspar, SOI Technology and Devices X, Electrochemical Society Proceedings, PV 01-03, 39, (2001)
4. C. Maleville, O. Rayssac, H. Moriceau, B. Biasse, L. Baroux, B. Aspar and M. Bruel, Semiconductor wafer bonding, Science Technology and Applications IV, Electrochemical Society Proceedings, PV 97-36, 46 (1998)
5. M. Nishida, M. Okuyama and Y. Hamakawa, Appl. Surf. Sc., 79/80, 409, (1994)
6. A. C. Adams, Solid State Technology, 26, 135, (1983)
7. C. Maleville, T. Barge, B. Ghyselen, A.J. Auberton, H. Moriceau, A.M. Cartier, 2000 IEEE Intern. SOI Conference, 134 (2000)

8. B. Ghyselen, Y. Bogumilowicz, C. Aulnette, A. Abbadie, B. Osternaud, P. Besson, N. Daval, F.Andrieu, I. Cayrefourq, H. Moriceau, T. Ernst, A. Tiberj, O. Rayssac, B. Blondeau, C. Mazure, C.Lagahe-Blanchard, S. Pocas, A.-M. Cartier, J.-M. Hartmann, P. Leduc, C. Di Nardo, J.-F. Lugand, F. Fournel, M.-N. Semeria, N. Kernevez, Y. Campidelli, O. Kermarrec, Y. Morand, M. Rivoire, D.Bensahel, V. Paillard, L. Vincent, A. Claverie, P. Boucaud. Submitted to 2004 MRS Spring meeting (2004)

9. B. Ghyselen, J.-M. Hartmann, T. Ernst, C. Aulnette, B. Osternaud, Y. Bogumilowicz, A. Abbadie, P.Besson, O. Rayssac, A. Tiberj, N. Daval, I. Cayrefourq, F. Fournel, H. Moriceau, C. Di Nardo, F.Andrieu, V. Paillard , M. Cabié, L. Vincent, E. Snoeck, F. Cristiano, A. Rocher, A. Ponchet, A. Claverie, P. Boucaud , M. -N. Semeria, D. Bensahel, N. Kernevez and C. Mazure. To be pulished in Solid State Electronics / Special issue on strained Si and heterostructures and Devices – August 2004

10. V. Paillard, B. Ghyselen, C. Aulnette, B. Osternaud, N. Daval, F. Fournel, H. Moriceau, T. Ernst, J.M. Hartmann, C. Lagahe-Blanchard, S. Pocas, P. Leduc, L. Vincent, F. Cristiano, Y. Campidelli, O. Kermarrec, P. Besson, Y. Morand, Microelectronic Engineering 72, 367–373, (2004)

11. F. Letertre, C. Deguet, C. Richtarch, B. Faure,J.M. Hartmann, F. Chieux, A. Beaumont, J. Dechamp, C. Morales, AM Cartier, F. Allibert, P. Perreau, S. Pocas, S. Personnic, C. Lagahe-Blanchard, B. Ghyselen, YM Le Vaillant, E. Jalaguier, N. Kernevez, C. Mazure. Submitted to 2004 MRS Spring meeting (2004).

POROUS SILICON BASED SOI: HISTORY AND PROSPECTS

V. Bondarenko[1], G. Troyanova[1], M. Balucani[2], and A. Ferrari[2]

[1]*Microelectronics Department, Laboratory of Porous Silicon, Belarussian State University of Informatics & Radioelectronics, P. Brovka Street 6, Minsk 220027, Belarus;* [2]*INFM Unit 6, Rome University "La Sapienza", Via Eudossiana 18, Rome 00184, Italy*

Abstract: This paper is a review of application of porous silicon to SOI technology. Three main approaches for fabricating the SOI structures based on PS are critically reviewed. We show that there exists renewed interest in all these approaches although SOI technology based on layer transfer with PS as the splitting layer is the dominated method. Patent analysis is presented to reveal new potentialities of porous silicon-based SOI technologies.

Key words: Porous Silicon (PS); Oxidised Porous Silicon (OPS); Silicon-on-Insulator (SOI).

1.   INTRODUCTION

Porous silicon (PS) is not a new material for microelectronics applications. It was discovered by Artur Uhlir from Bell Lab in 1956[1] and was first turned to practical use for dielectric isolation in 1969 by NTT[2] and Sony[3]. The extremely high chemical reactivity of PS was proposed as the means[2,3] to form thick oxidized PS with dielectric properties equivalent to those of conventional thermal silica.

Among the physical-chemical properties of PS, the following properties are of principal interest for SOI technology:

- fast oxidation;
- possibility of epitaxial growth;
- etching selectivity;
- possibility of stress relaxation;
- mechanical weakness.

D. Flandre et al. (eds.), Science and Technology of Semiconductor-On-Insulator Structures and Devices Operating in a Harsh Environment, 53-64.

The first idea to create SOI based on PS was patented by Watanabe and Sakai from NTT in 1969[2]. The idea was founded on the combination of porous anodization, deposition of a silicon layer, oxidation of PS, and accurate polishing. From 1969 to 1982 a variety of techniques for the formation of SOI based on PS were investigated. Even the idea of converting the whole of a silicon wafer into oxidised PS was proposed[4]. These ideas were attractive but difficult, if not impossible, to realize and they had no practical applications.

The situation radically changed in 1982 once Kazio Imai from NTT invented the method called FIPOS (Full Isolation by Porous Oxidised Silicon)[5]. Note that most investigators referred to all methods of full dielectric isolation by oxidised PS by this name. It is the FIPOS method that stimulated intensive research on application of PS to SOI. Many scientific research centers (Bell Lab., Sandia National Lab., CNET, RSRE) as well as famous microelectronics companies (NTT, IBM, Texas Instruments, etc) were involved in this research developing different methods for the realization of SOI structures based on PS[6].

All known SOI fabrication methods exploiting PS can be classified into three main groups[7]: a) selective lateral anodisation underneath a silicon layer followed by oxidation of the PS to form isolated islands; b) epitaxial growth of silicon on the surface of PS followed by oxidation of the PS layer under a silicon epitaxial layer and c) epitaxial growth of silicon on PS followed by silicon layer transfer from the PS to another silicon substrate.

2.   PS SOI FABRICATION METHODS

2.1   Selective anodization

The first group is based on the selectivity of the electrochemical reaction. This approach contains the following steps shown schematically in Fig. 1: a) formation of fast anodized regions inside the silicon substrate; b) formation of mesas structures; c) selective anodization of fast anodized regions; d) oxidation of the PS and e) filling the openings with a suitable dielectric material.

To provide the selective anodization several approaches can be used:
- Formation of diffused (ion-implanted) regions in the substrate or epitaxial structure[8,9], including also high-energy ion implantation to form buried layers[10];
- Local proton implantation[11];
- Selective epitaxy[12];
- Deposition of a SiGe strain layer[13].

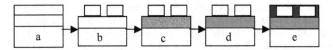

Figure 1. Schematic of a SOI technology based on selective anodization.

FIPOS[5,14] is the method based on selective anodization. The original FIPOS process relies on the fact that p-type silicon can be selectively converted into PS by anodization in HF solution. The convertion rate of n-type silicon is much lower. FIPOS process proceeds as follows: at first, a silicon nitride film is deposited on the top of initial p-type silicon wafer and is patterned by photolithography, then boron is implanted to control the porosity of the PS surface layer. The n-type regions are formed by conversion of the p-type silicon into n-type silicon by proton implantation. The p-type silicon is then converted into PS by anodization. Optimal regime yields a 56% porosity. In this way, the volume of the oxidised PS is equal to that of the porous silicon and stress in the layers can be minimized. FIPOS method was used at NTT to fabricate complete SOI circuits with very good characteristics gaining a 30% speed increase compared to bulk technology due to reduced parasitic capacitance[15]. However, the FIPOS was not followed up by further development because it has several limitations. Only small isolated silicon islands were practicable with FIPOS because the thickness of the PS layer increases when the dimensions of islands increase. The necessity of thick PS layers stems from the fact that PS formation front spreads both laterally and vertically from the exposed surface. Thus, the thickness of the PS layer must be equal to half the width of the island to form complete isolation of the island. When oxidised, such thick PS layers induce wafer warpage. Moreover, the PS layer is formed of non-uniform thickness. Indeed, because formation of the PS layer begins from the exposed surface between n-type islands, it has to be thicker between the islands in comparison with regions thereunder.

Different approaches for solving the problems of FIPOS were proposed. Nesbit showed[16] that the thickness of oxidised PS may be reduced by formation of a buried p+-layer located between a p-type substrate and n-type islands. This allows wide islands with a thinner layer of oxidised PS, because of enhanced anodization of heavily doped silicon, to be isolated. Electron mobilities of about 500 cm^2/V·s have been reported on MOS transistors fabricated with this method. However, the thickness uniformity problem has persisted[6].

The most promising approaches were those based on anodization of n/n+/n structures, where porous silicon is selectively formed in a thin heavily doped buried n+-type layer. The n/n+/n method was extensively studied in several laboratories because this method is free from the disadvantages peculiar to FIPOS. In this method, the thickness of the PS layer is uniform and easily controlled by the dopant concentration in the n+-layer[17]. The method is simple and cheap. Excellent MOS transistors characteristics have been reported[7,17] with this method by CNET, RSRE, and Texas Instruments.

SOI technology based on anodization of n/n+/n structures has been very well established by Integral Co.(Belarus). Digital radiation hardened CMOS IC's were manufactured in a pilot line during few years. Low cost and simplicity of n/n+/n SOI technology[18-20] were the driving Integral's considerations. However, the disintegration of the USSR interfered with Integral's plan and in 1992 the investments in SOI technology were stopped.

In 1995-1997 our group (Belarussian State University of Informatics & Radioelectronics and INFM Unit 6 of Rome University "La Sapienza") continued the advancement of SOI based on selective anodization of n/n+/n structures. Due to modification of the process, some improvement of the SOI parameters was gained. The thickness of the silicon layer was reduced from 0.3 μm to 0.1 μm, the range of the BOX thickness was expanded from 1.5-1.8 μm to 0.8-3.0 μm. Warpage of the wafers was reduced from 20 μm to 5 μm. The use of low temperature epitaxy and optimization of technological regimes made possible above-mentioned improvements[21-23].

All SOI fabrication methods based on selective anodization suffer from the disadvantage that they provide SOI structures of a predermined configuration dictated by a mask. That is why finished SOI structures are dedicated to fixed die designs and cannot be commercialised by a vendor. Nevertheless, renewed interest to SOI based on selective anodization has been recently reported. Some examples are presented in Section 3.

2.2   Epitaxy of silicon on PS

The second group includes SOI technologies which are based on direct silicon epitaxy on a PS layer previously formed on the surface of a silicon substrate [24, 25]. SOI process flow based on direct epitaxy of silicon on PS is shown schematically in Fig. 2.

The process consists of the following main steps: a) porous anodization; b) silicon epitaxy; c) pattern the epilayer by photolithography and plasma etching; d) oxidation of the PS and e) fill the openings with suitable dielectric material.

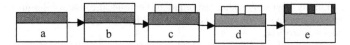

Figure 2. Schematic of a SOI technology based on direct silicon epitaxy on PS.

The epitaxial process must be at a reasonably low temperature to prevent the PS sintering and to retain the high reactivity of the PS to oxidation. Silicon epitaxy can be provided by MBE, PECVD and LPCVD techniques. Good electrical characteristics of thin-film MOS transistors fabricated in SOI based on molecular beam epitaxy of silicon on PS have been demonstrated[26]. However, the technique of MBE is still unique, complex, and high cost. SOI structures of satisfactory crystalline quality were reported with PECVD and LPCVD. Despite the presence of microtwins and dislocations in the silicon epitaxial layer, reasonably good characteristics of MOS transistors were shown. However, more complex electronic devices have not been demonstrated yet. Very interesting results were presented by Romanov and co-workers from ISP RAS[27]. They developed a method based on direct epitaxy of silicon on the surface of the PS layer, which is specially oxidised. The method allows the formation of the continuous SOI, in contrast to the methods of the first group (above).

New applications of direct epitaxy of silicon on PS for the creation of SOI structures are presented in Section 3.

2.3 SOI based on Bonding and Layer Transfer technology

In this group, PS layer is used as a sacrificial layer that is responsible for wafer splitting. In SOI structures of this group, the buried oxide (BOX) consists of a thermal oxide and is not oxidised PS. All SOI methods in this group are based on two principal properties of PS:

- PS remains crystalline even at the highest porosity ($p \sim 90\ \%$);
- mechanical properties of PS depend strongly on the porosity and in general are much worse than those of crystalline silicon. As an example, the Vickers hardness of 70-80% porosity PS is about 4 GPa which is 3 times less than that of monocrystalline silicon[28].

At first PS layer was proposed for application in bonding technology[29-31]. This process involves the formation of a PS layer on the top of a silicon wafer, epitaxy of silicon on the PS, bonding of this wafer to a second wafer with a previously formed thermal oxide layer, mechanical polishing of initial silicon wafer from the back side followed by chemical etching down to the PS. Due to the high chemical activity of PS, the etching process can be

controllably stopped providing uniformity of the residual epitaxial layer in
this SOI structure. In this process, both the PS layer and the initial silicon
wafer are sacrificial.

More recently modified Layer Transfer methods where splitting of the
bonded wafer along the PS sacrificial layer have been developed. These
methods avoid consuming the initial silicon wafer, which may be reused. A
basic Layer Transfer process with a sacrificial porous layer for fabricating
the SOI structures is shown schematically in Fig. 3. The process consists of
the following steps:

a) Formation of multi-layer PS (low-porous surface and high porous
interface with substrate) - by changing the current density[32] or by
changing the dopant concentration[33] or by ion (hydrogen, nitrogen or
noble gas) implantation into uniform porous layer[34]; stabilization of PS
structure by low temperature ($\sim$ 400 °C) oxidation in dry O_2[32];
modification of PS surface by wet treatment in HF followed by high
temperature (1100°C) annealing in H_2 and SiH_xCl_x pre-injection (in epi
chamber)[32];

b) Epitaxial Growth of silicon on PS sintered surface by Ion-assisted
Deposition (700°C)[35] or by CVD (SiH_4, 900-1000°C [36] or SiH_2Cl_2,
1100°C[32]) or by Liquid Phase Epitaxy[37];

c) Bonding to a supported substrate; splitting along the PS by mechanical
pulling force[35,36] by centrifugal force[38] or by water jet[32] or by
ultrasound [35, 36] or by cooling[38] or by freeze-thaw technique[39];

d) Removal of remaining PS by wet etching or H_2 annealing [32].

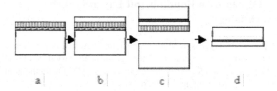

a   b   c   d

Figure 3. Schematic of a SOI technology based on layer transfer

Main developers of this technology are Canon Co; Sony Corporation,
Mitsubishi Denki KK (Japan); Institut fur Physikalische Elektronik (IPE) of
University Stuttgart; Bavarian Center of Applied Energy Research (ZAE
Bayern), Robert Bosch GmbH (Germany); IBM Corporation, Silicon
Genesis Corporation (USA); Samsung Electronics Co., (South Korea).

ELTRAN (Epitaxial Layer Tranfer) is a representative of the Layer
Transfer methods exploiting PS as a sacrificial layer. Basis of the ELTRAN
SOI technology developed by Takao Yonehara from CANON in 1990.

Among the SOI methods based on PS, ELTRAN is the first to reach the maturity required for VLSI CMOS applications. Many years of basic research have been necessary to reach this level. There were a lot of problems to be resolved prior to realizing high quality SOI structures by the ELTRAN process. However, the enthusiasm of inventors, advances in equipment and all contributed to improvements in the material quality and manufacturability. In 2000, Canon Co. announced that ELTRAN SOI was at an industrial level with production of about 100 000 wafers per year. In 2002, Takao Yonehara was awarded a Special Price at the International Conference PSST-2002[40] for the development of an industrial SOI technology based on PS. The availability of a wide range of silicon and buried oxide thicknesses gives this technology a large degree of flexibility in device design. To study the ELTRAN, we refer readers to the original review of Yonehara and Sakaguchi[32].

Patent analysis presented in Section 3 shows that the technological features of the ELTRAN process are actively patented by Canon and Sony.

3.   PATENT ANALYSIS AND PROSPECTS

Fig. 4 shows the number of US patents and patent applications published per year during the period from 1976 up to the present. Up to 1993, two to three patents were published per year. Starting in 1999, there has been a considerable increase in the number of patents dedicated to applications of PS for SOI. Currently almost 50 patents and patent applications are published per year. From January to June 2004, 10 Patents and 21 Patent applications were published.

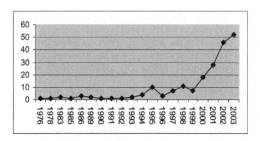

Figure 4. The number of US patents devoted to SOI based on PS and published per year.

Total number of US patents and patent applications devoted to SOI based on PS is about 250 and may reach 300 by the end of 2004. New developments include integration of PS based SOI, MEMS, and photonics based upon of standard silicon technology[7].

Canon Co. is the main assignee of SOI based on PS (more than 140 patents). Other assignees are Sony, Texas Instruments, IBM, Samsung Electronics, Kulite Semiconductor Products, Lucent Technologies, Robert Bosch, STMicroelectronics. The overwhelming majority of patents and applications (more than 170) are related to Bonding and Layer Transfer technology. Nevertheless, there is a definite renewed interest in SOI based on selective anodization and direct epitaxy of silicon on PS.

It is well known, that SOI structures present several advantages over bulk silicon. One of the principal disadvantage of standard SOI structures with continuous BOX layer is due to electrical isolation of the active region from the silicon substrate resulting in floating body effects. These phenomena occur when excess carriers are accumulated in the floated body during device operation. To solve these problems, a quasi-SOI structure has been proposed, in which a partially forming contact hole under the active region is formed for extracting excess carriers.

Quasi-SOI structures based on selective anodization may be formed by stopping the lateral anodization process before the lateral PS regions are connected under the silicon islands as proposed in IBM patent[41]. This process is shown in Fig. 5.

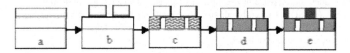

Figure 5. Schematic of a quasi-SOI technology based on selective anodization.

The process consists of the following steps: a) formation of local fast anodized regions inside Si substrate; b) formation of mesas; c) lateral partial anodization of fast anodized regions; d) oxidation of PS and e) fill the openings with a dielectric material.

Samsung Electronics developed another method[42] for producing the Quasi-SOI. The method is based on direct silicon epitaxy on PS. Main steps of this process are shown in Fig. 6: a) porous anodization; b) high temperature sintering of PS surface; c) silicon epitaxy, d) uncomplete etching of PS; e) filling the openings by dielectric material.

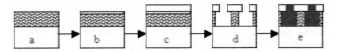

Figure 6. Schematic of a quasi-SOI technology based on direct epitaxy on PS.

Recently, it was proposed[43,44] to replace PS or OPS regions in SOI structures with another dielectric material. The main steps of this method are formation of local slow or fast anodized regions inside a Si substrate; formation of mesas; anodization of the fast anodized regions; oxidation of the PS (optional); removal of the PS or OPS local regions; oxidation and fill air gaps under the Si islands which form due to the volume increase.

In 1995-1998, a new process[45] based on vertical anodization and PS oxidation was patented by Kulite Semicon Products, Inc. and Lucent Technologies Inc. This process contains the following main steps as shown in Fig. 7: a) anodization through a mask; b) oxidation of the PS and c) densification of the OPS by heating in an oxygen ambient.

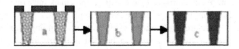

Figure 7. Schematic of the technological flow to form Si structure with full dielectric isolation.

This new approach allows fabrication of fully isolated silicon regions made of original monocrystalline silicon and separated by deep oxidised PS regions. It is more correct to name these structures Silicon-in-Insulator instead of SOI. The main application fields for these structure are multi-level integrated chips, high voltage devices, and mixed-signal devices with full isolation between digital and analog elements.

Up to now, there have been no publications on the design of circuits in direct silicon epitaxy on PS for industrial SOI. Nevertheless, recently a new interest in this approach has appeared. Possible reasons for the renewed interest in "Epi on PS" technology are:

- improvement of epitaxial technology in the context of ELTRAN;
- possibility of easy formation of contact to under-gate region of MOSFET;
- shortening the process flow in comparison with ELTRAN or other Layer Transfer approaches.

Texas Instruments patented a new method to create SOI structures with continuos nonocrystalline silicon layer on BOX[46]. This methods includes:

a) porous anodization; b) partial oxidation of PS and high temperature sintering of the PS surface; c) epitaxy and d) capping and annealing (~ 6 hours) for the PS transform into single crystal Si and SiO_2 layers. High temperature annealing at 1325 °C causes the porous layer to collapse forming a mixture of silicon and SiO_2 which are not soluble in each other. Thus, annealing results in the division of partial oxidized PS into separate layers. One more similar solution is to implant oxygen ions into PS before epitaxy [47].

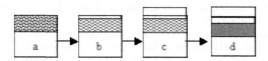

Figure 8. Schematic of the new SOI technology based on direct epitaxy on PS.

This method has the advantage of providing both high quality silicon for the active regions and a high quality uniform thickness of oxide for insulation. It minimizes warpage of the wafer, consumes only one wafer, is low cost and easily ramped into volume production.

In addition to standard applications of PS in SOI technology (as insulator in SOI structures; an etch-stop layer in bonded SOI structures and as the splitting layer in Layer Transfer technology) new application fields have recently been developed:

- PS for thinning the silicon layer in SOI structures[48] ;
- PS as stress relaxing layer in supporting substrate[49,50].

4.   SUMMARY

In this paper, SOI technology based on PS is reviewed. At the time of writing this paper, the history of applications of PS in SOI technology provides many interesting examples of major technological achivements. Many of these methods facilitate the fabrication of electronic devices with unique characteristics. Nevertheless, only ELTRAN SOI, based on a layer transfer process with PS as the sacrificial layert is incorporated in industrial scale circuit manufacture. To date there is a renewed interest to SOI based on PS due to new possibilites provided by PS for integration of different SOI based electronic and optoelectronic devices, MEMS, bio-sensors, microfuel cells in the framework of standard silicon technology.

ACKNOWLEDGEMENTS

The authors thank L. Dolgyi for fruitful discussion and the Ministry of Industry of Belarus for the financial support.

This work is a part of the research programs of Belarussian State University of Informatics and Radioelectronics and Rome University "La Sapienza".

REFERENCES

1. A. Uhlir, Electrolytic shaping of germanium and silicon, *Bell Syst. Tech. J.* **35** (9), 333-337 (1956).
2. US Patent 3,640,806.
3. Jap. Patent 49-19019.
4. US Patent 3,954,523.
5. US Patent 4,393,577.
6. S. Tsao, Porous silicon techniques for SOI structures, *IEEE Circuit & Device Magazine* **3**, 3-7 (1987).
7. V. Bondarenko et al. in: *Progress in SOI Structures and Devices Operating at Extreme Conditions*, edited by F. Balestra et al. (Kluwer NATO Science Series, 2000), pp.309-327.
8. US Patents 3,954,523; 4,532,700; 4,627,883; 4,628,591; 4,104,090; 4,628,591; 4,810,667; 5,597,738; 5,686,342; 5,767,561; 5,773,353; 6,506,658.
9. US Patent Application 20040048437.
10. US Patent 6,331,456.
11. US Patents 4,393,577; 4,532,700.
12. US Patents 5,110,755; 4,910,165.
13. US Patents 4,849,370; 4,982,263.
14. K. Imai et al, Crystalline quality of silicon layer formed by FIPOS technology, *J. Crystal Growth* **63**, 547-553 (1983).
15. K. Imai and H. Unno, FIPOS (Full Isolation by Porous Oxidized Silicon): Technology and Its application to LSI's, *IEEE Trans. Electron Dev.* **ED-31**(3), 297-303 (1984).
16. L. Nesbit, Advances in oxidized porous silicon for SOI, *Technical Digest, IEEE International Electron Devices Meeting*, **Cat. No 84CH2099-0**, 800-803 (1984).
17. K. Barla et al, SOI technology using buried layers of oxidized porous silicon, *IEEE Circuits & Device Magazine* **3**, 11-14 (1987).
18. V. Bondarenko et al, SOI structures based on oxidized porous silicon, *Rus. Microelectron.* **23**, 61-68 (1994).
19. V. Bondarenko et al. in: *Physical and Technical Problems of SOI structures and Devices*, edited by J.P.Colinge et al. (Kluwer NATO Science Series, 1994), pp.275-280.
20. V. Bondarenko et al, Total gamma dose characteristics of CMOS devices in SOI structures based on oxidized porous silicon, *IEEE Transactions on Nuclear Science* **44**(5), 1719-1723 (1997).
21. International Patent PCT/IT00/00329.
22. International Patent PCT/IT00/00330.
23. International Patent PCT/IT00/00331.
24. T. Unagami and M. Seki, Structure of porous silicon and heat-treatment effect, *J. Electrochem. Soc.* **125**(8), 1339-1344 (1978).

25. V. Labunov et al, Process of formation of porous silicon and autoepitaxy on its surface, *Rus. Microelektron.* **12**(1), 11-16 (1983).
26. H. Takai and T. Itoh, Porous silicon layers and its oxide for the SOI structure, *J. Appl. Phys.* **60** (1), 222-225 (1986).
27. S. Romanov et al. in: *Perspectives, Science and Technologies for Novel Silicon on Insulator Devices*, edited by P. Hemment et al. (Kluwer NATO Science Series, 2000), pp. 29-46.
28. S. Duttagupta and P. Fauchet. in: *Properties of Porous Silicon*, edited by L. Canham (EMIS datareviews series No18, INSPEC, 1997), pp.132-137.
29. US Patents 5,250,460; 5,277,748.
30. N. Sato et al. Epitaxial growth on porous Si for a newbond and etch back silicon-on-insulator, *J. Electrochem. Soc.* **142**, 3116-3122 (1995).
31. K. Sakaguchi et al. Extremely High selective Etching of Porous Si for Single Etch-Stop Bond-and-Etch-Back Silicon-on-Insulator, *Jpn. J. Appl. Phys.* **34**, 842-847 (1995)
32. T. Yonehara and K.Sakaguchi. in: *Progress in SOI Structures and Devices Operating at Extreme Conditions*, edited by F. Balestra et al. (Kluwer NATO Science Series, 2000), pp.39-86.
33. US Patent Application 20020153595.
34. US Patent Application 20030008477.
35. R. Brendel, A novel process for ultrathin monocrystalline silicon solar cells on glass, *Proc. 14th European Photovoltaic Solar Energy Conf.*, 1354-1358 (1997).
36. http://solar.anu.edu.au/pages/pdfs/review.pdf.
37. T. Yonehara, Eltran SOI-Epi and SCLIPS by epitaxial layer transfer from porous Si, *Extended abstracts Second International Conf. Porous Semiconductors-Science and Technology*, 14 (2000).
38. US Patent Application 20020000242.
39. US Patent Application 20020096717.
40. http://www.upv.es/psst_2002.
41. US Patent: 6,429,091.
42. US Patent 6, 448,115; 6, 657,258.
43. US Patents 6,277,703; 6,469,350.
44. US Patent Application 20030080383.
45. US Patents 5,455,445; 5,461,001; 5,789,793; 5,767,561.
46. US Patent 6,376,285; 6,376,859.
47. US Patent Application 20020086463.
48. US Patents 5,556,503 and 5,650,042.
49. US Patents 6,271,101; 6,602,761; 6,037,634.
50. US Patent Application 20040023448.

ACHIEVEMENT OF SiGe-ON-INSULATOR TECHNOLOGY

Yukari Ishikawa[1], N. Shibata[1] and S. Fukatsu[2]
[1]Japan Fine Ceramics Center, 2-4-1, Mutsuno, Atsuta, Nagoya, Japan; [2]The University of Tokyo, 3-8-1, Komaba, Meguro, Tokyo,Japan

Abstract: Fabrication technologies of SiGe-OI substrate, using SIMOX, bonding and growth of SiGe on top of Si-OI, are described. Attention was paid to the mechanism how SiGe-OI develops, and how the Ge concentration ceiling can be lifted. It is also indicated the achieved level of lattice relaxation and dislocation density of the SiGe layer in SiGe-OI. Remaining issues, which would be solved in order to develop devices based on SiGe-OI, are pointed out.

Key words: SiGe-on-insulator, SIMOX, Bonding, Etch back, Smart-cut®, ITOX, oxidation, condensation, strain, dislocation.

1.   INTRODUCTION

Silicon-on-insulator (SOI) technology has been widely accepted as being of unparalleled importance not only for electrical isolation but also in the context of single electron and optoelectronics devices. Recently, SiGe-on-insulator (SiGe-OI) has become the focus of attention since it offers more advantages than conventional SOI substrates, such as low energy consumption, radiation hardness, low parasitic capacitance, and reducing of short channel effects.[1] In particular, the predicted high electron mobility in a tensile strained Si channel lattice-matched to strain-relaxed SiGe and SiGe-OI holds greater promise in the field of electronics.[2] One example of a novel electronic device that uses SiGe-OI is shown in Fig.1. As such, controlling the concentration of Ge in the top SiGe layer and the flatness of the SiGe layer surface are a primary concern, as electron mobility is strongly dependent on the Ge concentration in the top SiGe layer and also on interface morphology of the strained Si channel.[3]

D. Flandre et al. (eds.), Science and Technology of Semiconductor-On-Insulator Structures and Devices Operating in a Harsh Environment, 65-75.
© 2005 *Kluwer Academic Publishers. Printed in the Netherlands.*

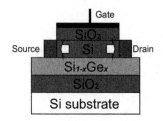

Figure 1. Cross-section of strained-Si/SiGe-OI MOSFET

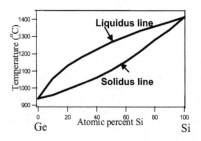

Figure 2. Phase diagram of the Si-Ge binary alloy system[15]

The SiGe-OI substrates have been realized by a variety of methods: SIMOX,[3-6] bonding,[7-9] and deposition of SiGe or Ge on SOI.[10-14] However, lowering of the melting point (Fig.2)[15] of SiGe-OI as compared to SOI raises such serious problems as Ge concentration ceiling and morphological degradation when applying standard Si-OI technology to the fabrication of SiGe-OI.[5,16]

In this article, fabrication technologies of SiGe-OI substrates will be reviewed from the standpoint of device applications. Focus will be placed upon the mechanism enabling SiGe-OI structures to be formed, and how the Ge concentration ceiling can be lifted. Also other issues will be addressed.

2.   SIMOX

SIMOX technology was the first to produce useful SiGe-OI material.[4,5] The mechanism of the formation of the buried oxide layer during SiGe-OI is much the same as that for conventional SIMOX. Oxygen implantation into the starting substrate, which was SiGe grown on top of Si substrate, and high temperature annealing were performed as shown in Fig.3 (a-c). Oxygen concentration depth profiles of the as oxygen implanted SiGe substrate were almost showed close agreement with profiles calculated by the SRIM2000 code.[17] The range of oxygen ions in SiGe became a little shallower than in Si consistent with the Ge content of the substrate. Small precipitates were distributed around the oxygen ion range in the as implanted SiGe substrate (Fig.4).[18] A continuous buried oxide layer was formed by growth and/or coalescence of precipitates during high temperature annealing.[18] In this process, Si was preferentially oxidized and Ge was swept out from the peak of the oxygen profile. Thus, Ge around the oxygen ion range diffused both into the top SiGe layer and the substrate. It was confirmed that the buried

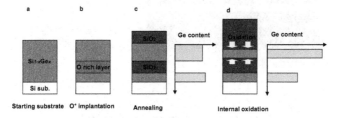

Figure 3. SIMOX process combined with ITOX

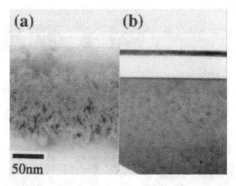

Figure 4. Cross-section of SiGe substrate: a) as O⁺ implanted and b) after high temperature annealing[18]

oxide was SiO_2, with the concentration of Ge in the buried oxide being less than the sensitivity of EDS, XPS and AES. Interfaces of SiGe/buried oxide and SiGe/surface oxide were atomically flat as shown in Fig.5.[4] The existence of an optimum dose window (Fig.6)[4] for the formation of a nearly perfect SOI structure was demonstrated and found to be energy 25keV and dose $2\text{-}2.5\times10^{17}cm^{-2}$. In contrast, doses below and above the dose window resulted in SiO_2 precipitates running parallel to the substrate interface and a nonuniform top SiGe layer along with parasitic islands in the buried oxide, respectively (Fig.6).

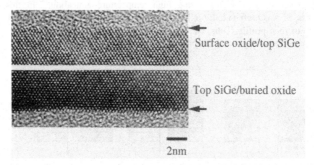

Figure 5. Lattice images of the interfaces of SiO_2/SiGe[4]

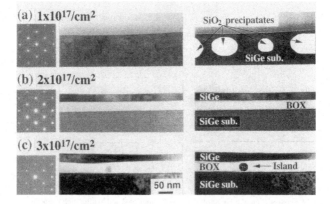

Figure 6. Variation of SIMOX structure with increasing oxygen dose[4]

The presence of Ge in the substrate sets the SIMOX window. The morphological degradation during oxygen implantation into SiGe with increasing Ge content by the desorption of Ge and O in the form of volatile GeO during oxygen implantation was observed. For example the large amount of oxygen lost from the oxygen-implanted region of $Si_{0.7}Ge_{0.3}$ resulted in a disrupted or thinner buried oxide layer. The desorption of GeO was more pronounced for $Si_{0.49}Ge_{0.51}$. Thus, $Si_{1-x}Ge_x$ (x<0.3) substrates should be used for oxygen implantation. The melting point of SiGe is a major factor, which lowers the ceiling of the SIMOX window. Annealing at higher temperature than the melting point of SiGe results in destroying the SiGe layer, however the temperature for post-implant annealing should be set higher than 1250°C for the 25keV-SIMOX in order to form a continuous buried oxide layer. It has been found that ~20% Ge concentration is the ceiling obtained for 25kV-SiGe-SIMOX, as the melting point of $Si_{0.8}Ge_{0.2}$[15] set the lower limit temperature for the post implant annealing (Fig2).

Interestingly, the internal oxidation (ITOX) process was found to be effective in increasing the otherwise bounded Ge concentration in the SiGe-SIMOX.[3] ITOX process is performed at 1050°C under oxygen gas flow after SiGe-SIMOX as illustrated in Fig.3 (d). As pointed out above, Si was preferentially oxidized and Ge was swept out from the SiO_2 to the interface of SiO_2/SiGe. The rejected Ge diffuses in the SiGe layer, however very few Ge can diffuse into SiO_2 layer. Thus, the Ge content in the SiGe layer formed by the ITOX process was controlled by SiO_2 layers, which were formed on both sides of SiGe layer, to act as diffusion barrier layers for Ge, as shown in Fig.3. Ge content of SiGe layer after ITOX is estimated by simple calculation as follows,

$$C_{ITOX} = (t_{SIMOX} \times C_{SIMOX})/t_{ITOX} \tag{1}$$

where C_{SIMOX}, t_{SIMOX} and t_{ITOX} are the concentration of Ge in SiGe layer of the starting substrate, thicknesses of SiGe layer of the starting substrate and after ITOX, respectively.

Creation of a fully relaxed 57% SiGe-OI substrate was demonstrated using this technique.[18]

3. BONDING

SiGe-OI was also demonstrated using bonding method (bonding and etch back or smart-cut®) by optimizing conditions of bonding, H-implantation and etching.[7-9]

3.1   Bonding and etch back

The bonding and etch back process[8,9] is shown in Fig.7. Firstly, strain-relaxed SiGe was grown on a compositionally graded SiGe buffer layer on Si substrate and an SiO_2 layer was grown on a handle wafer by dry oxidation. The SiGe was polished by CMP in order to eliminate cross-hatch roughness. Epitaxialgrowth of SiGe (the bonding layer) upon a strained Si epitaxial layer was carried out on the flat surface of the SiGe substrate. The deposited wafer was bonded to the handle Si wafer following an HF dip after a piranha clean (1:3=H_2O_2:H_2SO_4). The bonded wafers were annealed at 800°C in N_2 for 2 h to strengthen the bonds. Si etching by a KOH solution followed by grinding of backside Si was performed on the bonded wafers. The etching stopped at the SiGe buffer layer, when the Ge content of the buffer layer is in excess of 20%. Finally, the remaining SiGe was etched by a solution of HF:H_2O_2:CH_3COOH (1:2:3), which preferentially etches SiGe. Therefore, a SiGe-OI substrate capped with by strained Si was obtained.

Control the etching is a key component for the creation of SiGe-OI substrates using this method. Especially, relaxed $Si_{0.8}Ge_{0.2}$, which acts as an etch-stopper, avoids the necessity of an undesirable P^{++} etch-stop layer. The threading dislocation density was strongly dependent on the quality of the virtual SiGe layer deposited on a SiGe graded buffer and was typically 10^5-10^6 cm^{-2}.[8] Unfortunately, it was not possible to determine the limit of the Ge content in the SiGe layer, however it is expected that annealing at a low temperature will increase to push up the limit of Ge content in SiGe-OI.

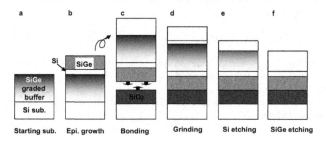

Figure 7. Bonding and etch back process

3.2   Bonding and smart-cut®

The process of bonding and layer transfer (smart-cut®) method is as follows:[7,9] 1) an SiO_2 layer was thermally grown on a Si handle wafer and a strain-relaxed SiGe layer was grown on Si substrate, 2) relaxed SiGe grown on Si substrate was implanted with H ions, 3) the H-implanted substrate and the handle wafer were polished by a chemical-mechanical polishing (CMP) process and cleaned, 4) the H-implanted substrate was bonded to the handle wafer at room temperature, 5) thermal treatment at 500-600 °C was performed to induce layer splitting in the SiGe layer of the bonded wafer followed by annealing of the SiGe-OI substrate at 800-900 °C to strengthen the bonded interface, as shown in Fig.8. The essential point in this process is to determine the optimal condition for the H-induced surface blistering. The optimum implantation and annealing conditions are $120keV-H_2^+$ implantation to the dose of $3.5\times10^{16}cm^{-2}$ followed by annealing at 500-600°C for 1hr[7] and $100keV-H^+$ implantation to the dose of $5\times10^{16}cm^{-2}$ followed by annealing at 600°C for 3hr.[9] Strain-relaxed SiGe layers, were directly deposited on Si with a uniform Ge content and also on a compositionally graded SiGe buffer layer on Si. However, setting the H ion ranges shallower than the interface of SiGe/Si and SiGe/SiGe buffer, respectively, is important to reduce the threading dislocation density in the transferred SiGe layer. Surface roughness of the as-split SiGe layer was the same as the Si layer, while CMP process improved it down to 0.5nm rms.[7] The upper limit of the Ge content was also identified for this method. The limit in the top SiGe layer was 25%, due to a tendency of islanding associated with the SiGe layers.[7] Presumably, the rather high Ge content causes agglomeration or precipitation of Ge during annealing.

To the authors' knowledge, there are no articles, which describe lattice relaxation of the SiGe layer in SiGe-OI by bonding and smart-cut[®].

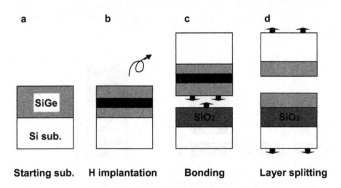

Figure 8. Bonding and smart-cut[®] process

However, it is shown that the Si layer, which is deposited on SiGe-OI, is strained and acts as a high quality Si channel as determined from measurements of the electron mobility and density.

4.   GROWTH OF SiGe ON TOP OF SOI

Epitaxial growth of SiGe on top of a SOI wafer was found to yield SiGe-OI structures. In the early phase of this process, a Si layer was found to remain in between the top SiGe and buried oxide layer.[10] Recently, a complete SiGe-OI substrate has been obtained by combining an annealing process after deposition of SiGe or Ge on SOI.[11-14] The details of this method are shown in Fig.9. Firstly, a SiGe or Ge layer is deposited on the SOI substrate. Next, a SiO_2 cap layer was formed on the top of the substrate. Finally, high temperature annealing of this sample is performed. There are several methods to form the SiO_2 layer, however the SiO_2 layer, which caps the deposited layer, is important. Degradation of the SiGe layer by Ge desorption is less pronounced, since the SiO_2 cap suppresses out-diffusion of Ge. In addition, the suppression of out-diffusion of Ge by the SiO_2 layer enables intermixing of Ge and Si between the cap and buried oxide layer during annealing. Thus, high temperature annealing produces a homogeneous top SiGe layer (Fig.9). Ge condensation in the SiGe layer by high temperature annealing under oxygen gas flow, as described in SIMOX section, is also applicable.[11] Moreover, this method also facilitates the formation of a Ge-OI substrate.[14] Atomically flat interfaces at the cap SiO_2/ Ge and Ge/buried SiO_2 (Fig.10)[14] and low surface roughness (Fig.11),[14] with an rms value of 0.4nm in a 10-μm square, indicate that this method has an ability to produce high quality SiGe and Ge -OI substrates.

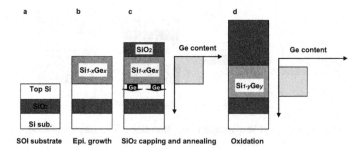

Figure 9. Epitaxial growth of SiGe (Ge) on SOI followed by annealing

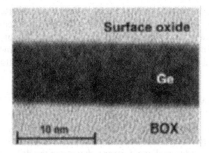

Figure 10. Lattice images of the 7nm-Ge layer of Ge-OI substrate[14]

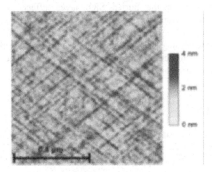

Figure 11. AFM image of Ge-OI layer surface[14]

It should be noted that the annealing temperature was chosen to lie below the solidus line in the phase diagram (Fig.2) of Si-Ge binary alloy system, thereby avoiding the melting of the SiGe layer. Higher temperature annealing, above the solidus line, resulted in disorder and creation of a microstructure in the SiGe layer.[12]

Fabrication of structures with fully relaxed layer of several hundred nm of $Si_{1-x}Ge_x$ (0.2<x<0.4)-OI, and $Si_{0.7}Ge_{0.3}$-OI substrates with a defect density less than $1000cm^{-2}$ and a surface roughness of 0.39nm rms were achieved.[13] However, thin (<100nm) SiGe-OI substrates were partially-relaxed, and the dislocation density was higher than 10^4cm^{-2}.[11] Ge-condensation combined with mesa isolation was proposed for reducing the strain in the SiGe layer.[19,20] Mesa etching was performed on a SiGe capped SOI substrate prior to the high temperature oxidation. The mesa structure caused the compressively strained SiGe to relax by lateral expansion, as the underlying SiO_2 layer can be deformed by viscous flow at the oxidation temperature. It is noted that strain relaxation resulted in the suppression of dislocation formation. This method enables $Si_{1-x}Ge_x$ (x<0.35)-OI to be formed, which is thin, fully relaxed and free from surface undulation and dislocations.[19,20]

5. SUMMARY

SiGe-OI substrates with a flat surface/interface and top SiGe layer with a well-controlled Ge content have been fabricated. However, a more dedicated study will be required to realize a further reduction in the dislocation density of the top SiGe layer and to achieve a better control over the strain in the top SiGe layer. Especially, we will have to know how small amounts of Ge, presumably contained in the buried SiO_2 layer, affects the electrical properties (breakdown voltage, parasitic capacitance and resistance, etc.) so that we can exploit of the SiGe-OI materials. In addition, an appropriate surface treatment technology needs to be established for SiGe with a high Ge concentration as the surface instability under exposure to oxygen and water hampers deposition of high quality Si layers on SiGe-OI.

REFERENCES

1. J. P. Colinge, Silicon-On-Insulator Technology: Materials to VLSI (Kluwer, Boston, 1997).
2. K. Ismail, M. Arafa, K. L. Saenger, J. O.Chu, and B. S. Meyerson, Extremely high electron mobility in Si/SiGe modulation-doped heterostructures, Appl. Phys. Lett. 66(9), 1077-1079 (1995).
3. T. Tezuka, N. Sugiyama, T. Mizuno, M. Suzuki and S. Takagi, A Novel Fabrication Technique of Ultrathin and Relaxed SiGe Buffer Layers with High Ge Fraction for Sub-100nm Strained Silicon-on-Insulator MOSFETs, Jpn. J. Appl. Phys. 40(4B), 2866-2874 (2001).
4. S. Fukatsu, Y, Ishikawa, T. Saito, N. Shibata, SiGe-based semiconductor-on-insulator substrate created by low-energy separation-by-implanted-oxygen, Appl. Phys. Lett., 72(26), 3485-3487 (1998).
5. Y. Ishikawa, N. Shibata, S. Fukatsu, SiGe-on-insulator substrate using SiGe alloy grown Si(001), Appl. Phys. Lett., 75(7), 983-985 (1999).
6. Z. An, Y. Wu, M. Zhang, Z. Di, C. Lin, R. K. Y. Fu, P. Chen, P. K. Chu, W. Y. Cheung and S. P. Wong, Relaxed silicon-germanium-on-insulator substrate by oxygen implantation into pseudomorphic silicon germanium/silicon heterostructures, Appl. Phys. Lett., 82(15), 2452-2454 (2003).
7. L. J. Huang, J. O. Chu, D. F. Canaperi, C. P. D'Emic, R. M. Anderson, S. J. Koester, and H. S. P. Wong, SiGe-on-insulator prepared by wafer bonding and layer transfer for high-performance field-effect transistors, Appl. Phys. Lett., 78(9), 1267-1269 (2001).

8. G. Taraschi, T. A. Langdo, M. T. Currie, E. A. Fitzgerald and D. A. Antoniadis, Relaxed SiGe-on-insulator fabricated via wafer bonding and etch back, J. Vac. Sci. Technol., B20(2), 725-727 (2002).

9. G. Taraschi, Z. Y. Cheng, M. T. Currie, C. W. Leitz, T. A. Langdo, M. L. Lee, A. Pitera, E. A. Fitzgerald, Relaxed SiGe on Insulator Fabricated via Wafer Bonding and Layer Transfer: Etch-Back and Smart-Cut Alternatives, Electrochemical Society Proceedings Vol.2001-3, 27-32(2001).

10. A. R. Powell, S. S. Iyer and F. K. Legoues, New approach to the growth of low dislocation relaxed SiGe material, Appl. Phys. Lett. 64(14), 1856-1858 (1994).

11. T. Tezuka, N. Sugiyama, and S. Takagi, Fabrication of strained Si on an ultarathin SiGe-on-insulator virtual sustrate with a high-Ge fraction, Appl. Phys. Lett., 79(12), 1798-1800 (2001).

12. N. Sugii, S. Yamaguchi and K. Washio, SiGe-on-insulator substrate fabricated by melt solidification for a strained-silicon complementary metal-oxide-semiconductor, J. Vac. Sci. Technol. B20(5), 1891-1896 (2002).

13. K. Kutsukale, N. Usami, K. Fujiwara, T. Ujihara, G. Sazaki, B. Zhang, Y. Segawa, and K. Nakajima, Fabrication of SiGe-on-Insulator through Thermal Diffusion of Ge on Si-on-Insulator Substrate, Jpn. J. Appl. Phys., 42(3A), L232-L234 (2003).

14. S. Nakaharai, T. Tezuka, N. Sugiyama, Y. Moriyama and S. Takagi, Characterization of 7-nm-thick strained Ge-on-insulator layer fabricated by Ge-condensation technique, Appl. Phys. Lett., 83(17), 3516-3518 (2003).

15. R. W. Olesinski and G. J. Abbaschian, Bull. Alloy Phase Diagram 5, 180 (1984).

16. Y. Ishikawa, N. Shibata and S. Fukatsu, Factors limiting the composition window for fabrication of SiGe-on-insulator substrate by low-energy oxygen implantation, Thin Solid Films, 369, 213-216(2000).

17. http://www.srim.org/

18. N. Sugiyama, T. Mizuno, M. Suzuki and S. Takagi, Formation of SiGe on Insulator Structure and Approach to Obtain Highly Strained Si layer for MOSFETs, Jpn. J. Appl. Phys. 40 (4B) 2875-2880 (2001).

19. T. Tezuka, N. Sugiyama, S. Takagi, and T. Kawakubo, Dislocation-free formation of relaxed SiGe-on-insulator layers, Appl.Phys. Lett. 80(19), 3560-3562(2002).

20. T. Tezuka, N. Sugiyama, and S. Takagi, Dislocation-free relaxed SiGe-on-insulator mesa structures fabricated by high-temperature oxidation, J. Appl. Phys., 94(12), 7553-7559 (2003).

CVD DIAMOND FILMS FOR SOI TECHNOLOGIES

V. Ralchenko[1], T. Galkina[2], A. Klokov[2], A. Sharkov[2], S. Chernook[2] and V. Martovitsky[2]

[1]*A.M. Prokhorov General Physics Institute RAS, Vavilov str. 38, Moscow 119991, Russia;*
[2]*P.N. Lebedev Physical Institute RAS, Leninsky prospekt 53, 119991 Moscow, Russia*

Abstract: Diamond films is an interesting material for silicon-on-insulator technologies due to a unique combination of properties that can enhance to electronic device performance. Current CVD techniques for diamond growth allow the production of diamond layers with a thickness from less than 1 micron to a few mm on Si substrates of large area. As the direct diamond-to-silicon bonding requires high temperature (>1150°C) and still is problematic, the use of a Si substrate as the device layer after diamond deposition could be an alternative. No dramatic degradation of silicon after diamond deposition at 750°C by microwave plasma assisted CVD is found as revealed from resistivity, X-ray diffraction and acoustic phonon scattering measurements. Further studies of electronic properties of the diamond-coated Si are required to evaluate the feasibility of Si-on-diamond (SOD) device fabrication.

Key words: SOI, diamond film, thermal conductivity, stress, silicon, interface, ballistic phonons.

1. INTRODUCTION

Conventional silicon-on-insulator (SOI) technology uses a buried silicon dioxide SiO_2 isolation layer to allow device fabrication operating at lower voltage, at higher frequencies, in harsh environment. However, because of low thermal conductivity of SiO_2 any increased integration density and power dissipation may cause problems. Diamond can be investigated as a novel dielectric material for SOI systems, as it possesses a unique combination of properties useful to electronic device performance (Table 1). In particular, its thermal conductivity, up to 22 W/cmK, is three orders of magnitude higher that that of SiO_2 and five times better than for copper.

D. Flandre et al. (eds.), Science and Technology of Semiconductor-On-Insulator Structures and Devices Operating in a Harsh Environment, 77-84.
© 2005 *Kluwer Academic Publishers. Printed in the Netherlands.*

High breakdown field strength, low dielectric constant, high radiation resistance, extreme hardness make diamond very attractive for high temperature, high power, radiation-hard devices.

Currently, thin coatings (one micrometer) and thick (up to 2 mm), polycrystalline diamond wafers can be grown at low pressure using a chemical vapor deposition (CVD) technique[1-4]. The synthesis process is based on the decomposition of a methane-hydrogen gas mixture by a variety of methods (plasma, flame, hot filament and some others), which resulting in the deposition of diamond on a hot (typically 600-1000°C) substrate, in most cases silicon. The most pure and high quality diamond films are produced by microwave plasma enhanced CVD (MPCVD) technique. The MPCVD reactor (5 kW, 2.45 GHz) developed at General Physics Institute RAS for diamond synthesis, and a selection of produced diamond wafers and thin films on Si substrates of up to 100 mm diameter, are shown in Fig. 1.

Table 1. Properties of Si and some dielectrics important for SOI devices @300K.

Property	Si	SiO_2	sapphire	diamond
Thermal conductivity, W/cmK	1.51	0.014	0.34	10-22
Resistivity, Ohm_cm	10^{-3}-10^4 (*)	10^{13}-10^{19}	10^{16}	10^{12}-10^{15}
Dielectric constant, ε	11.7	3.8	9.4	5.7
Breakdown field, V/cm	3.7×10^5	10^7 (**)	6.4×10^5	10^7
Loss tangent, 10^{-6} @145 GHz	3*	300	200	3-8
Thermal expansion coef., 10^{-6} K^{-1}	2.5	0.5	8.2	0.8
Young's modulus, GPa	160	73	380	1050

*doped with Au; **~10 nm thick film

Figure 1. MPCVD reactor for diamond deposition (left) and a selection of produced diamond films and wafers (right).

The diamond can be separated from the substrate by chemically removing the silicon in HF-HNO₃ acid to produce a free-standing film. The material is polycrystalline with randomly oriented grains, with a very rough top surface. A surface roughness less than 10 nm can be achieved by mechanical polishing, which, is a difficult and slow procedure because of the extreme hardness of diamond. The bottom (nucleation) side is smooth, being the replica of the mirror-polished Si substrate. Typically a surface roughness R_a < 15 nm is measured on the nucleation side provided high-density diamond seeding of the substrate before deposition[5]. The machining of diamond films, such as cutting, drilling, patterning, rough polishing can conveniently be performed using lasers[6].

The thermal conductivity of translucent thick (several hundreds microns) films is k=16-18 W/cm K at room temperature. With the temperature is increased to T=200°C the thermal conductivity decreases down to 10-12 W/cmK (Fig. 2), mainly due to phonon-phonon scattering mechanisms following the approximate relationship $k \sim T^{-1}$ [7]. Thin films consist of small grains laced with defects and impurities, primarily Si and H, and the thermal conductivity of the films with thickness up to 10 μm typically is <10 W/cmK.[8] As the crystallite size and perfection increase with film thickness there is a gradient in thermal conductivity across the film, the difference between the conductivity at the top and bottom surfaces can amount to factor of three[9]. The thermal conductivity displays an anisotropy because the grains have columnar shape with their axes directed perpendicular to the film surface. Perpendicular conductivity $k_\perp$ is typically larger by 10-20% than the in-plane values k_{II} [7].

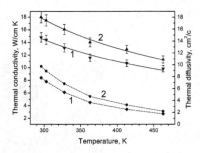

Figure 2. Temperature dependence of in-plane thermal diffusivity D (dotted lines) and thermal conductivity k (solid lines) for two diamond films (Nos. 1&2) of approx. 0.3 mm thickness[7]. The data are obtained using the laser-induced dynamic grating technique.

The device layer may be placed on the diamond substrate in two ways: (i) by a bonding process, or (ii) via direct deposition of diamond film on silicon, followed by Si thinning. Yushin *et al.*[10,11] found that the complete bonding of diamond to Si is possible at temperatures above 1150°C, however, cracks always appear due to thermal stress caused by the high bonding temperature. The thermal treatment of CVD diamond in vacuum should be limited to 1200-1300°C, at which temperatures the internal graphitization takes place at the grain boundaries (GB)[12,13], and the diamond becomes electrically conducting along the transformed GBs. The de-bonding of hydrogen which passivates the grain boundaries and extended defects in as-grown material, is believed to trigger the GB graphitization process.

In case of the diamond deposition a little is known about changes in Si substrate structure induced by high temperature (700-900°C), and its interaction with hydrogen and carbon during the deposition process. When the radial temperature gradient in the substrate takes place a plastic deformation was observed in silicon near the film-substrate interface[14]. Here we address the issue of possible defects formation in Si substrate as a result of diamond growth on it.

2.   EXPERIMENTAL DETAILS AND RESULTS

Diamond films were deposited on Si wafers at T=720-750°C by microwave plasma CVD using CH_4 (2.5%)/H_2 (97.5%) mixture as the source gas[3]. Dislocation-free (dislocation density <10 cm^{-2}) polished (100) Si wafers 10 mm in diameter and 3 mm thick were used as the substrates. Typically, 15 μm thick films with grain size of 3-5 μm were produced for deposition runs of 9 hours at 5 kW MW power and 90 Torr pressure (Fig. 3).

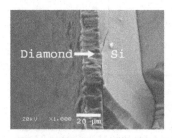 

Figure 3. SEM pictures of 15 μm thick polycrystalline diamond film on Si: (left) cross-section, (right) typical diamond surface morphology.

MicroRaman spectroscopy (ISA, S3000 model) was used to identify the diamond structure and stress in the film and substrate. The scattering was excited at wavelength 488 nm of an Ar⁺ laser. The laser spot diameter on the analyzed surface was either 2 or 20 μm. A sharp peak at 1333.5 cm⁻¹ with width (FWHM) 6.7 cm⁻¹ in the spectrum (Fig. 4) confirms the cubic diamond structure with a negligible inclusion of amorphous carbon (wide weak band around 1550 cm⁻¹). The peak is shifted to higher frequencies respective to the peak position of 1332.5 cm⁻¹ characteristic of unstressed diamond. This indicates a compressive stress σ= 0.6 GPa in the film, as estimated from the correlation[15] between the stress σ and the shift Δv in Raman peak position: σ [GPa] = - 0.61 Δv [cm⁻¹]. This stress is very close to the thermal stress value of 0.7 GPa for deposition temperature T=830°C [14], which is caused by a mismatch between the thermal expansion coefficients of diamond and Si, being generated upon sample cooling after the process finish.

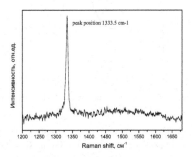

Figure 4. Raman spectrum of diamond film.

A stress in silicon induced by diamond growth process has also been evaluated by X-ray diffraction measured on the sample cross-section with spatial resolution 200 μm. The width (FWHM) of the (400) signal in the rocking curve was 8 sec. for virgin Si. After the diamond deposition it increased to 20 sec. only in a layer (<200μm) adjacent to the film (Fig. 5), but remained narrow (~8 sec.) across the rest thickness except the layer, adjacent to the diamond film. This indicates the presence of some stress in the silicon.

As an additional characterization of silicon substrate quality the propagation of non-equilibrium acoustic phonons across the diamond/Si structure was studied in transmission geometry at cryogenic temperature T= 1.8 K (Fig. 6). The phonons were generated either by direct photoexcitation of silicon or diamond by the UV pulses of a nitrogen laser (τ=7.5 ns, λ=337 nm)[16]. The temporal evolution of the bolometer signal from phonons propagating through the virgin and exposed to diamond deposition silicon is shown in Fig. 7. The insert in Fig. 7 shows the spatially resolved phonon flux at the time t=700 ns after the laser pulse. The degree of darkness corresponds to phonon flux intensity. The observed anisotropy is the proof of low phonon scattering in silicon substrate after CVD deposition.

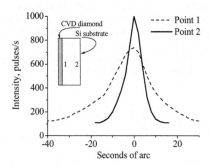

Figure 5. Rocking curves of (400) signal for the monocrystalline Si substrate measured at Si/diamond interface (point 1) and from the back side (point 2).

The only scattering source in the virgin sample is the scattering by isotopes. No increase of acoustic phonon scattering in Si was observed after diamond deposition, so the generation of thermally activated defects by the deposition process was insignificant. No change in the resistivity of Si substrate compared to the virgin Si (3.3 kOhm·cm) has been found after the deposition process. Of course, a more detailed analysis is needed to better characterize the Si modifications, yet there is a hope that the potential degradation of Si in diamond growth environment can be minimized by the careful choice of CVD process conditions. This would make possible to obtain an adherent device layer on diamond substrate after for SOD electronic devices fabrication.

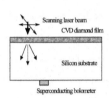
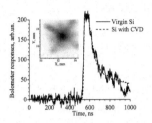

Figure 6. Schematics of laser induced phonon generation

Figure 7. Bolometric responses to laser-generated phonons propagating in virgin Si and diamond-on-silicon sample. The insert shows the spatially resolved phonon flux at t=700 ns.

3.   CONCLUSIONS

Polycrystalline CVD diamond is a promising dielectric material for high power, high-frequency, radiation-hard SOI systems, since it possesses extremely high thermal conductivity, can be produced on large area Si substrates in form of thin films as well as wafers. Potentially, silicon-on-diamond should have advantage over silicon-on sapphire devices, if the problem of Si/diamond bonding or direct diamond deposition on Si in mild conditions could be solved. The present experiments reveal that the Si substrates can withstand the MPCVD diamond growth environment without significant degradation. Although currently some technological hurdles (stress generation, polishing, high cost) retard the SOD development, the impressive progress of CVD diamond technologies during the last decade in growth of high quality, large size, uniform diamond films gives a hope that the SOD systems will attract the adequate attention in near future.

ACKNOWLEDGEMENTS

The authors are thankful to K. Anisimov for preparation of diamond film samples, I. Vlasov and S. Terekhov for Raman spectra measurements. This work was supported in part by Russian Foundation for Basic Research, Nos. 02-02-17392 and 04-02-17060.

REFERENCES

1. B. Dischler, C. Wild (Eds), Low-Pressure Synthetic Diamond: Manufacturing and Applications, Springer, Berlin, 1998.
2. S.E. Coe, R.S. Sussmann, Optical, thermal and mechanical properties of CVD diamond, Diamond Relat. Mater. **9**, 1726-1729 (2000).
3. V.G. Ralchenko, A.A. Smolin, et al. Large-area diamond deposition by microwave plasma, Diamond Relat.Mater. **6**, 417-421 (1997).
4. D.M. Gruen, Nanocrystalline diamond films, Ann. Rev. Mater. Sci. **29**, 211-259 (1999).
5. V. Ralchenko, Nano-and microcrystalline CVD diamond films on surfaces with intricate shape, in Nanostructured Thin Films and Nanodispersion Strengthened Coatings, NATO Science Series, II: Mathematics, Physics and Chemistry, Kluwer (in press).
6. V. Ralchenko, V. Migulin, et al. Treatment of diamond films with lasers, in Diamond Films, 9th CIMTEC'98 - Forum on New Materials, Symp. IV-Diamond Films, P. Vincenzini (Ed), Techna, (1999), pp.109-118.
7. E.V. Ivakin, A.V. Suhkodolov, et al. Measurement of thermal conductivity of polycrystalline CVD diamond by laser-induced transient grating technique, Quantum Electronics (Moscow), **32**, 367-372 (2002).
8. J.E. Graebner, V.G. Ralchenko. Thermal conductivity of thin diamond films grown from d.c. discharge, Diamond Relat. Mater. **5**, 693-698 (1996).
9. V.G. Ralchenko, A.V. Vlasov, et al. Measurements of thermal conductivity of undoped and boron-doped CVD diamond by transient grating and laser flash techniques, Proc. ADC/FCT 2003, NASA/CP-2003-212319 (2003), pp. 309-314.
10. G.N. Youshin, S.D.Walter, et al. Study of fusion bonding of diamond to silicon for silicon-on-diamond technology, Appl. Phys. Lett. **81**, 3275-3277 (2002).
11. S.D.Walter, G.N. Youshin, et al. Direct fusion bonding of silicon diamond to polycrystalline diamond, Diamond Relat. Mater. **11**, 482-486 (2002).
12. V. Ralchenko, L. Nistor, et al. Structure and properties of high-temperature annealed CVD diamond, Diamond Relat. Mater. **12**, 1964-1970 (2003).
13. A.V. Khomich, V.G. Ralchenko, et al. Effect of high temperature annealing on optical and thermal properties of CVD diamond, *Diamond Relat. Mater.* **10**, 546-551 (2001).
14. J. Michler, M. Mermoux, et al. Residual stress in diamond films: origins and modeling, Thin Solid Films, **357**, 189-201 (1999).
15. J.W. Ager, M.D. Drory, Quantitative measurement of residual biaxial stress by Raman spectroscopy in diamond grown on a Ti alloy by chemical vapor deposition, Phys. Rev B, **48**, 2601-2607 (1993).
16. T.I. Galkina, A.Yu. Klokov, et al. Propagation of acoustic phonons across the interface in CdTe and Si/CVD diamond and quasi-two-dimensional phonon wind in CdTe/ZnTe quantum wells, Physica B, **316-317**, 243-246 (2002).

RADICAL BEAM QUASIEPITAXY TECHNOLOGY FOR FABRICATION OF WIDE-GAP SEMICONDUCTORS ON INSULATOR

Giorgi Natsvlishvili, Tamaz Butkhuzi, Maia Sharvashidze, Lia Trapaidze and Dimitri Peikrishvili
Iv. Javakhishvili Tbilisi State University, Laboratory of Semiconducting Materials, 3 Chavchavadze Ave., 380028 Tbilisi, Georgia

Abstract: Current work presents a new method – Radical Beam Quasiepitaxy (RBQE) -- of forming monocrystalline layers of ZnO and ZnS on basic crystals of ZnO or ZnS. Basic crystals were stoichiometric insulating samples with minimal concentration of residual impurities. Studies were made of the electrical and optical properties of grown layers as well as basic crystals, i.e. Photoconductivity (PC), Photoluminescence (PL), Thermo-electromotive force (EMF) and Hall Effect. In PL spectra of $10^{11} – 10^{13}$ Ωcm resistivity ZnO newly grown layers obtained at 650 K free A, B, C excitons emission and polariton emission have been observed. After removing the grown layers, surface layers of the basic crystals always demonstrated *p*-type conductivity. Defects responsible for the observed peaks in the PL spectra are identified.

Key words: RBQE; *p*-type ZnO; polaritons; free excitons.

1. INTRODUCTION

The fabrication of optoelectronic devices operating in the visible or ultraviolet regions is one of the major tasks in optical electronics. Semiconductors having wide direct gaps and high efficiency of radiative recombination are essential for optoelectronic devices[1]. The fabrication of such semiconductors and the investigation of their electrical and optical properties is a very important issue. ZnS and ZnO are of particular interest.

Radical Beam Quasiepitaxy enables the growth of new quasiepitaxial layers on the surface of binary compounds like the basic crystals. In this case the RBQE method enables monocrystalline layers of ZnO and ZnS to be

D. Flandre et al. (eds.), Science and Technology of Semiconductor-On-Insulator Structures and Devices Operating in a Harsh Environment, 85-90.
© 2005 *Kluwer Academic Publishers. Printed in the Netherlands.*

created on the basic crystal of ZnO and ZnS, but generally it might be of use for other binary compounds. It should be noted that the formation of new layers takes place in a wide temperature range (400 K to 1500 K). Basic crystals were stoichiometric insulating samples with minimal concentration of residual impurities. This is necessary to inhibit the transition of residual impurities to the newly grown layers during the growth process.

2.   RADICAL BEAM QUASIEPITAXY

Figure 1 shows the RBQE scheme.

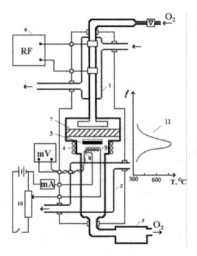

Figure 1. 1) and 2) cooling systems, 3) substrate, 4) heater, 5) vacuum pump, 6) RF generator, 7) filter which separates only oxygen radicals, from oxygen plasma, 8) thermocouple, 9) platinum coil, 10) potentiometer, 11) temperature distribution in the reactor.

During RBQE, the basic crystal undergoes treatment in an atmosphere of singlet non-metal radicals created by a RF discharge. The RF oscillator converts the molecular non-metal stream into plasma containing ions, electrons etc. in addition to non-metal radicals. The applied electric and magnetic fields cause "filtration", after which only non-metal radicals reach the basic crystal. Due to their activity such radicals interact with the basic crystal and the growth of new layers is carried out by non-metals absorbed from the gas phase and metals extracted from the basic crystal. The radicals are highly active when they reach the surface of the basic crystal with which they react. During the process the radicals stimulate extraction of Zn atoms from the bulk and the creation of a new ZnO (or ZnS) quasiepitaxial layer on the surface. Meanwhile, a number of Zn vacancies are created on the surface area of the basic crystal. The narrow profile of temperature distribution required is maintained with a water-cooling system in the upper and lower parts of the reactor.

Control of the electrical properties of the created layers as well as the surface layers of the basic crystals is possible with the correct choice of the temperature regime. The magnetic and electric fields control the distance between the plasma and basic crystal. The experiment shows that the effective distance is d≈4 to 7 mm. At d<4 mm the crystal evaporation (ion etching) takes place; at d>7 mm, radicals can unify as molecules reducing the efficiency of layer growth. Concentration of non-metal radicals can be measured using several different techniques, which provide almost similar results. The most convenient technique will be discussed, ie - Elias method of catalytic sample heating[2]. A platinum coil was used as a catalyst. Measurements were conducted in an isothermal state with a given coil temperature maintained by a DC supply. The constant temperature of the coil, with and without oxygen radicals in the process, was maintained using a potentiometer. The temperature was measured by the thermocouple connected to the coil. When oxygen radicals were present, the partial pressure of oxygen radicals was estimated using the following formula:

$$P_O = 7.00 \times 10^{-3} \frac{R(I_O^2 - I^2)}{A}$$

R – is the resistance of the coil at a set temperature, I_O and I – are current strengths measured with and without oxygen atoms, A – is the surface area of the coil. The measurements showed that the concentration of oxygen radicals is close to $n_o \approx 10^{13}$ cm^{-3}. If account is also taken of the coefficient of platinum accommodation (~0.1), then the actual concentration of oxygen radicals should be: $n_o \approx 1 \times 10^{14} - 5 \times 10^{15}$ cm^{-3}. It should be noted that it is practically impossible to obtain such a concentration in the equilibrium condition.

3. EXCITON PHOTOLUMINESCENCE OF ZnO LAYERS WITH $\rho=10^{11}$ Ωcm AND $\rho=10^{13}$ Ωcm

The dielectric layers of ZnO grown by RBQE (of μm thickness and ρ= 10^{13} Ωcm resistivity) on the ZnO basic crystal were excited by the nitrogen laser (λ=337.1 nm) at T=77 K temperature. It is important to note that under the complete absence of visible luminescence the bands λ_{max}=367.7 nm, λ_{max}=366.8 nm and λ_{max}=362.7 nm are observed in the PL spectrum of these layers. The band λ_{max}=366.8 nm was characterized by sharply defined dependence of intensity on the polarization of the falling beam. The spectral position of these bands coincides with the exciton model of ZnO and their position in the reflection spectra[3]. So the radiative recombination of free excitons A (λ_{max}=367.7 nm), B (λ_{max}=366.8 nm) and C (λ_{max}=362.7 nm) is observed.

At high levels of excitation the samples with ρ=10^{11} Ωcm resistivity are characterized by the complete absence of visible luminescence and by the pronounced bands λ_{max}=363.1 nm, λ_{max}=366.8 nm and λ_{max}=370.6 nm. As mentioned above the band λ_{max}=366.8 nm is due to the radiative recombination of free B exciton. The band λ_{max}=370.6 nm is known as the M band and there is not a one-valued interpretation. At the relatively low level of excitation the complicated structure of this band with the maxima λ_1=368.5 nm (M_1), λ_2=370 nm (M_L) and λ_{max}=372.5 nm (M_2) are observed. The band λ_{max}=363.1 nm (T band) is first observed (Fig. 2).

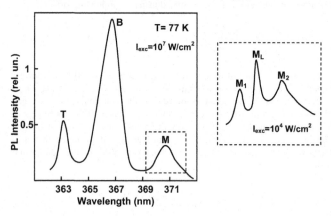

Figure 2. The PL spectrum of ZnO layers of ρ= 10^{11} Ωcm resistivity

The experimental spectra obtained can be explained by the presence of the following three channels of exciton molecule decay (Fig. 3):

1. As the result of exciton molecule (EM) decay a light-like state M_1 of the low polariton branch (LPB) and exciton-like state of the LPB are formed. Reaching the crystal surface the light-like polariton escapes as a luminescence photon M_1 (λ=368.5 nm; 3.362 eV).

2. EM causes a light-like state of the LPB and a longitudinal exciton state B_L. Reaching the crystal surface the light-like polariton escapes as a luminescence photon M_L (λ=370.0 nm; 3.348 eV).

3. The EM decay results in the appearance of two light-like polaritons of LPB and upper polariton branch (UPB). Both polaritons reach the crystal surface and escape as photons M_2 (λ=373.0 nm; 3.322 eV) and T (λ=363.1 nm; 3.412 eV).

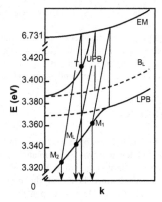

Figure 3. Dispersion curves of excitons (dashed lines), biexcitons, and exciton-polaritons in orientation **k** $\perp$ **C** (solid lines)

4.   OPTICAL PROPERTIES of *p*-TYPE ZnO

Studies were made of PL and Photoconductivity (PC) spectra of the ZnO basic crystal after removing newly grown layers. In the PL spectra of ZnO basic crystals treated at 900 to 1000 °C (ρ=10^4 to 10^5 Ωcm; μ=11 cm^{2}Vs; p=10^{13} cm^{-3}), peaks with maxima 395 nm, 460 nm and 510 nm were observed. Disintegration of the line using Alentsev-Fock method[4] shows two elementary peaks with maxima 385 nm and 400 nm (Fig. 4).

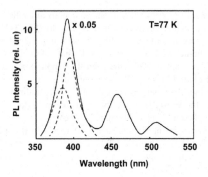

Figure 4. The PL spectra of the basic ZnO crystal after removing newly grown layers

It is considered that the 385 nm band corresponds to the complex (V^{-}_{Zn} – V^{++}_{O}). 400 nm band is connected with electron transition from the conductivity band to the V^{++}_{O} center. In the PL spectrum 460 nm band is connected with V^{-}_{Zn} center[5]. The 510 nm band is connected with V^{+}_{O} center[6].

5.   CONCLUSION

An original method for RBQE has been developed. In PL spectra of 10^{11} – 10^{13} Ωcm resistivity ZnO newly grown layers obtained at 650 K, free A, B, C excitons emission and polariton emission have been observed. After removing the grown layers, surface layers of the basic crystals always demonstrated *p*-type conductivity. Defects responsible for the observed peaks in the PL spectra have been identified.

REFERENCES

1. A. Nurmikko and R. L. Gunshor, *Semicond. Sci. Technol.*, **12**, 1337-1344 (1997)
2. L. Elias, E. A. Orgyzlo and H. J. Schiff, *Can. Chem.*, **37**, N 10, 1680-1683 (1959)
3. D. G. Thomas, *J. Phys. and Chem. Solids,* **15**, 86 (1960)
4. M. V. Fock, *Proc. Lebedev Physics Institute Acad. Science USSR,* **59**, 3 (1972)
5. F. Kroeger, *Khimia Nesovershennix Kristallov,* Mir, M (1969) Russian
6. W. F. Wei, *Phys. Rev. B* **15**, 2250 (1977)

IMPACT OF HYDROSTATIC PRESSURE DURING ANNEALING OF Si:O ON CREATION OF SIMOX - LIKE STRUCTURES

A.Misiuk, J.Ratajczak, J.Kątcki, I.V.Antonova[1]
Iinstitute of Electron Technology, Warsaw, Poland; [1]Institute of Semiconductor Physics, RAS, Novosibirsk, Russia

Abstract: The new and earlier reported results on the effect of hydrostatic pressure (HP, up to 1.23 GPa) and annealing at $\leq$ 1570 K of oxygen implanted silicon (Si:O) are discussed. HP affects the misfit at the SiO$_x$/Si boundary, solubility and diffusivity of oxygen, the number and dimensions of SiO$_x$ precipitates.

Key words: silicon, oxygen, implantation, hydrostatic pressure, annealing

1.   INTRODUCTION

High hydrostatic pressure (HP) of gaseous ambient during annealing at high temperature (HT) of Si implanted with oxygen (Si:O) exerts pronounced effects on the structure of oxygen-enriched (for low oxygen dose, D) or a continuous (for $D \geq 6 \times 10^{17}$cm^{-2}) buried layer composed of SiO$_2$ [1-6]. The HT-HP treatment results in a changed misfit at the SiO$_2$ / Si boundary, in a creation of nucleation centers for oxygen precipitation, affects oxygen diffusivity and dissolution of SiO$_{2-x}$ clusters, especially at temperatures $\geq$ 1400 K [5]. New results concerning the effect of HP on oxygen implanted Si at HT are presented and discussed in the light of earlier published data.

D. Flandre et al. (eds.), Science and Technology of Semiconductor-On-Insulator Structures and Devices Operating in a Harsh Environment, 91-96.

2.   EXPERIMENTAL DETAILS

Czochralski (Cz-Si) and floating zone (Fz-Si) grown, (001) and (111) oriented Si wafers were implanted with O^+ (oxygen doses, $D = 1x10^{15} - 2x10^{18}$ cm^{-2}, energies, $E = 50 - 200$ keV) and treated at HT up to 1570 K under HP up to 1.23 GPa for up to 5 h. The structure of HT–HP treated Si:O was determined using Transmission Electron Microscopy (XTEM) and photoluminescence (PL) measurements at 10 K (Ar laser, $\lambda = 488$ nm).

3.   RESULTS AND DISCUSSION

The impact of HT-HP on Si:O is dependent on numerous implantation - (mainly D and E) and treatment - related parameters. This effect is, however, dependent first of all on the implanted oxygen dose; it is the reason why the results presented below are listed accordingly to the specified ranges of D.

3.1   Si:O prepared by implantation with $D \le 1x10^{17}$ cm^{-2}

Annealing of Si:O ($D \le 1x10^{17}$ cm^{-2}) at 1230–1470 K under 10^5 Pa results in the creation of SiO$_x$ precipitates embedded in Si with dislocations formed if the misfit, ε, at the SiO$_x$/Si boundary exceeds a critical value. HP results in a decreased ε and so dislocations are not created. The 250 nm thick zone, containing precipitates and dislocations, has been detected in Si:O annealed at 1400 K – 10^5 Pa (Fig. 1A); however no dislocations were detected after the HP treatment at 1400 / 1470 K (Fig. 1B, C).

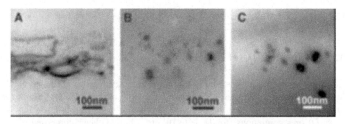

Figure 1. XTEM images of Cz-Si:O ($D = 1x10^{16}$ cm^{-2}, $E = 200$ keV), treated for 5 h at 1400 K under 10^5 Pa (A), under 1.2 GPa (B) and for 1 h at 1470 K under 1.5 GPa (C).

Dislocations were not created also in the Si:O samples treated at 1570 K under 10^7 Pa but were formed under 1.23 GPa (Figs 2A, B). Their creation under HP is caused by overcompensation of the volume – related misfit (the different relative volumes of Si and SiO_x) by the HP – induced misfit of the opposite sign. The nature and density of defects formed are dependent also on treatment time, t (Fig. 2B, C).

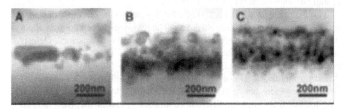

Figure 2. XTEM images of Cz-Si:O ($D = 1\times10^{17}$ cm^{-2}, $E = 200$ keV) treated for 5 h at 1570 K under 10^7 Pa (A), 1.23 GPa (B) and for 10 min. at 1550 K – 1.5 GPa (C).

The intensity of the dislocation-related D1 peaks at 0.81 eV decreases with HP[1], especially for Si:O prepared by implantation with low D (10^{16} – 10^{17} cm^{-2}) and treated at 1400 – 1470 K under HP for 5–10 h (Fig. 3). The treatments at $\leq$ 1400 K – HP, as well as at $\geq$ 1400 K – HP but of short duration, result in the presence of dislocations (Fig. 2C) as also evidenced by the increased intensity of the D1 peaks (Fig. 3).

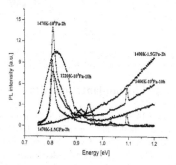

Figure 3. PL spectra of 001 oriented Cz-Si:O ($D = 1\times10^{16}$ cm^{-2}, $E = 200$ keV), HT – HP treated at up 1470 K under HP up to 1.5 GPa.

3.2   Si:O prepared by implantation with $D = 6\times10^{17}$ cm^{-2}

Continuous or semi-continuous SiO_2 layers, containing crystalline Si inclusions, are formed at HT-HP in Si:O prepared by implantation with

1×10^{17} cm^{-2} < D ≤ 1×10^{18} cm^{-2}. For example, the structure prepared by implantation with $D = 6\times10^{17}$cm^{-2}, $E = 170$ keV and treated at 1400 – 1570 K under HP ≥ 1 GPa, contains BOX layers. The structure of such samples is strongly dependent on HT, HP and t (Fig. 4).

The treatment at 1570 K – 10^7 Pa for 5 h results in the formation of a buried oxide of about 130 nm thickness. The highly disturbed zone of about 100 nm thickness has been created just above the SiO$_2$/Si interface. Defects observed are mostly dislocations and micro-twins (Fig. 4A). The same treatment but under 1.23 GPa produces a buried SiO$_2$ layer containing almost no inclusions. However, the 100 nm thick defected zone containing mostly micro-twins has been created near the sample top (Fig. 4B). The short time treatment at slightly different conditions produces the well defined BOX layer but still with numerous Si inclusions and micro-twins in the interfacial zone (Fig. 4C). The distribution of defects is strongly asymmetric (Fig. 4), with more numerous defects created near the sample surface[6].

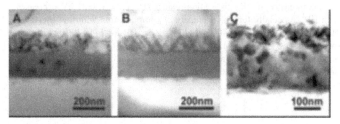

Figure 4. XTEM images of Fz-Si:O ($D = 6\times10^{17}$cm^{-2}, $E = 170$ keV) treated for 5 h at 1570 K under 10^7 Pa (A), 1.23 GPa (B) and for 1 h at 1470 K – 1.5 GPa (C).

3.3   Si:O prepared by implantation with D ≥ 1×10^{18}cm^{-2}

The treatment at 1230 / 1400 K – 1.2 GPa of Si:O prepared by high dose oxygen implantation ($D = 1\text{-}2\times10^{18}$ cm^{-2}, $E = 50$ keV) leads to the formation of a continuous near-surface SiO$_2$ layer. Numerous dislocations are, however, formed in such samples; their density decreases with HT while the dependence of this density on HP is not so straightforward (Fig. 5).

Figure 5. XTEM images of Cz-Si:O ($D = 1.1 \times 10^{18}$ cm^{-2}, $E = 50$ keV) treated for 5 h at 1400 K under 1.1 GPa (A), at 1570 K under 10^7 Pa (B) and under 1.23 GPa (C).

The treatment of Si:O prepared by oxygen implantation with $D = 1.1 \times 10^{18}$ cm^{-2} at 1400 K under 1.2 GPa produces numerous small (of about 20 nm diameter) precipitates and dislocations placed just below a BOX layer (Fig. 5A). The treatment at the highest applied temperature (1570 K) under comparatively low pressure (10^7 Pa) results in the formation of the perfect BOX layer (Fig. 5B) while, in the case of treatment under 1.23 GPa, the oxide layer is of irregular thickness and oxide precipitates can be observed above it (Fig. 5C).

Enhanced HP adversely affects the re-crystallization of amorphous Si produced by heavy oxygen implantation. At the first stage of treatment, well defined SiO$_x$ precipitates are formed while much less numerous extended defects are created (Figs 1B, C). Under the highest HP these precipitates remain isolated from one another even after prolonged annealing and so the polycrystalline material between these precipitates creates extended defects.

3.4 Qualitative explanation of observed effects

The oxygen-enriched layer in Si:O is strongly disturbed just after implantation. The SiO$_{2-x}$ agglomerates are created during annealing, serving as nucleation centers for further growth of a BOX-like SiO$_2$ layer in the case of sufficiently high dose of implanted oxygen.

The following treatment parameters are most important, contributing in the HP induced effects in Si:O (see also[1-6]):
- Mobility and solubility of implanted oxygen as well as of silicon interstitials are dependent on HP and HT;
- Stability of oxygen agglomerates is dependent on HP;
- Most probably, oxygen diffusivity decreases with HP;
- The misfit at the SiO$_2$/Si boundary is tuned by HP.

In terms of the misfit at the SiO_2/Si boundary, it exists an equivalence of the effects of annealing (of HT) and of HP. Equation [1] describes equivalence of HT and HP in the respect discussed for growing precipitate:

$$\Delta HT \sim \Delta HP \, (1/K_m - 1/K_p) \, / \, (\beta_p - \beta_m) \qquad\qquad [1]$$

In this equation: K_m and K_p - the bulk modulus of the matrix and precipitate material, respectively; β_p and β_m - the volume thermal expansion coefficients of Si and of precipitate material (SiO_2). In the case of Si with embedded SiO_2 precipitates the effect of $\Delta HP = 1$ GPa would be equivalent (in terms of the HP - induced misfit and so of the misfit-induced shear stress) to that of $\Delta HT \approx 1300$ K This estimation is oversimplified but still proves that the HT – HP treatment exerts very strong effect on the misfit and so on a creation (or its prohibiting) of defects in Si:O.

The experiments reported here are some of the first to consider the role of very high pressure during the annealing of Si:O samples. Further investigations are, however, needed to understand the mechanisms involved.

ACKNOWLEDGEMENTS

The authors thank Dr G.Gawlik from IEMT, Warsaw for sample preparation and Dr L.Bryja from Technical University, Wroclaw for PL data.

REFERENCES

1. Misiuk, A. Barcz, J. Ratajczak, J. Katcki, J. Bak-Misiuk, L. Bryja, B. Surma, G. Gawlik, Structure of oxygen-implanted silicon single crystals treated at ≥ 1400 K under high argon pressure, *Cryst. Res. Technol.* **36**, 933-941 (2001).
2. Misiuk, J. Katcki, J. Ratajczak, V. Raineri, J. Bak-Misiuk, L. Gawlik, L. Bryja, J. Jun, in: *Atomic Aspects of Epitaxial Growth*, edited by M. Kotrla et al., (Kluver Academic Publishers, The Netherlands, 2002), pp. 457-466.
3. Misiuk, J. Bak-Misiuk, L. Bryja, J. Katcki, J. Ratajczak, J. Jun, B. Surma, Oxygen precipitation in Si:O annealed under high hydrostatic pressure, *Acta Phys. Polon. A*, **101**, 719-727 (2002).
4. Misiuk, L. Bryja, J. Katcki, J. Ratajczak, Effect of uniform stress on SiO_2/Si interface in oxygen implanted Si and SIMOX structures, *Optica Applicata*, **32**, 397-407 (2002).
5. Misiuk, A. Barcz, J. Ratajczak, I.V. Antonova, J. Jun, Hydrostatic pressure effect on the redistribution of oxygen atoms in oxygen-implanted silicon, *Solid State Phen.* **82-84**, 115-120 (2002).
6. Misiuk, A. Barcz, J. Ratajczak, L .Bryja, Effect of high hydrostatic pressure during annealing on silicon implanted with oxygen, *J. Mater. Sci.: Mater. Electr.* **14**, 295-298 (2003).

SIO$_2$ AND SI$_3$N$_4$ PHASE FORMATION BY ION IMPLANTATION WITH IN-SITU ULTRASOUND TREATMENT

O. Martinyuk[2], D. Mazunov[1], V. Melnik[1], Ya. Olikh[1], V. Popov[1], B. Romanyuk[1] and I. Lisovskii[1]

[1]V. Lashkarev Institute of Semiconductor Physics of the National Academy of Sciences of Ukraine, 41 Prospect Nauki , Kyiv 03028, Ukraine; [2]National Technical University of Ukraine "Kiev Politechnikal Institute", 37 Prospect Peremogy, Kyev 03056, Ukraine

Abstract: For stimulation of ultra-thin dielectric layer formation we used *in-situ* ultrasound (US) excitation of the silicon wafer during N$^+$ or O$^+$ ion implantation. ToF-SIMS dopant profiling and infrared transmission spectroscopy has been used to analyze the buried film structure and composition.. The US treatment during an implantation gives in more effective SiO$_2$ phase growth in the area of R$_p$-ΔR$_p$. The thickness of the buried layer is ~5 nm less in comparison with the case of implantation without US. For the samples implanted by nitrogen, ultrasonic treatment leads to shrinkage of nitrogen distribution profile and its shift to a surface.

Key words: Silicon-on-Insulator; Ion Implantation; Ultrasound Treatment; Point Defects; Buried Layer; Oxygen; Nitrogen.

1.   INTRODUCTION

Silicon-On-Insulator (SOI) material synthesized by Separation by Implanted Oxygen (SIMOX) remains a leading candidate for advanced, large-scale, integrated circuit applications due to thickness uniformity and moderate defect density.[1] Presently, the most widely used SIMOX is material, typically produced by high-dose implantation of 1.8·10^{18} cm^{-2} of O^{16} ions at ~600°C, which results in a buried oxide (BOX) layer thickness of ~300 nm. However, in the recent years, there has been a growing interest in low-dose SIMOX (dose <1.0·10^{18} cm^{-2}), with BOX layer thicknesses from 80 to 200 nm, because of potential technological and economic advantages over

D. Flandre et al. (eds.), Science and Technology of Semiconductor-On-Insulator Structures and Devices Operating in a Harsh Environment, 97-102.

high-dose SIMOX. At lower dose of $2 \cdot 10^{17}$ cm^{-2} a layer did not form, but only disjointed, isolated oxide precipitates are developed.[2]

To promote the formation of ultra-thin (<0.1 μm) BOX, the implantation process was modified to produce a microstructure, which promotes coalescence of the oxygen into a continuous layer.[3] The method which yielded the best results consists of a standard implantation at $3 \cdot 10^{17}$ cm^{-2} followed by a dose of 10^{15} cm^{-2} at room temperature.

Nucleation and growth of the SiO$_2$ precipitates depends on the set of factors, especially on the point defect concentration in the precipitate formation region. For example, it was shown in [4] that carbon enhances oxide precipitation via C_s-Si_i interaction. Reducing of the supersaturation of Si_i and the strain fields around precipitates implies the enhancement effect on oxide-precipitate growth.

Increase of the vacancy concentration in the region of precipitate nucleation and growth also stimulates these processes. The defect concentration may be changed by different ways: annealing in the various ambients, additional ion implantation, deposition of layers on the sample surface.

It is determined nowadays that ultrasound treatment (UST) influences the defect structure and electro-physical characteristics of silicon and silicon-based structures.[5] We had shown[4] that the use of in situ UST during the implantation process leads to the separation of point defects and accumulation of the vacancy clusters within the R_p - ΔR_p region. In the given paper we describe the investigations devoted to the UST influence on the buried SiO$_2$ and Si$_3$N$_4$ layer formation.

2.   EXPERIMENTAL

(100) boron-doped Cz-silicon samples (10 Ω·cm) were mounted inside the implantation chamber on piezoelectric transducers via acoustic binders. The cell with a sample and US transducer in shown in Fig.1.

Ultrasound vibrations were generated in the wafer by operating the transducer in a resonance mode. The basic resonance frequency was of 6 MHz. The amplitude of the generated deformations did not exceed 10^{-5} of the lattice constant, corresponding to an acoustic power of 1 W·cm^{-2}.

We used N$^+$ or O$^+$ ion implantation with the energy of 25 keV and dose of $2 \cdot 10^{17}$ cm^{-2}. The ion flux was of $3 \cdot 10^{12}$ ions·cm^{-2}·s. After implantation we used the furnace annealing in Ar ambient at the temperature of 1200°C (1 hour).

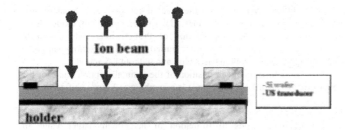

Figure 1. Schema of the cell for *in-situ* UST.

IR-transmission spectra have been measured within the range of 700 – 1400cm^{-1} using spectrometer IKS-25. Initial silicon wafer was used as a reference sample.

ToF-SIMS dopant profiling was performed with a ToF-SIMS IV system in the dual beam mode with 0.5 kV Cs^{+} sputtering and 10 kV Ar^{+} primary ion beam for secondary ion generation.

3.   RESULTS AND DISCUSSION

IR spectra (Fig. 2a) had typical shape for the case of absorption by silicon-oxygen phase with two well pronounced bands (maximum positions at ~800 and ~ 1070 cm^{-1}), which are known to be connected with symmetrical and asymmetrical vibrations of oxygen atoms in Si-O-Si bridges, correspondingly.[6]

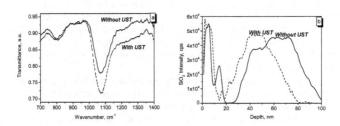

Figure 2. Influence of UST on SiO₂ phase formation after thermal annealing at 1200° C, 1h.; a- IR spectra, b-ToF-SIMS depth profiles.

A long high-frequency tail on the spectral curves is also usual for this case, but was observed only for rather thick (more than 40 nm) oxide films.[7] These facts evident that implantation of the O^+ ions and subsequent annealing resulted in appearance of the buried layer of silicon-oxygen phase (SiO_x) in silicon wafer both in cases with and without US action.

At the same time, the peak position of the main absorption band for the samples treated with US (1078 cm^{-1}) is something (~ 7 cm^{-1}) shifted to high frequency range, if compare with US-untreated crystal, and integral absorbance for these samples is also higher ($\sim 20\%$). These facts should mean that US treatment leads to formation of layer with larger value of stoichiometry index x, which is close to 2.

Fig.2b shows the SIMS profiles for an implanted sample with and without UST after annealing. The US-untreated samples show a broad oxygen distribution, which is caused by a sub-stoichiometric SiO_2. The distribution separates into two peaks, one of which is at $R_p = 65$ nm, and other at the damage peak at $R_p\text{-}\Delta R_p = 50$ nm. The oxygen distribution profile for the sample implanted with UST has one maximum in the range of $R_p\text{-}\Delta R_p \sim 45$ nm. Oxygen concentration in a maximum is substantially higher then for the sample implanted without UST and corresponds to stoichiometrical one for the SiO_2 film of 15 nm thickness.

IR spectra of the nitrogen-implanted samples (Fig.3a) had some pronounced bands (peak positions were ~ 840, 905, 940 and near 1060 cm^{-1}), which are characteristic for absorption on Si-N and Si-O bonds. According to literature data [8,9] the 830 cm^{-1} band is connected with absorption on Si-N bonds in amorphous S_3N_4 films, the band 900 cm^{-1} characterizes Si-N bonds in amorphous SiO_xN_y films, whereas the band 935 cm^{-1} is the main absorption band in crystalline $\beta\text{-}S_3N_4$ phase.

For the samples untreated with US the band 940 cm^{-1} was rather weak, this fact means that in this case nitrogen implantation leads to formation of mainly amorphous silicon-nitrogen phase. If ion implantation was combined with US treatment this band markedly (1.5 times) increased, whereas the 940cm^{-1} band increased drastically (about order of magnitude). Hence, US action leads not only to growth of silicon-nitride buried layer, but this layer is to a high extent crystalline.

Absorption on Si-O vibrational mode changed due to US treatment, but the maximum position of the corresponding band changed remarkably (1070 cm^{-1} in comparison with 1055 cm^{-1} for US-untreated sample), and absorption decrease also takes place. It may be caused by the increase of the surface oxide thickness during annealing the samples, implanted without UST, whereas implantation with UST leads to the enrichment of the surface by nitrogen, and it blocks Si oxidation during annealing process.

Nitrogen distribution profiles for the samples with and without UST are shown in Fig 3b. One can see that UST leads to some shift of the distribution profile toward sample surface and to its narrowing. Nitrogen concentration increases in the sub-surface region in the samples implanted with the UST.

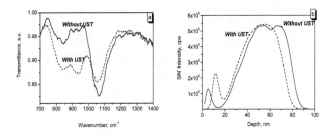

Figure 3. Influence of UST on Si₃N₄ phase formation after thermal annealing at 1200° C, 1h. a- IR spectra, b-ToF-SIMS depth profiles.

It is known that the kinetics of Si₃N₄ and SiO₂ dielectric phases is influenced by point defects. So, the presence of vacancies stimulates the SiO₂ phase precipitation and growth, whereas synthesis of a Si₃N₄ phase needs an excess concentration of interstitial atoms. We have shown[4] that the excitation of ultrasonic waves in a wafer during an implantation gives to spatial separation of the point defects. The interstitial silicon atoms under US wave action diffuse deep into the wafer, whereas the vacancies collect in complexes and remain in an area of an implantation. Such redistribution of the point defects does gives in the change of the buried layer formation kinetics in silicon.

4. CONCLUSIONS

It is shown that using the US treatment the ion-induced morphology near the R_p range can be manipulated to facilitate or promote the formation of higher quality buried layers, and gives the possibility to create super-thin stoicheometrical SiO₂ buried layers at doses $< 4 \cdot 10^{17}$ cm⁻².

ACKNOWLEDGEMENT

This work was partly supported by the STCU (Project #2367).

REFERENCES

1. A. Proßl et al., Silicon on insulator: materials aspects and applications, *Solid State Electron.*, **44**, 775-788 (2000).
2. S. Bagehi et al., Dose dependence of microstructural development of buried oxide in oxygen implanted SOI material, *Appl. Phys. Lett.*, **71**(15), 2136-2138 (1997).
3. O.W. Hollond et al., Formation of ultrathin, buried oxides in Si by O^+ ion implantation, *Appl. Phys. Lett.*, **69**(5), 674-676 (1996).
4. B. Romanyuk et al., Modification of the Si amorphization process by in situ ultrasonic trestment during ion implantation, *Semicond. Sci. Technol.*, **16**, 1-5 (2001).
5. S. Ostapenko et al., Change of minority carrier diffusion length in policrystalline silicon by ultrasound treatment, *Semicond. Sci. Technol.*, **10**, 1494-1500 (1995).
6. A. Lehmann et al., Optical phonons in amorphous silicon oxides. I. Calculation of the density of states and interpretation of LO-TO splittings of amorphous SiO_2, *Phys. Stat. Sol. (b)*, **117**(2), 689-698 (1983).
7. I.P. Lisovskyy et al., IR spectroscopic investigation of SiO_2 film structure, *Thin Solid Films*. **213**, 164-169 (1992).
8. T.A. Kruse et al., Peculiarities of the IR absorption on Si-N bonds in silicon implanted by nitrogen, *Non-Organic Materials (in Russian)*, **26**(12), 2457-2460 (1990).
9. B.N. Romanjuk et al., Processes of formation of buried insulating layer in Si at implantation of N^+ and O^+ ions, *Ukr. Journ. Phys. (in Russian)*, **37**(3), 389-394 (1992).

FABRICATION AND CHARACTERISATION OF SILICON ON INSULATOR SUBSTRATES INCORPORATING THERMAL VIAS

M. F. Bain, P. Baine, D. W. McNeill, G. Srinivasan, N. Jankovic, J. McCartney*, R. A. Moore*, B. M. Armstrong and H. S. Gamble
Northern Ireland Semiconductor Research Centre, School of Electrical and Electronic Engineering. The Queen's University Belfast, Northern Ireland. Belfast BT9 5AH
Microelectronics and Electrical Engineering, Trinity College, University of Dublin. Dublin 2. Ireland

Abstract: SOI substrates incorporating a thermal via (TV) in the buried oxide layer (BOX) were successfully produced. Various via refill were attempted and all were found to bond successfully. Inspection of the bond interface showed no micro void formation due to the thermal via. Raman analysis verified that the introduction of a buried TV structure did not cause stress in the adjacent SOI. Electrical measurements confirmed the breakdown voltage of the TV structure was similar to that of the blanket depositions and the integrity of the layers were not adversely affected by the fabrication process.

Key words: SOI, Thermal Via, planarisation

1.   INTRODUCTION

Self-heating in high performance electronic circuits results in the degradation of the device performance, limiting both speed and reliability.[1, 2] Devices on SOI substrates are particularly susceptible to these heating effects due to the poor thermal conductivity of the buried oxide layer (BOX). A possible solution to this is the incorporation of a localised thermal via (TV) to act as a heat conduit through the BOX between the SOI and the bulk handle silicon. Thus rather than replacing the entire BOX with a more thermally conductive material or stack, the work has concentrated on producing vias through the BOX refilled with suitable materials.

The main advantage of using tailored micro columns as thermal vias rather than substituting the entire BOX is that due to their dimensions (<20 microns) the vias have a negligible effect on the substrate parasitic

D. Flandre et al. (eds.), Science and Technology of Semiconductor-On-Insulator Structures and Devices Operating in a Harsh Environment, 103-108.

capacitance. Consideration of the materials to be used in the TV is also an important factor. To improve the thermal conductivity it was understood early on that the major contributor to the TV stack would be LPCVD polysilicon due to its relatively high thermal conductivity. At high frequencies the undoped polysilicon will behave as a dielectric helping to reduce capacitance (despite its relatively high permitivity). A further advantage of this process is that the via can be positioned to dissipate heat at a required location whilst the isolation properties of the SOI are maintained elsewhere.

The aim of this work was to incorporate a thermal via structure into an SOI substrate (TVSOI), which can offer an improvement in thermal conductivity over thermally grown silicon dioxide. Then to characterise any mechanical effect the buried TV structure has on the SOI and to investigate the electrical properties of the as fabricated via.

2.   EXPERIMENTAL

Initial experiments to identify potential thermal via multi layers were carried out on blanket depositions. The thermal and electrical properties were determined using thermal resistor and capacitor structures respectively. The technique used to measure thermal conductivity is described in greater detail in other publications.[3, 4]

The materials used had to be IC compatible and capable of withstanding the post bond processing temperatures if they were to be incorporated into the SOI substrate. These criteria limited the materials to: LPCVD polysilicon, silicon nitride, amorphous silicon and thermally grown silicon dioxide. Various combinations of these materials were investigated, a summary of the thermal and electrical results are presented in Table 1. The selection criteria required the structure be capable of supporting 30V.

Table 1. Electrical/Thermal properties of thermal via stack combinations

Thermal Via Material	Breakdown Voltage (V)	Thermal Cond (W/mK)
1. Thermal Oxide (0.1microns)	98	0.6
2. Silicon Nitride (0.1microns)	97	1.44
3. Ox/poly/Ox (20nm/1000nm/20nm)	18-24	2.4
4.Ox/AmorphSi/PolySi/Ox (20nm/200nm/800nm/20nm)	38-44	2.8 - 3.5

Table 1 shows that silicon nitride offers over a two fold increase in the thermal conductivity for layers 0.1 microns thick (1&2). The breakdown voltage exhibited by sample 3 (ox/poly/ox) did not meet the requirement. It is proposed that this is due to secondary grain growth compromising the oxide polysilicon interface. In order to reduce the roughness and improve the oxide/silicon interface, the polysilicon layer was replaced by a combination of amorphous and polysilicon depositions (4). The breakdown voltage improved significantly from 24 to 44 volts. Potentially the best results would be obtained from a nitride/amorphousSi/polySi/nit multi layer.

3. FABRICATION OF SOI INCORPORATING THERMAL VIAS

Trenches in a 1micron thick oxidised handle wafer were patterned by reactive ion etching (RIE) using CF_4/CHF_3 chemistry.

Various via refill materials were attempted to establish any possible processing issues: Ox/PolySi/Ox, Ox/AmorphSi/PolySi/Ox, Nit/PolySi/Nit, Nit/AmorphSi/PolySi/Nit. A 2.5 micron polysilicon layer is deposited over all the substrates, and used as a planarisation layer. The thickness of the polysilicon layer must be greater than that of the trench depth to allow planarisation. The oxide layer was used as polish stop. Planarisation of the polysilicon was achieved by using a combination of slurries Nalco 2355 and 2350 with a Suba 500 stock removal pad.[5]

The dishing in the polysilicon after planarisation was measured using a tencor α-step surface profilometer. A scan taken after planarisation across a 10 microns wide structure showed a 10nm dishing in the polysilicon when compared to the oxide surface. Once planarisation was complete the final layer of the TV structure was deposited. Finally a polysilicon layer was then deposited over the entire substrate and polished to produce a bondable surface. After a high temperature bond anneal (1000C 2hrs) the bond interface was examined by scanning acoustic microscopy (SAM).

If any micro voids formed above the TV the performance of the via would be catastrophically compromised. SAM analysis showed no void areas due to trench patterns in the handle wafer. The lack of voids was further verified by tape test after grinding, which showed no delamination of the SOI above the patterned regions. SEM cross sectional analysis showed no void regions at the bond interface due to the patterned handle Fig. 1.

All substrates bonded successfully and the quality of the bond was found to be independent of the refill used. The TVSOI substrates were then polished back to the required thickness of 2 microns. It was found that mask alignment could be performed using the buried thermal via alignment marks.

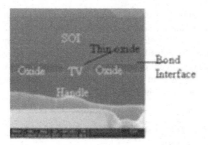

Figure 1. Cross section of Thermal Via structure incorporated in to an SOI substrate

4.   CHARACTERISATION OF TVSOI

Raman Characterisation

Raman analysis was employed to establish what effect the buried TV structure has on the SOI properties. Raman spectra were registered in backscattering geometry using a RENISHAW 1000 micro-Raman system equipped with a Leica microscope. To prevent sample heating the laser power was kept below 10mW. The measurements were performed at room temperature with 514.5 nm line of an Ar+ laser. An 1800 lines/mm grating was used in all measurements, which corresponds to a spectral resolution of ~ 1cm-1. The 100 times magnifying objectives of the Leica microscope focuses the beam into spot of about 1 micrometer in diameter. In order to define the position of the phonon lines with a higher accuracy, the spectral lines, used for the analysis, were fitted with Lorentzian functions.

A typical Raman spectrum for this type of structure will have phonon bands in the range of 520cm which belong to the Si-Si stretching mode. The presence of Si-Si peak associated with Si substrate (or Si buffer layer) will depend on how deep the laser probe beam will penetrate the sample during Raman measurements. This will depend on the absorption coefficient of the material at the wavelength of the exciting laser line and was calculated to be approximately 763nm.

By etching a series of mesas into the SOI above a long 8μm wide trench analysis was carried out at various thickness of SOI, as shown in Fig. 2. The mesas were defined using wet etching and yielded SOI thicknesses, greater than the depth of penetration, of approximately, 1, 1.5 and 3 microns. Raman analysis was performed using a line mapping technique, where measurements were taken across the whole via. The resulting Raman spectrum is shown in Fig.2. It can be seen that the shift from the unstressed

Si-Si position is negligible, the spectra were found to be constant through the SOI layer for the thickness values prepared, confirming that there is no stress introduced through the creation of the thermal vias.

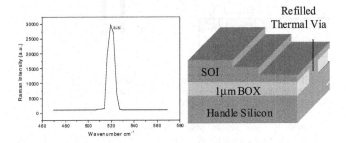

Figure 2. Raman analysis of TVSOI substrates at various SOI thicknesses over a buried trench structure.

Electrical Characterisation

Previous work, as shown in table 1, investigated the electrical properties of the blanket thermal via stack. In order to investigate the electrical properties of the as fabricated TVSOI substrates test substrates were produced in the same way but remained unbonded. Capacitors were fabricated on these substrates to investigate the breakdown characteristics of the TV structures.

Two types of TV refills were to be investigated: nitride/polysilicon/oxide and nitride/amorphous silicon/polysilicon/oxide. Oxide was formed at the top of the TV structure for ease of fabrication. Unfortunately this means the results cannot be compared directly with those obtained for blanket deposition. In order to fabricate the capacitor test structures all back depositions were removed and a polysilicon layer was deposited over the TVs to prevent aluminium punch through. Aluminium was then deposition front and back of the wafer. The aluminium and polysilicon were patterned over the largest structures creating the capacitors, as shown in Fig. 3.

The breakdown voltage value was taken when the leakage current reached 60 micro amps a summary of the results is shown in Fig. 3(b). From Fig. 3(b) it is clear that for the vias refilled with a combination of amorphous and polysilicon the breakdown voltage is significantly higher. Again this is attributed to a cleaner interface between the silicon and the oxide due to the presence of the amorphous silicon.

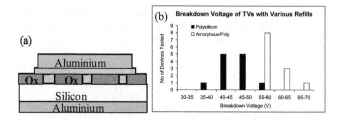

Figure 3. Breakdown characterisation of as fabricated TV structures (a) the test structure (b) breakdown voltages on two substrate (nitride/polysilicon/oxide and nitride/amorphous silicon/polysilicon/oxide)

Unfortunately these results cannot be compared directly to any bulk values as the via consisted of a nit/Si/ox structure. The breakdown characteristic however are very close to the values which would be expected, strongly suggesting the fabrication process does not compromise the electrical isolation of the via. Processing of the materials may even improve the breakdown characteristics, as the planarisation step will reduce the roughness of the polysilicon prior to the final dielectric deposition.

5.   CONCLUSIONS

Thermal vias have been successfully incorporated into SOI substrates it was found that of the via refill materials investigated all could be successfully bonded. No micro voids were detected that were attributable to the via structure. Raman analysis showed the buried TV structures had no measurable effect on the adjacent SOI. Electrical analysis confirmed the fabrication process did not compromise the integrity of the stack and the isolation provided by the stack was well within the specification.

REFERENCES

1. B.M. Tenbroek et al. IEEE Trans. Electron Devices. Vol. 43, No. 12, (1996), 2240 2248
2. L.T. Su, IEEE Trans. Electron Devices, Vol. 41, 1,(1994), 69-75
3. P.Baine, K.Y. Choon, H.S. Gamble, B.M. Armstrong and S.J.N. Mitchell. Journal of Material Science: Materials in Electronics 12 (2001)
4. P.T. Baine, D.L. Gay, B.M. Armstrong and H.S. Gamble J. Electrochem. Soc., Vol. 145, No. 5, Mat 1998 pp1738-1743.
5. B.M. Tenbroek, G.Whiting, M.S.L. Lee and C.F. Edwards. IEEE Trans. Electron Devices 46 (1999)

RELIABILITY AND ELECTRICAL FLUCTUATIONS IN ADVANCED SOI CMOS DEVICES

Jalal Jomaah and Francis Balestra
IMEP, ENSERG, 23 rue des Martyrs, BP 257, 38016 Grenoble Cedex 1, France

Abstract: With SOI to be adapted as a mainstream technology in the forthcoming years, two issues are still of a critical interest regarding circuit applications: low frequency noise behavior and hot-carrier reliability, in state-of-the art SOI MOSFETs. Both are thoroughly investigated in this paper and a set of recent results in advanced CMOS SOI devices is given.

Key words: Silicon-on-Insulator (SOI); Partially-depleted (PD); Floating body effect (FBE), Fully-depleted (FD); Low frequency noise (LFN); Hot-carrier reliability

1. INTRODUCTION

Thanks to their structure, the SOI technologies present several intrinsic properties for analog and RF applications. For instance, as it is well established now, these interesting devices allow the reduction of the power consumption at a given operating frequency. Moreover, the high insulating properties of SOI substrates, in particular with the use of high resistivity material, leads to high performance mixed-signal circuits.[1-3] However, in order to use this kind of device in such applications, low-frequency noise, which can directly impact RF or analog integrated circuits, needs to be thoroughly evaluated. Following the specifications of the ITRS Roadmap, the 1/f-noise amplitude is predicted to decrease in modern technologies. But, the maximum signal is also lowered with decreasing operation voltage leading to a deterioration of the signal to noise ratio. Therefore, accurate characterization of the noise behavior has to be established.

D. Flandre et al. (eds.), Science and Technology of Semiconductor-On-Insulator Structures and Devices Operating in a Harsh Environment, 109-120.
© 2005 *Kluwer Academic Publishers. Printed in the Netherlands.*

In this paper, LFN in N- and P-channel SOI MOSFETs is investigated. An overview of the LFN behavior for two different structures, i.e., Fully- and Partially-Depleted Si films, is also given. Furthermore, the impact on the electrical noise of the shrinking between 0.25 and 0.13µm SOI CMOS technology nodes will be shown, with different types of layout: floating body (FB) and body-contacted (BC) devices. A particular attention will be paid to the floating body effect that induces a kink-related excess noise, which superimposes a Lorentzian spectrum on the flicker noise. We also consider the gate-to-body tunneling current giving birth to the GIFBE as a new noise source whose impact on the LFN behavior of SOI MOSFETs is analyzed.

From another aspect, with Partially Depleted SOI is becoming a major trend among all the candidates to further improve VLSI applications, hot-carrier induced degradation is still an increasingly important issue since device dimensions continue to be scaled down. This has been a subject of interest over the last past years, but became a novel interest with the development of ultra-thin oxides since changes in the commonly admitted worst-case aging scenarios were observed. Recently, the first studies appeared on HCI in ultra-thin gate oxides[4-8] PD SOI MOSFETs, but they were either performed on SIMOX substrates or no special information was given concerning the buried oxide, which plays a role on the hot-carrier degradation. They report a worst-case aging condition for $V_g = V_d$, instead of $V_g = V_t$.[9] The worst-case (WC) conditions for HCI degradation of sub-100 nm PD SOI MOSFETs were also discussed,[6] and the authors concluded that both $V_g = V_d$ and $V_g = V_d/2$ were important regimes to consider for the determination of the WC scenario depending on the chosen drain bias during stress. In this work, we present results obtained on Unibond SOI devices. Since FB MOSFETs may exhibit a different degradation behavior compared to BC[8] transistors depending on the applied drain bias, we also address some comparisons between the different device architectures.

2.   DEVICES USED

A 0.13µm SOI CMOS PD technology, processed on conventional Unibond wafers and provided by STMicroelectronics (Crolles, France), is used in this study. Floating-Body (FB), Body-Contacted (BC) and Body-Tied (BT) devices with the following characteristics are considered: front nitrided gate oxide thickness $t_{ox1} = 2nm$, Silicon film thickness $t_{Si} = 160nm$ with channel doping close to 1.10^{18} cm^{-3} and back oxide thickness $t_{ox2} = 400nm$. N$^-$ LDD zones are implanted on both sides of the channel, as well as P$^+$ HALO pockets. In BC devices, the Silicon film is externally grounded using an additional control whereas in BT devices the body is directly connected to the source. As regards Fully Depleted devices, front oxide thickness is t_{ox1}= 4.5nm for 0.25μm CMOS technology and Silicon film thickness t_{Si}= 40nm. For both FD technologies, back oxide thickness t_{ox2} is 400nm. Advanced SOI FD transistors with channel length down to 0.1μm fabricated in CEA/LETI have the following characteristics: 10-18nm film thickness (t_{Si}), 2nm thick pure SiO_2 gate oxide (t_{ox1}).

3.   LOW FREQUENCY NOISE

Noise measurements at low and high drain voltages were carried out in the 1 Hz to 100 kHz frequency range for both N- and P-MOSFETs. The DC bias is supplied to the device under test by the Berkeley Technology Associate BTA 9812B noise analyzer. The noise current of the tested device is amplified by the low-noise amplifiers of the BTA 9812B system before being applied to the Stanford Research SR 780 dynamic signal analyzer for FFT (fast Fourier transform). A personal computer installed with *Noise Pro 2000.2.8*[TM] allows an automation of the whole measurement process.

3.1   Noise at low drain bias

We will now consider low frequency noise measurements carried out for PD and FD devices in the linear regime to identify the main noise sources at the front gate oxide interface. Figure 1 shows, in linear operation (V_d=50mV), the normalized drain current power spectral density S_{Id}/I_d^2 plotted as a function of the drain current for N-channel PD SOI MOSFETs (Fig. 1.a) and for N-channel FD SOI MOSFETs (Fig. 1.b) for two channel lengths: L=0.25 and 0.12μm (0.10μm in Fig. 1.b). In this plot, the straight line represents the front gate power spectral density S_{Vg} multiplied by the ratio $(G_m/I_d)^2$ where G_m stands for the gate transconductance. A good correlation is obtained between these two amounts, confirming results predicted by the McWhorter model which associates the 1/f noise to carrier number fluctuations.[21]

In this model, the fluctuations of the drain current are due to those of the inversion charge near the Si-SiO_2 interface caused by the dynamic trapping/detrapping of free carriers onto traps located in the oxide near the interface. The normalized drain current spectral density is given by:[10]

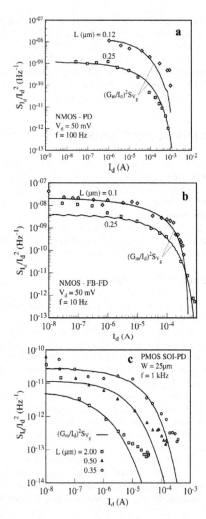

Figure 1. Normalized drain current power spectral density S_{Id}/I_d^2 at V_d=50 mV at different channel lengths for (a) FB Partially Depleted N-MOS, (b) Fully Depleted N-MOS and (c) FB Partially Depleted P-MOS SOI devices. Full lines : $S_{Vg}*(G_m/I_d)^2$.

$$\frac{S_{Id}}{I_d^2} = \left(\frac{G_m}{I_d}\right)^2 S_{Vg} \tag{1}$$

Moreover, some difference in strong inversion can be observed (case of P-channel) (Fig. 1.c) which is attributed to correlated mobility fluctuations. Indeed, by taking into account the dependence of the carrier mobility on the insulator charge (Coulomb scattering), the fluctuations of the insulator charge give rise to a supplementary change of the mobility, which induces an extra drain current fluctuation. The drain current fluctuations can then be evaluated as:

$$\frac{S_{Id}}{I_d^2} = \left(1 \pm \alpha\mu_{eff} C_{ox} \frac{I_d}{G_m}\right)^2 \frac{G_m^2}{I_d^2} S_{Vg} \tag{2}$$

where μ_{eff} is the effective mobility, C_{ox} is the gate oxide capacitance and α is a parameter in the order of a few 10^4 V/s.

3.2   Excess noise due to kink effect at high drain bias

Low frequency noise characterization of SOI devices also needs to be carried out in the saturation mode whose bias conditions give birth to the kink-related excess noise observed in SOI devices. The drain current spectral density S_{Id}, normalized by W, is plotted (Fig. 2) for a fixed gate overdrive voltage V_{gt} versus the applied drain voltage V_d for 0.12μm FB PD SOI and for L=0.25μm with FB and BC PD structures.

A substantial kink effect is clearly observed in the case of FB PD devices, and for a drain bias superior to the kink onset drain voltage a low-frequency excess noise occurs. The noise peak shifts towards higher V_d with increasing frequency. Indeed, the floating body effect induces a kink-related excess noise, which superimposes a Lorentzian spectrum. Nevertheless, the kink effect disappears when the SOI film is contacted to the ground.

Several mechanisms have been proposed to explain this excess noise, such as trap-assisted generation-recombination noise[11] or body-source shot noise amplified by floating body effect.[12] Note also that the magnitude for L=0.12μm is only one decade high, contrary to the 0.25μm SOI CMOS technology for which two orders of magnitude were usually obtained.[13] This results from a better control of FBEs in shorter lengths devices for which the kink excess drain current is also reduced because of enhanced short-channel effects.

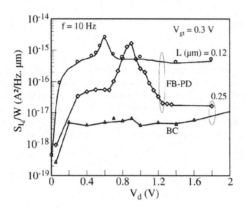

Figure 2. Drain current power spectral density, normalized by the width W, at f=10Hz for FB PD devices with two channel lengths (L=0.25 and 0.12μm) and for a BC device with L=0.25μm.

3.3   Excess noise due to Gate-Induced Floating Body Effect at low drain bias

Low frequency excess noise associated with the effect of gate-induced floating body is reported in Partially Depleted SOI MOSFETs with ultra-thin gate oxide.[14] This is investigated with respect to floating body devices biased in linear regime. Due to a body charging from the gate, a Lorentzian-like noise component superimposes to the conventional 1/F noise spectrum. This excess noise exhibits the same behavior as the Kink-related excess noise previously observed in Partially Depleted devices in saturation regime. Indeed, scaling metal-oxide-semiconductor (MOS) devices to very-deep submicron dimensions has resulted in an aggressive shrinking of the gate oxide thickness.

In this ultra-thin gate oxide range, direct tunneling from the gate clearly appears, and increases exponentially with decreasing oxide thickness.[15] These currents strongly affect the body potential causing the Gate Induced Floating Body Effects (GIFBE) observed even in linear regime.[16] Figure 3.a illustrates the drain current and transconductance measured in linear regime for V_D= 10 up to 50 mV for a W/L=10/0.12 μm FB-PD N-MOSFET. A sudden increase of the drain current occurs close to V_G=1.1-1.2 V whatever the drain bias is, and this results in an unforeseen second hump in the transconductance characteristic, whose value exceeds the normal peak obtained at lower gate voltage. The same feature is observed for P-

MOSFETs. Then, we considered the drain current power spectral density versus the applied front gate bias for two different frequencies (F=10 and 80 Hz). The results are plotted in Figure 3.b for a W/L=10/1 μm FB-PD SOI MOSFET biased with a drain voltage V_D=50 mV. The noise overshoot magnitude attributed to the GIFBE is almost two decades for this device. The shift of the noise peak towards higher gate biases with increasing the measurement frequency is clearly shown in Fig. 3.b.

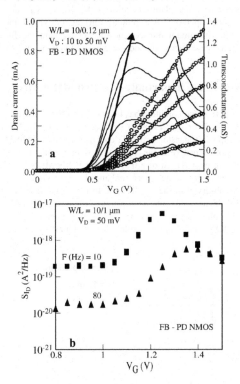

Figure 3. (a) Transfer characteristics I_D and $G_m(V_G)$ of a W/L=10/0.12 μm FB-PD N-MOSFET at low drain voltage (V_D=10 to 50 mV) and (b) Drain current power spectral density of a W/L=10/1 μm FB-PD N-MOSFET with V_D=50 mV. The two different frequencies are F=10 and 80 Hz.

4.   HOT CARRIER RELIABILITY

Both N- and P-MOSFETs were stressed during approximately 27500 seconds, i.e. 7,5 hours. Device stressing is performed on HP4155 semiconductor parameter analyzers, static measurements in linear region (V_d=50 mV) with the back gate and/or the body grounded are carried out at each intermediate stress point. Various drain and gate biases are considered to cover the different degradation conditions (V_g=V_t, V_g=V_d/2 and V_g=V_d). During the stress operation, we monitor on the one hand the relative variations of the maximum transconductance G_{mmax}, and the linear drain current I_d at fixed gate overdrive, and the threshold voltage (V_t) shift on the other hand.

4.1   Worst-case in partially-depleted SOI devices

The worst-case aging condition has been debated for years to be whether the I_{submax} (V_g=V_d/2) or I_{gmax} (V_g=V_d) stress condition, or for V_g=V_t where the parasitic bipolar transistor action is enhanced. SOI N-MOSFETs were known to be more sensitive to the V_g=V_t stress condition.[17] Figure 4 shows the G_{mmax} relative degradation of a 0.25 µm FB PD N-MOSFET with three degradation conditions. The more enhanced sensitivity to hot-carrier degradation is observed in the V_g=V_t case. In contrast, for 2nm gate oxide 10/0.12µm FB devices on Unibond substrates, Figure 5 shows a worst-case aging scenario obtained for V_g=V_d which was also recently reported in PD devices on SIMOX with ultra-thin oxide.[18] The width effect was also studied

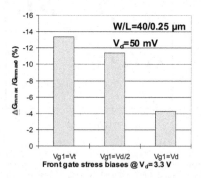

Figure 4. Maximal transconductance relative degradation in 40/0.25µm FB N-MOSFETs for $V_{dstress}$=3.3 V under 3 different front gate bias conditions.

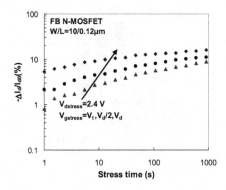

Figure 5. Drain current degradation measured at V_{gt}=0.5V with V_d=50 mV for the three bias conditions during the stress operation.

with W ranging from 10 down to 0.4μm in the worst-case aging condition, and no enhancement of the degradation is observed contrary to the case of bulk Silicon N-MOSFETs from various technologies.[19-20]

In the case of P-MOSFETs, we reported a noticeable degradation only for the V_g=V_d case, confirming that the hot-carrier reliability has now to be considered for the ultra-thin gate oxides, and not only the Negative Bias Temperature Injection (NBTI) which is usually studied for these devices. We address the hot-carrier issue in P-MOSFETs in a next section of this paragraph.

4.2　Hot-carrier degradation in various device architectures

In Figure 6.a is reported the relative degradation of the drain current at constant gate overdrive (V_{gt}=0.5 V) for 5/0.12μm FB, BC & BT N-MOSFETs for the I_{gmax} (V_g=V_d) degradation condition. BC devices exhibit an enhanced hot-carrier immunity because of the collected holes coming from impact ionization at the drain edge. A similar trend is also observed at V_g=V_d/2 (not shown here) where BC devices are also less degraded than FB ones. The main difference between the three considered architectures comes from the shift of the threshold voltage, which is stronger in FB devices, associated with a more pronounced electron injection as outlined in Figure 6.b where the gate current is also represented both before and after stress for BC and FB devices.

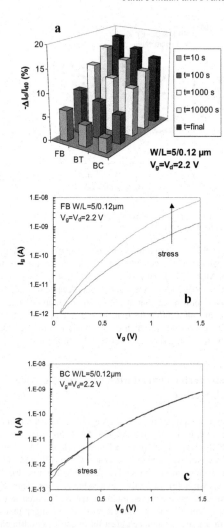

Figure 6. (a) Drain current degradation measured at V_{gt}=0.5 V with V_d=50 mV for various device architectures of W/L=5/0.12 µm N-MOSFETs in the worst-case aging scenario. (b) Gate current generation measured at V_{gt}=0.5 V for both FB and BC N-MOSFETs.

5.   CONCLUSION

Low-frequency noise behavior of Partially and Fully Depleted SOI N- and P-MOSFETs was shown, with the identification of the noise source in each case. The impact of the shrinking between 0.25 and 0.13µm SOI CMOS technology nodes on the LFN was also demonstrated. The emerging GIFBE can also be considered as a new noise source, and consequently give birth to a noise overshoot in the LFN. The kink-related noise overshoot and its suppression in BC devices was reported in saturation mode. In a last part, we paid attention to hot-carrier effects in N-MOSFETs for various device architectures (FB, BC and BT). Contrary to the case of 0.25µm SOI devices, the worst-case aging scenario was in the $V_g=V_d$ stress condition as recently reported on Silicon bulk or SIMOX SOI MOSFETs. Comparing hot-carrier immunity between various device architectures, BC devices exhibited an increased resistance to hot-carrier effects because of the suppression of floating-body effects.

REFERENCES

1. A. K. Agarwal, M. C. Driver, M. H. Hanes, H. M. Hodgood, P. G. McMullin, H. C. Nathanson, T. W. O'Keeffe, T. J. Smith, J. R. Szedon and R. N. Thomas, "MICROX™ - An advanced silicon technology for microwave circuits up to X-band", in Proc. IEDM Tech. Dig., pp. 687-690, Dec. 91.
2. D. Eggert, P. Huebler, A. Huerrich, H. Kuerck, W. Budde and M. Vorwerk, "A SOI-RF-CMOS technology on high resistivity SIMOX substrates for microwave applications to 5 GHz", IEEE Trans. on Electron Devices, vol. 44, no. 11, pp. 1981-1989, Nov. 97.
3. O. Rozeau, J. Jomaah, S. Haendler, J. Boussey, and F. Balestra, "SOI Technologies Overview for Low-Power Low-Voltage Radio-Frequency Applications", Analog Integrated Circuits and Signal Processing, 25, 2000, Kluwer Academic Publishers – Special issue of SOI.
4. C.W. Tsai et al., "Valence-band tunneling enhanced hot carrier degradation in ultra-thin oxide nMOSFETs", IEDM Tech Digest, 2000.
5. B.W. Min et al., "Hot carrier enhanced gate current and its impact in short channel nMOSFET reliability with ultra-thin gate oxides", IEDM Tech Digest, 2001.
6. E-X Zhao et al., "Worst case conditions for Hot-Carrier Induced Degradation of sub-100 nm Partially Depleted SOI MOSFETs", 2001 IEEE SOI Conf., pp.121-122.
7. W-K Yeh et al., "Hot-carrier induced degradation on 0.1µm SOI CMOSFET", 2002 IEEE SOI Conf., pp.107-108.
8. J. Chan et al, "Charge pumping study of hot-carrier induced degradation of sub-100 nm Partially Depleted SOI MOSFETs", 2002 IEEE SOI Conf., pp.43-44.
9. S.-H. Renn et al., "Hot-carrier effect and reliable lifetime prediction in deep submicron N- and P-channel SOI MOSFET's", IEEE Trans. El. Dev., Vol. 45, N°11, pp. 2335-2342, Nov. 1998.

10. G. Ghibaudo, O. Roux, Ch. Nguyen-Duc, F. Balestra, and J. Brini, "Improved Analysis of Low Frequency Noise in Field-Effect MOS Transistors", phys. stat. sol. (a) 124, p. 571, 1991.

11. E. Simoen, U. Magnusson, Antonio L. P. Rotondaro and Cor Claeys, "The Kink-related excess low frequency noise in silicon-on-insulator MOST's", IEEE Trans. Electron Devices., vol. 41, pp.330-339, March 1994.

12. Y.-C. Tseng, W.M. Huang, P.J. Welch, J.M. Ford, and J.C.S. Woo., "Empirical correlation between AC kink and low frequency noise overshoot in SOI MOSFETs", IEEE EDL, vol. 19, no 5, pp. 157-159, May 1998.

13. F. Dieudonné, S. Haendler, J. Jomaah, C. Raynaud, K. De Meyer, H. Van Meer, and F. Balestra "Shrinking from 0.25 down to 0.12µm SOI CMOS technology node : a contribution to 1/f noise in Partially Depleted N-MOSFETs", in Proc. 3rd European Workshop on Ultimate Integration of Silicon, Munich, March 2002, pp. 33-36.

14. F. Dieudonné et al., Gate-induced floating body effect excess noise in ultra-thin gate oxide Partially Depleted SOI MOSFETs, IEEE El. Dev. Lett., Vol. 23, N°12, pp. 737-739, Dec. 2002.

15. W-C Lee and C. Hu, "Modeling CMOS tunneling currents through ultra-thin gate oxide due to conduction- and valence-band electron and hole tunneling" IEEE Trans. Electron Devices, vol. 48, pp. 1366-1373, July 2001.

16. J. Pretet, T. Matsumoto, T. Poiroux, S. Cristoloveanu, R. Gwoziecki, C. Raynaud, A. Roveda and H. Brut, "New mechanism of body charging in Partially Depleted SOI-MOSFETs with ultra-thin gate oxides", in Proc. 32nd ESSDERC, Firenze, Italy, September 2002, pp. 515-518.

17. S.-H. Renn et al., "Hot-carrier effect and reliable lifetime prediction in deep submicron N- and P-channel SOI MOSFET's", IEEE Trans. El. Dev., Vol. 45, N°11, pp. 2335-2342, November 1998.

18. W-K Yeh et al., "Hot-carrier induced degradation on 0.1µm SOI CMOSFET", 2002 IEEE SOI Conf., pp.107-108.

19. W. Lee and H. Hwang, "Hot carrier degradation for narrow width MOSFET with shallow trench isolation", Mic. Relia., Vol. 40, pp. 49-56, 2000.

20. E. Li and S. Prasad, "Channel width dependence of NMOSFET hot carrier degradation", IEEE Trans. El. Dev., Vol. 50, N°6, pp. 1545-1548, June 2003.

21. A.L. McWhorter, Semiconductor Surface Physics, University of Pennsylvania Press, Philadelphia 1957.

HYDROGEN AND HIGH-TEMPERATURE CHARGE INSTABILITY OF SOI STRUCTURES AND MOSFETS

A.N. Nazarov
Lashkaryov Institute of Semiconductor Physics, National Academy of Sciences of Ukraine, Prospekt Nauki 41, 03028 Kiev-28, Ukraine

Abstract: This paper surveys the hydrogen behavior (trapping, diffusion and generation) in gate and buried oxides in Metal-Oxide-Silicon (MOS) and Silicon-On-Insulator (SOI) structures and MOSFETs. New phenomena relating to proton generation in buried oxide (BOX) of SOI structures and MOSFETs during high-temperature hydrogen annealing and bias-temperature (BT) stress are considered. Analysis of the high-temperature charge instability, which is created at BT stress in the BOX is performed. The model of proton defect-assisted generation in the BOX due to temperature and negative bias stress is presented.

Key words: Hydrogen, charge instability, Silicon-on-Insulator, SiO_2-Si structures, buried oxide, gate oxide, SOI MOSFET

1.   INTRODUCTION

One of the fastest diffused impurity in SiO_2 is hydrogen. Hydrogen imbedding into SiO_2 network during the dielectric growth or deposition on silicon wafer or hydrogen annealing of the SiO_2-Si structure can result in a decrease of both interface state density and oxide positive charge with subsequent influence on a decrease of radiation hardness at stable operation at harsh conditions. In order to understand the hydrogen and proton behavior in both amorphous gate oxide in metal-oxide-silicon (MOS) structures and buried oxide in silicon-on-insulator (SOI) structures we have to consider the oxide structure and defects associated with hydrogen in these oxides.

121

D. Flandre et al. (eds.), Science and Technology of Semiconductor-On-Insulator Structures and Devices Operating in a Harsh Environment, 121-132.
© 2005 *Kluwer Academic Publishers. Printed in the Netherlands.*

2.   STRUCTURE OF A-SIO₂ NETWORK IN GATE
AND BURIED OXIDES

It's well known that main electronic and optical properties of amorphous SiO₂ (a-SiO₂) is controlled by existing of short-range order in this material, which is connected with main component of dioxide amorphous network, that is SiO₄ tetrahedral blocks[1] (Fig.1a). Bond angles and distances between atoms in these blocks are exactly determined and are not strongly changed for different silicon dioxides. However, the mechanical and electrical properties of a-SiO₂ are associated with long-range order, which is determined by the Si-O-Si bridging bond and the bond angle Φ, linking the neighboring SiO₄ tetrahedrons (Fig. 1b).

It was shown[2] that the minimal bond energy for the bringing bond is obtained at 144^0 of this angle (Fig. 2). Such value of the angle is observed

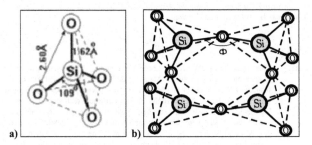

Figure 1. A schematic structure of SiO₄ tetrahedral block (a) and a-SiO₂ network (b).

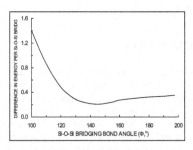

Figure 2. The calculated energy for Si-O-Si bridging bond as a function of the bond angle.[2]

for the mechanical relaxed dioxide. For such kind of oxides the SiO_4 tetrahedrons are arranged into rings with 6 members. If the a-SiO_2 possesses a decreased density the average angle of the Si-O-Si bridging bonds is more then 144^0, and the SiO_4 tetrahedrons have to be arranged into rings with 7 and 8 members. If the a-SiO_2 is high dense oxide, Φ is less then 144^0, and the SiO_4 tetrahedrons are arranged preferably into rings with 4 members. The last "strained" layers are usually observed in the SiO_2-Si interfaces, especially after high-temperature annealing.[3,4] Additionally buried oxides (BOX) obtained by SIMOX technique with supplemental oxygen implantation are preferably low dense, since the single implant SIMOX BOXs are preferably high dense.[5] Thus in the BOXs the Φ angle can be changed in wide range in dependence on the fabrication technology.

3.   HYDROGEN ASSOCIATED DEFECTS IN A-SIO₂

Silicon oxidation in presence of the water vapor is widely used in silicon technology[6] that leads to hydrogen and hydroxyl group incorporation into silicon dioxide network. During SiO_2 growth two kinds of reactions may occur.[7] First one is oxide growth:

$$Si+OH (H_2O) \rightarrow SiO_2 + H_2\uparrow, \tag{1}$$

and the second reaction is interaction of water with the dioxide network

$$O_3\equiv Si\text{-}O\text{-}Si\equiv O_3 + H_2O \rightarrow O_3\equiv Si\text{-}OH + HO\text{-}Si\equiv O_3 . \tag{2}$$

Besides, hydrogen will be incorporated into the SiO_2 network due to annealing in hydrogen ambient at elevated temperature of SiO_2-Si structure. Thus, a third reaction between hydrogen and SiO_2 can take place:

$$O_3\equiv Si\text{-}O\text{-}Si\equiv O_3 + H_2 \rightarrow O_3\equiv Si\text{-}OH + H\text{-}Si\equiv O_3 . \tag{3}$$

The Si-OH and Si-H bonds are suggested[8,9] to be candidates for electron traps with capture cross-section from 10^{-18} to 10^{-17} cm^2. The electron traps with capture cross-section of 2×10^{-18} cm^2 can be associated with H_2O molecule located inside of the dioxide network.[9]

In papers[10,11] the E'-centre or the positively charged oxygen-vacancy centre ($O_3\equiv Si^+$ $^\bullet Si\equiv O_3$), studying with EPR technique, is considered as a main positively charged centre in dioxide network, which concentration distribution correlates with one of the positive charge in gate oxide after irradiation.[11] However in many cases appearance of the positive charge in the gate[12] and buried[13] oxides was not associated with a presence of the paramagnetic E'-centres in the oxides. Recently, EPR study of changing concentration of the E'-centres and atomic hydrogen generation by investigation of boron passivation in Si space charge region of the SiO_2-Si structure due to hole injection shown, that the $O_3\equiv Si$-H defect can be associated with hole traps with capture cross-section of 3×10^{-14} cm^2. [12,14]

In papers[15,16] it was shown that bridging oxygen bond can be exibited as a hydrogen trap, and increase of the Φ angle more than 144^0 results in an increase of the hydrogen bond energy. Becides, calculations performed by Yokozawa and Migamoto have demonstrated,[17] that the Si-OH-Si complex is preferently positively charged in the dioxide matrix, whereas Si-OH bond is negatively charged one. Additionally, the large fixed positive charge can be introduced into oxide (especially buried oxide) by high-temperature (near 600-700^0C) hydrogen annealing,[13,18] that is the hydrogen associated defects can operate like hole traps.

After low-temperature (77K) x-ray irradiation of high-OH silica some other defects associated with hydrogen were exibited with EPR technique.[19,20] These defects were marked as 74G and 10.6G duplets and, as it was supposed in[20] represent themselves E'-centre where one bonding oxygen was replaced by hydrogen ($O_2=Si^{\cdot}$-H) or hydroxil group ($O_2=Si^{\cdot}$-OH), respectively. Additionally to above mentioned defects some other diamagnetic defects relating to oxygen vacancy and hydrogen can be exited in the dioxide. Totally hydrogen passivated oxygen vacancy can be formed after hydrogen annealing of irradiated dioxide.[21] Two possible structures of the defect and their binding energies with hydrogen are presented in Table 1. In neutral hydrogen bridge defect ($O_3\equiv Si$-H-$Si\equiv O_3$) hydrogen possesses the smalest binding energy in compare with other defects considered in Table 1. However the energy is sufficiently high to keep hydrogen at room temperature.

Table 1. Binding energy of hydrogen for various defects calculated by first-principles density functional theory[21]

Defect	Binding energy (eV)
(Si-OH)	6.3
(Si-H)	5.7
(Si-H H-Si)	4.4
(Si-H-Si)	2.0

Thus hydrogen concentration in dioxide can reflect existing of the "strained" and dangling bonds in the dioxide network. Indeed, after thermal annealing of single implant (SI) SIMOX SOI structures at 700-900^0C in deuterium ambient the deuterium concentration in the BOX exceeds $10^{20}cm^{-3}$ that was considered as evidence for a large concentration of "strained" bonds in SI SIMOX BOX.[22] Distribution of deuterium concentration in BOX determined from SIMS analysis of plasma deuterated SI SIMOX SOI structures[23] shown an increase of the deuterium concentration near the BOX/Si wafer interface that correlated with the increase of Si concentration in this region of the BOX (see Fig.3a) and with the height of protrusions, observed by AFM technique after HF etching of the BOX.[24] The SOI structures fabricated by UNIBOND technique, which bears better quality of the BOX then SI SIMOX BOX, demonstrate after deuterium treatment narrow deuterium peak near the BOX/Si wafer interface (Fig.3b).

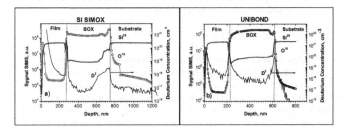

Figure 3. Distribution of Si, O and D atoms through the deuterated SOI structures fabricated by SI SIMOX (a) and UNIBOND (b) technique.

4. HYDROGEN DIFFUSION IN A-SIO$_2$

Investigation of proton diffusion in BOX of SOI structures subjected by high-temperature (600-700°C) hydrogen annealing[25] attests that the diffused ion current (I_{H^+}) can be better described like space-charge-limited one,[26] that is

$$I_{H^+} \propto \frac{V_G^2}{d^3} \mu_{H^+} ,$$ (4)

where V$_G$ is the bias applied to the BOX during the diffusion, d is the BOX thickness and μ_{H^+} is the ionic mobility. To first order, the ionic mobility can be described by thermally activated hopping process as following:

$$\mu_{H^+} = (\frac{Q_i a^2 v}{6kT}) \exp(-\frac{E_a}{kT}) \equiv \frac{Q_i D_o}{kT} \exp(-\frac{E_a}{kT}) ,$$ (5)

where Q_i is the ionic charge, a is the intersite hopping distance, v is the attempt frequency, D_0 is the diffusion coefficient at T=0K, other amounts are common used ones. For these suggestions the activation energy for H^+ and D^+ diffusion was found to be 0.80 eV.[25] The similar approach employed in paper[27] has shown that the activation energy of proton diffusion in BOX depends on the fabrication technology of SOI wafers and changes in the range from 0.6 to 0.7 eV.

In Hofstain's papers,[26,28] where effect of water absorption in gate oxide has been studied, the minimal values of diffusion coefficient for fast migrate positively charged ions has been estimated as

$$D \geq 2 \times 10^{-2} \exp(-0.68/kT) \ [cm^2/sec].$$

It was suggested that the positively charged moving ions are protons.

Theoretical calculations using first-principles density functional theory[29] have shown that the most stable state of proton in amorphous dioxide network is proton bonded with bridging oxygen, and rout for proton diffusion is hop between nearest bridging oxygens. The energy barrier which proton has to overcome for such jump was calculated to be near 1.3 eV that is higher then the experimental values for proton diffusion activation energy. In paper[30] the hopping model was extended. Introduction of zero-point vibration energy in the theory allows to obtain a value for activation energy for proton motion in model ring system (see Fig. 4) very similar to experimental data (near 0.90 eV). It's very interesting to note that the ring structure of a-SiO$_2$ is very flexible and in frame of this model it is possible to obtain hydrogen-bonded structure for proton motion across the ring with very small activation energy (0.1-0.2 eV).

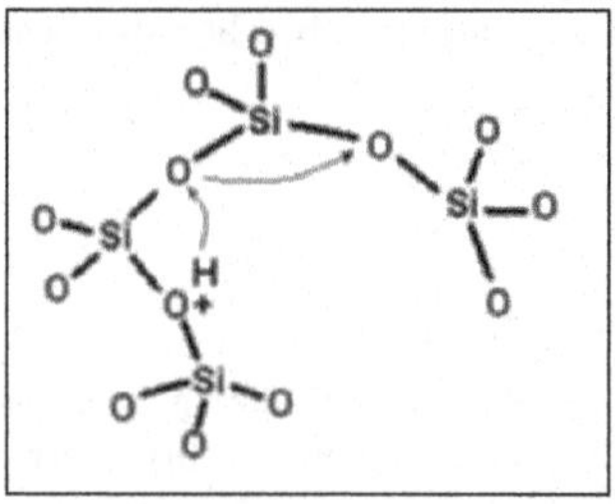

Figure 4. The model ring system for proton motion in SiO$_2$.[30]

5. HYDROGEN GENERATION DUE TO ACTIVE TREATMENTS OF A-SIO₂

Existing of bonded hydrogen in SiO_2 can result in hydrogen release during different active treatments. Historically first researches of hydrogen release in dioxide were connected with study of surface state generation in the SiO_2/Si interface during radiation effect on the SiO_2-Si structures. Careful investigations of this phenomenon demonstrated that time dependence of the interface state generation could be explained only by employing the fast diffusing particles such as holes or protons.[31] The subsequent study of the surface state generation kinetics in post-irradiation hydrogen annealed SiO_2-Si structures totally supported idea of hydrogen participation in this process.[32] Besides, in Stahlbush's papers[32,33] it was demonstrated that molecular hydrogen (H_2) can be cracked by positively charged defects in dioxide with creation of protons, which drifts to the SiO_2-Si interface and is involved in the surface state generation process.

Hydrogen embedding in the SiO_2-Si interface due to irradiation at positive voltage applied to gate of the Al-SiO_2-Si structure was confirmed by nuclear reactions method (NRM).[34] Hydrogen accumulation in the SiO_2-Si interface was also observed by SIMS[35] and NRM[36] after avalanche and Fowler-Nordheim electron injection into dioxide, during which the molecular hydrogen was also introduces into subsurface silicon layer.[37]

In last ten years in papers[25,38,39] it was demonstrated that hydrogen annealing (at 600-700⁰C) of SOI structures fabricated both SIMOX and UNIBOND techniques leads to generation of positive fixed and mobile charges in the buried dielectric. The positively charged moving charge was associated with protons,[25] which are generated only in the case of defect existing in dioxide.[40]

Thus, the previous research has established a proton generation on defect sites, which are often positively charged and usually located near the SiO_2-Si interface.

6. HIGH-TEMPERATURE CHARGE INSTABILITY OF SOI STRUCTURE AND MOSFETS

6.1 SOI structures

So, the existing of defects and "strained" bonds in BOX of the SOI structures can results in proton generation in the oxide near the BOX/silicon substrate interface after high-temperature hydrogen annealing. Such effect leads to fast movement of this positive charge due to bias applied that can be used for non-volatile memory devices fabrication.[25]

However in some cases such kind of charge instability can be created in the BOX at considerably lower temperature ($\sim$250^{0}C) resulting in intolerant device operation at high temperature.[41] In paper[42] the processes of charge generation and drift in the BOX of SIMOX SOI capacitors using thermally stimulated polarization/depolarization (TSP/TSD) method have been considered. It was shown (see Fig.5a) that virgin samples demonstrate no existing of ion current peaks for both negative and positive voltage applied to the Si substrate up to 200^{0}C during linear heating. If a positive voltage is applied no current peaks is observed up to 400^{0}C. However, if we apply a negative voltage to the silicon substrate the strong current peak with maximum near 270^{0}C is observed. Activation energy of this process is near 1.2 eV.

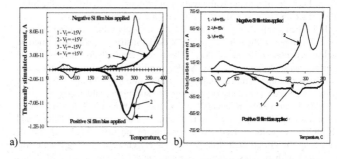

Figure 5. Typical TSP current spectra for SI SIMOX (a) and UNIBOND (b) SOI structures.[42]

After observation of this high-temperature current peak on next step of measurement, when positive voltage is applied to the substrate the low-temperature current peak in temperature range from room temperature to 100^{0}C is observed (Fig.5a). Activation energy of this low-temperature process depends on applied electric field and at 0.8MV/cm equals to 0.3 eV. A taking into account Pool-Frankel effect results in the activation energy of this process at small electric field in the BOX near 0.6 eV. Similar phenomenon is also observed in UNIBOND SOI structures (Fig.5b). However activation energy of the high-temperature process in this case is not monoenergetic one and is distributed in energy range from 0.8 to 1.3 eV.

It should be noted that negative bias applied to silicon substrate results also in creation of positive charge located in the BOX/substrate interface. Additionally, this positive charge is embedded in the interface from edges of the SOI capacitor,[42] that is very similar to generation of the fixed positive charge during high-temperature hydrogen annealing.[43]

6.2   SOI FD MOSFETs

In next part the high-temperature charge instability processes that occurs in SOI devices fabricated with full CMOS process sequence are considered. In papers[41,44-46] fully depleted (FD) SOI MOSFETs fabricated on basis of SI SIMOX and UNIBOND SOI wafers were studied. For study the charge instability processes in the BOX a sweep voltage was applied to the silicon substrate. When a negative voltage was applied to the substrate at temperature above 200[0]C and then was changed into side of positive voltages a significant drain current jump at bias voltage near zero has been observed. Arising drain current jump increased with an increase of hold time and temperature and was observed both SIMOX and UNIBOND wafer (Fig.6). The observed phenomenon was not abolished by grounding of the silicon film,[46] that attested the charge drift process in the BOX could be play considerable role in observed instability but not floating body effects.

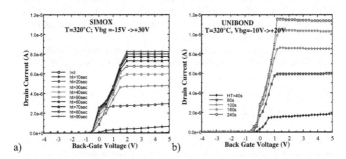

Figure 6. Dependence of the high-temperature drain current jump in FD SOI MOSFETs on hold time of the negative voltage applied to the substrate of SI SIMOX (a) and UNIBOND (b) SOI structures.

It should be noted that an origin of the drain current jump correlates with creation of current peak in high-temperature dynamic current-voltage (DIV) characteristics of MOS SOI capacitor fabricated in the same chip where the MOSFET was located.[46] It's well known that creation of current peaks in the high-temperature DIV characteristic is associated with ion movement inside of dielectric of MOS structure.[47] Furthermore, the observed hysteresis of the subthreshold region of the MOSFET drain current vs. back-gate voltage characteristics and current peak in the DIV characteristic corresponds to positively charged ion movement through the BOX.

Measurement of amount of the dynamic current peak square and amount of drain current jump of the $I_D V_{BG}$ characteristic allows to determine an amount of generated moving positive charge.[46] Dependence of this amount from hold time and measurement temperature gives a possibility to estimate an activation energy of the generation process.[46] The activation energy, calculated with using of the considered methods, comprised 1.1±0.1 eV for SIMOX wafer and from 0.9 to 1.5 eV for UNIBOND wafer, that are in good agreement with ones obtained by TSD technique.

Also in paper[46] it was shown that after moving positive charge generation at temperature above 200^0C the positively charged ions can be remained in motion state in the BOX at low temperature (up to 50^0C) for quite a long time. At these conditions in SOI capacitors considerable shift of maximum of the dynamic current peak at different temperatures and back-gate scan speeds was observed. Using this shift a diffusion coefficient and its dependence on temperature for positively charged ions has been estimated as $D=4x10^{-4}exp(-0.55/kT)$ (in cm^2/s). Obtained activation energy of the moving ions is similar to ones for protons presented in papers.[27,28] Diffusion coefficient at 140^0C for sodium ions, which often occurs in dioxide is about $1x10^{-15}$ cm^2/s,[28] that is considerable lower than for observed ions ($6x10^{-11}$ cm^2/s), and it is close to the amount obtained for protons[26] ($\geq 5x10^{-11}$ cm^2/s).

Thus it can be concluded that positively charged ions generated by electric field and high temperature in the BOX of UNIBOND and SIMOX SOI structures are protons.

7.   CONCLUSION

Defects and "strained" bonds in the BOX near the BOX/substrate interface of SOI structures can be saturated by hydrogen during fabrication technological processes, that results in release of moving protons at some extreme conditions (radiation, high temperature and electric field).

Negative voltage applied to Si substrate of SOI structure at temperature above 200^0C leads to generation of moving positive charge near the BOX/Si substrate interface in the BOX, that produces the high-temperature charge instability effects in FD IM MOSFET, which appears as drain current jump near zero back-gate voltage. Process of positive charge generation in the BOX appears activation energy near 1eV and can be associated with proton generation during hydrogen interaction with defect sited located near the BOX/substrate interface. Average activation energy of positive mobile charge diffusion determined from dynamic currents is 0.55 eV and corresponds to proton diffusion in dioxide.

ACKNOWLEDGEMENTS

The author thank the technical stuff of the UCL Microelectronics Lab for device fabrication using in this paper and Prof. J.P.Colinge (UC Davis, USA), Prof. D.Flandre (UCL, DICE, Belgium) and Dr. D.Ballutaud (CNRS, Maidon, France) for fruitful discussions. This work was supported partially by the NATO Linkage Grant PST. CLG. 979999.

REFERENCES

1. A.G.Revesz, *Phys. Stat. Sol. (a),* **57**, 235-243 (1980)
2. A.G.Revesz and G.V.Gibbs, *The Physics of MOS Insulators,* edited by G.Lukovsky et. al. (Pergamon, New York, 1980) pp. 92-96.
3. F.J.Grunthaner, B.F.Levis, N.Zamini at. al., *IEEE Trans. Nucl. Sci.,* **NS-27**, 1640-1646 (1980)
4. F.J.Grunthaner, P.J.Grunthaner and J.Maserjian, *IEEE Trans. Nucl. Sci.,* **NS-29**, 1462-1466 (1982)
5. B.J.Mrstik, V.V.Afanas'ev, A.Stesmans et. al., *J. Appl. Phys.,* **85**, 6577-6588 (1999) VLSI Technology, edited by S.M.Sze (McGraw-Hill Book Comp., New York, 1983)
6. A.G.Revesz, *J. Electrochem. Soc.,* **126**, 121-130 (1979).
7. T.Sugano, *Acta Polytechn. Semicond. Electr. Eng. Sci.,* **64**, 220-241 (1989)
8. A.Hortstein and D.R.Young, *Appl. Phys. Lett.,* **38**, 631-633 (1981).
9. J.F.Conley, P.M.Lenahan and P.Roitman, *IEEE Trans. Nucl. Sci.,* **NS-38**, 1247-1252 (1991)
10. J.F.Conley and P.M.Lenahan, *The Physics and Chemistry of SiO$_2$ and the Si-SiO$_2$ nterface-3,* edited by H.Z.Massoud et. al., (ECS Inc., NJ, 96-1) pp.214-249.
11. V.V.Afanas'ev and A.Stesmans, *J.Phys: Condens. Matter.,* **12**, 2285-2290 (2000)
12. K.Vanheusden and A.Stesmans, *Microelectronic Engineering,* **22**, 371-374 (1993).
13. A.Rivera, A.van Veen, H.Schut et. al., *Sol. St. Electron.,* **46**, 1775-1785 (2002)
14. A.H.Edwards and G.Germann, *Nuclea Instr. Methods Phys. Res.,* **B32**, 238-247 (1988)
15. K.Vanheusden, P.P.Korambath, H.A.Kurtz et.al., *IEEE Trans. Nucl. Sci.,* **46**, 1562-1567 (1999)
16. A.Yokozawa and Y.Miyamoto, *Phys. Rev.,* **B55**, 13783-13788 (1997)
17. K.Vanheusden, A.Stesmans and V.V.Afanas'ev, *J. Non-Cryst. Solids.,* **187**, 253-256 (1995)
18. D.L.Griscom, *Nucl. Instr. and Meth.,* **B1**, 481- (1984)
19. T.-E.Tsai and D.L.Griscom, *J.Non-Cryst. Solids,* **91**, 170- (1987)
20. P.E.Bumson, M.DiVentra, S.T.Pantelides et. al., *IEEE Trans. Nucl. Sci.,* **47**, 2289-2296 (2000)
21. S.M.Myers, G.A.Brawn, A.G.Revesz and H.L.Hughes, *J. Appl. Phys.,* **73**, 2196-2206 (1993)

22. A.Boutry-Forveille, A.Nazarov and D.Ballutaud, *Hydrogen in Semiconductors and Metals*, v.**513**, edited by N.H.Nickel et al. (MRS, Pennsylvania, 1998) pp. 319-324.
23. V.V.Afanas'ev, A.Stesmans, A.G.Revesz and H.L.Hughes, *J. Appl. Phys.*, **82**, 2184-2199 (1997)
24. K.Vanheusden, W.L.Warren, R.A.B.Devine et. al., *Nature*, **386**, 587-589 (1997)
25. S.R.Hofstein, *IEEE Trans. Electron. Devices*, **ED-14**, 749-759 (1967)
26. R.E.Stahlbush, R.K.Lawrence and H.L.Hughes, *IEEE Trans. Nucl. Sci.*, **45**, 2398-2407 (1998)
27. S.R.Hofstein, *IEEE Trans. Electron. Devices*, **ED-13**, 222-237 (1966)
28. P.E.Bunson, M.diVentra, S.T.Pantelides et. al., *IEEE Trans. Nucl. Sci.*, **46**, 1568-1573 (1999)
29. H.A.Kurtz and S.P.Karna, *IEEE Trans. Nucl. Sci.*, **46**, 1574-1577 (1998)
30. F.B.McLean, A framework for understanding radiation-induced interface states in SiO_2 MOS structures, *IEEE Trans. Nucl. Sci.*, **NS-27**, 1651-1657 (1980)
31. R.E.Stahlbush and G.A.Brawn, *IEEE Trans. Nucl. Sci.*, **NS-42**, 1708-1715 (1995)
32. R.E.Stahlbush, A.H.Edwards, D.L.Griscom and B.J.Mrstik, *J. Appl. Phys.*, **73**, 658-667 (1993)
33. J.Krauser, F.Wulf, M.A.Biere at. al., *Microelectronic Engineering*, **22**, 65-68 (1993)
34. R.Gale, F.J.Feigl, C.W.Magee and D.R.Young, *J. Appl. Phys.*, **54**, 6938-6942 (1983)
35. D.A.Buchanan, A.D.Marwick, D.J.DiMaria and L.Dori, *J. Appl. Phys.*, **76**, 3595-3608 (1994)
36. C.T.Sah, J.Y.-C.Sun and J.J.T.Tzou, *J. Appl. Phys.*, **54**, 944-956 (1983)
37. K.Vanheusden and A.Stesmans, *Appl. Phys. Lett.*, **64**, 2575-2577 (1994)
38. K.Vanheusden, A.Stesmans and V.V.Afanas'ev, *J. Appl. Phys.*, **77**, 2419-2422 (1995)
39. K.Vanheusden and R.A.B.Devine, *Appl. Phys. Lett.*, **76**, 3109-3111 (2000)
40. A.N.Nazarov, J.P.Colinge and I.P.Barchuk, *Microelectronic Engineering*, **36**, 363-366 (1997)
41. I.Barchuk, V.Kilchytska and A.Nazarov, *Microelectronic Reliability*, **40**, 811-814 (2000)
42. K.Vanheusden, W.L.Warren and R.A.B.Devine, *J. Non-Cryst. Solids*, **216**, 116-123 (1997)
43. A.N.Nazarov, I.P.Barchuk, V.S.Lysenko and J.P.Colinge, *Microelectronic Engineering*, **48**, 379-282 (1999)
44. A.N.Nazarov, I.P.Barchuk, V.S.Lysenko and J.P.Colinge, in *Silicon-On-Insulator Technology and Devices IX*, ed. by P.L.F.Hemment (ECS Inc. v.99-3) pp.299-304.
45. A.N.Nazarov, V.S.Lysenko, J.P.Colinge and D.Flandre, in *Silicon-On-Insulator Technology and Devices XI*, ed. by S.Cristoloveanu (ECS Inc. v.2003-05) pp.455-460.
46. M.W.Hillen, G,Greeuw and J.F.Verweij, *J. Appl. Phys.*, **50**, 4834-4837 (1979)

RECENT ADVANCES IN SOI MOSFET DEVICES AND CIRCUITS FOR ULTRA-LOW POWER / HIGH TEMPERATURE APPLICATIONS

David Levacq[1], Vincent Dessard[2] and Denis Flandre[1]

[1]DICE, Université catholique de Louvain, 3 place du Levant, 1348 Louvain-la-Neuve, Belgium; [2]CISSOID, 4 bte 7 place des Sciences, 1348 Louvain-la-Neuve, Belgium.

Abstract: This work focuses on new structures with ultra-low power dissipation. They are based on a new diode architecture that features ultra-low leakage current and a negative impedance in its reverse bias mode. This characteristic allows to realize an ultra-low power (ULP) latch with strongly reduced static current at stable states. We propose to implement this latch in new MTCMOS flip-flops where the supply voltage can be gated by high threshold voltage transistors to reduce the static power dissipation in the sleep mode. The presence of the ULP latch avoids data loss during standby. The work is supported by simulations in 0.12μm PD CMOS/SOI.

Key words: SOI, Ultra-Low power, diode, MTCMOS, flip-flops

1. INTRODUCTION

With the scaling of supply voltage (V_{DD}) and threshold voltage (V_{th}) levels in advanced CMOS, standby power contribution to the total power dissipation of VLSI circuits becomes more and more important: leakage is beginning to approach 50% of a digital circuit's operation power[1]. Operation in a high temperature environment further increases the leakage currents. In applications like cell phones, pagers or X-terminals that spend upwards of 90% of their time in standby mode, reduction of leakage currents during sleep periods is of prime importance[2]. Future wireless sensor networks will spend even more time in standby mode. Since these systems are intended to be self-powered, harvesting energy from their environment (vibrations, heat, light...), their requirement for ultra-low power dissipation in standby mode is stringent. Multi-threshold CMOS

D. Flandre et al. (eds.), Science and Technology of Semiconductor-On-Insulator Structures and Devices Operating in a Harsh Environment, 133-144.

(MTCMOS) techniques allow to strongly reduce static leakage currents. A first class of MTCMOS circuits uses low-V_{th} logic gates in the critical paths to speed up the circuit while other circuit blocks are realized with high-V_{th} transistors to reduce the static leakage[3,4]. A second class of MTCMOS circuits introduces high-V_{th} sleep devices to gate the power supplies of low-V_{th} logic blocks during standby (or 'sleep') periods[2,5]. The presence of floating nodes during sleep mode may lead to some problems since information on these nodes will be lost, which can be unacceptable in some circuits with memory functions such as latches and flip-flops. Solutions are then needed towards the retention of stored values during the sleep mode.

This paper presents different cells aimed at solving the problem of the static power dissipation in low voltage circuits. We first propose a new diode architecture with strongly reduced reverse current (section 2). The new diode features an interesting characteristic in reverse bias mode that allows to implement latches with ultra-low static current dissipation as shown in section 3. These latches are particularly suited for MTCMOS flip-flops with retention capability during sleep mode. In section 4 we propose two possible structures and present simulation results. All simulations were performed on a 0.12μm Partially-Depleted (PD) SOI process (t_{ox}=2.5nm).

2.   ULTRA-LOW LEAKAGE DIODE

A standard CMOS diode is classically implemented by connecting the gate of a MOSFET to its drain (Fig. 1a). In reverse bias mode, it is the source that appears connected to the gate. The reverse current is then equal to the MOS current I_S at zero gate-to-source voltage (V_{gs}=0V, see Fig. 1b), which becomes much higher than the junction leakage current for low V_{th}. We propose a new diode structure (ULPD for Ultra-Low Power diode)[6] combining an nMOS and a pMOSFET (Fig. 1a) that strongly reduces reverse current. When the ULPD is reverse biased, the n and pMOSFET sources appear connected and both transistors operate with negative V_{gs} in very weak inversion. When, starting from a 0V bias, reverse bias is progressively increased, the ULPD current first increases because drain-to-source voltage (V_{ds}) of both n and pMOSFETs increase from zero to positive values. Reverse current increases up to a maximum value, and then strongly decreases due to the more and more negative V_{gs} that is applied to the MOSFETs. This realizes ultra-low leakage. Simulations on a 0.12μm PD SOI technology (with floating body) show a reduction of the reverse current by more than three orders of magnitude when compared to a standard MOS diode for a reverse bias voltage of 0.5V (see Fig. 1b).

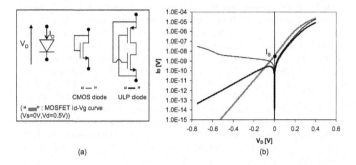

Figure 1. Architectures (a) and simulated characteristics (b) of standard CMOS and ULP diodes (W=0.5μm, L=0.12μm).

The negative resistance effect that appears in the current-voltage characteristic of the ULPD in reverse bias mode allows to realize memory cells, as described in next section.

3. LATCHES AND MEMORIES

By connecting two reverse-biased ULP diodes in series, a digital latch is obtained (Fig. 2a). Indeed, thanks to the negative impedance of each diode, two stable states appear, at "0" and "V_{DD}", in the plot of the current difference between the lower and the upper ULPD versus the voltage level at the memory node (V_x) (Fig. 2b). Subsequently, when a V_x between 0 and $V_{DD}/2$ (respectively, $V_{DD}/2$ and V_{DD}) is imposed and then left floating, I_{D2}-I_{D1} is positive or negative, respectively, and eventually drives V_x to "0" or V_{DD}, respectively. Such an architecture is therefore auto-regenerative for both logical levels "0" and "1". The value of the peak current in the regenerative characteristic depends on device doping levels and temperature[6]. Asymmetry in the threshold voltages of n- and p-MOSFETs of the diodes results in an increase of the regenerative current, thereby enlarging noise margins. A wider voltage range for the regeneration is also obtained when the temperature increases, allowing high temperature functionality of the cell which was demonstrated up to 280°C[7].

At each stable state, one diode is reverse-biased and the other one features almost zero bias voltage, leading to ultra-low power consumption when compared with standard static CMOS latch made of two inverters. The theoretical static consumption of the ULP latch is given by the point where

David Levacq, Vincent Dessard and Denis Flandre

the I_d-V_{gs} characteristics in weak inversion of the n- and p-MOSFETS show a voltage difference of V_{DD} (Fig. 3a). The improvement compared to standard static latch architectures based on inverters is related to the fact that in a logic stable state, the transistors of our cells are biased with low V_{ds} and negative gate to source voltages (V_{gs}) in very weak inversion, while V_{ds} is high and the minimum V_{gs} is 0 in CMOS inverters. Figure 3b shows the static current vs. V_{DD} for a ULP and a standard CMOS latches. For V_{DD}=0.5V, the ULP latch demonstrates a reduction of the dissipation of more than 3 orders of magnitude. As V_{DD} increases, its static current further decreases exponentially, the minimum dissipation being practically bounded by the junction leakage current of the MOSFETs. In the case of the standard CMOS latch, the static current increases with V_{DD} due to Drain-Induced Barrier Lowering.

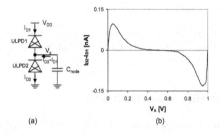

(a)   (b)

Figure 2. ULP latch architecture (a) and simulated characteristic (b) (V_{DD}=1V, L=0.13µm, W=0.6µm).

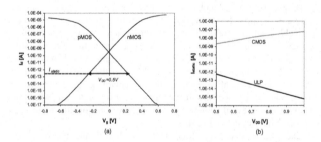

(a)   (b)

Figure 3. (a) Simulated I_d-V_g curves of n- and pMOSFETs and theoretical static current of an ULP latch; (b) Simulated static current of a classical CMOS latch and an ULP latch as function of V_{DD}.

4.   MTCMOS FLIP-FLOPS

4.1   Architecture

As stated in the introduction, one major problem of MTCMOS circuits based on the supply gating during standby by low leakage (high-V_{th}) transistors is the apparition of floating nodes[2,5]. When woken up, the circuit has lost its previous state and all the related data. The presence of floating nodes and thus undetermined levels also creates a problem when interfacing MTCMOS and CMOS gates since it leads to short circuits currents in the CMOS gates that degrade the power dissipation[8]. Based on the new latch architecture presented in section 3, we propose two new dynamic flip-flops (FF) structures (ULPFF1 and ULPFF2, Fig. 4 a and b). Besides their fast circuit performance during active mode, these FF can be put in a standby state with ultra-low static power dissipation while retaining their last state. It is thus particular interesting for applications that stay long periods in standby mode and need high performance (high speed) in active mode. Their functionality is similar to the Leakage Feedback Dynamic Flip Flop (LFDFF) proposed by J. Kao and A. Chandrakasan[8] (Fig. 4 c).

4.2   Description of operation

4.2.1   ULPFF1

ULPFF1 is obtained by directly connecting a ULP latch on the internal node of a standard dynamic FF. In the active mode, sleep signals S_{master} and S_{slave} are set to logical low voltage level and the flip-flop operates like a standard dynamic flip-flop. Information storing is made possible by charge storage on parasitic capacitors associated with each node. This requires having non-overlapping clock signals on both transmission gates and a clock frequency high enough to keep stable node voltages over one clock period. Almost no speed penalty is observed in the active mode due to the addition of the ULP latch since its input capacitance is about $1.25*C_{ox}*W*L$ only (W and L being the widths and lengths of one MOS device of the ULPD). For comparison, the input capacitance of an inverter is 2.6 times larger. Moreover, the positive feedback introduced by the memory cell is easily fought by the preceding low-V_{th} logic stage to change node x logic state. For example, the ULP latch of Fig. 2 presents regenerative current peaks around 0.1nA for $V_{DD}=1$V. These currents are however sufficient to memorize the logic state of node x during standby state when high-V_{th} sleep transistors are switched off by sleep signals and Ck1 (Ck2) is high (low).

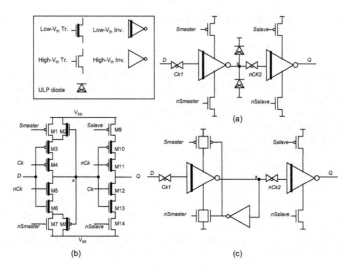

Figure 4. Flip-Flops structures: (a) ULPFF1 (b) ULPPFF2 (c) Leakage Feedback Dynamic Flip-Flop (Ck1 and Ck2 are non-overlapping clock signals).

When the FF has to be woken up, we proceed as follows. Because the input voltage of the first inverter can float, the timing to exit of the sleep state requires that the phase of the master latch sleep signals, S_{master} and nS_{master}, lag the slave sleep signals, S_{slave} and nS_{slave}, by one half cycle. The master slave can exit the sleep condition only after the clock goes low and the slave stage latches the stored data.

4.2.2   ULPFF2

This architecture integrates the ULP latch inside a C^2MOS type flip-flop[9]. It presents the usual advantages particular to the C^2MOS FF: clock skew insensitivity, no need to have clock non-overlapping schemes, less contacts area requirement. Moreover, the sleep storing scheme adds only 2 transistors while ULPFF1 and LFDFF need 4 additional transistors. Another advantage of this solution is that it presents only 2 current branches between the supply lines which is favorable to reduce the static power consumption. In sleep mode, when clock (Ck) is 'high' and sleep transistors are cut-off, the equivalent circuit of the FF is given in Fig. 5. The first branch behaves like the basic ULP latch of Fig. 2, memorizing the last state on node x. Because the input voltage of the first inverter can float, similarly to ULPFF1, the

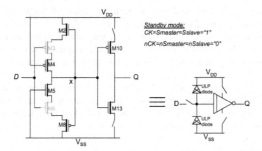

Figure 5. Equivalent circuit of ULPFF2 in sleep mode.

timing to exit of the sleep state requires that the phase of the master latch sleep signals, S_{master} and nS_{master} lag the slave stage sleep signals, S_{slave} and nS_{slave}, by one half cycle. The master slave can exit the sleep condition only after the clock goes low and the slave stage latches the stored data.

4.3   Simulations

4.3.1   Test bench

ULPFF1 and ULPFF2 were both simulated on a 0.12µm PD SOI technology under V_{DD}=1V. LFDFF was simulated as well for comparison. Our analysis was based on the approach suggested by V. Stojanovic[10]. The simulation test bench is given in Fig 6.

Capacitive load at the data input simulates the fanout signal degradation from previous stages. Capacitive load at the output simulates the fanout signal degradation caused by the succeeding stages. They were both set equal to 50fF, corresponding to the input capacitance of ~15 minimum-sized inverters.

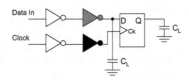

Figure 6. Simulation test bench.

The simulations were made for a clock frequency of 100MHz, with Data/Clock slopes of ideal signals equal to 100ps.

Dynamic power dissipation in active mode was calculated by taking the average power dissipation using a pseudo-random sequence at data input with a data activity rate α=0.5. Three kinds of power dissipation were included:

1. Local data power dissipation (portion of the gray inverter's power consumption (Fig. 6) dissipated on switching the data input capacitance).
2. Local clock power dissipation (portion of the black inverter's power consumption (Fig. 6) dissipated on switching the clock input capacitance).
3. Internal power dissipation (intrinsic power dissipated on switching the internal nodes of the circuit, excluding the power dissipated on switching the output load capacitances C_L since this one is independent of the FF structure).

Minimum D-Q delay was the parameter chosen as delay meter since it represents the effective portion of time that the flip-flop structure takes out of the clock cycle[10].

4.3.2 Results

In all simulated MTCMOS circuits, sleep transistors were sized following the same procedure[11] considering that we could tolerate 10% degradation in speed performance due to the presence of the sleep transistors (see Appendix). Minimum widths of the low-V_{th} were set to 0.6μm. The low and high V_{th} of the n(p)MOSFETs were equal to 0.3V(-0.25V) and 0.6V(-0.5V) respectively. Main simulation results are given in Table 1. Figure 7 allows easier comparison of results. The low-V_{th} MOSFETs of the ULP latch of ULPFF1 must be sized in order to ensure proper memorization of the state during sleep mode despite leakage of sleep transistors. This means that the regeneration current peak of the latch (Fig. 2b) must be large enough. The large channel widths that were needed (2.5μm) leads to a degradation of speed performance and active mode power dissipation when compared to LFDFF (Fig. 7). The size of the latch transistors could be reduced if transistors with lower V_{th} were available. For example, an additional V_{th} of 0.2V (-0.15V) for the n(p)MOSFETs allows to implement a minimum-sized latch. Simulations results in this case (ULPFF1b in Table 1 and Fig. 7) demonstrate improved speed performance (ULPFF1b featuring the minimum delay) and reduced power dissipation.

ULPFF2 features the lowest power dissipation in active mode. Despite slight degradation of the speed performance, it presents better Power-Delay product (PDP) than ULPFF1 and LFDFF. Active area surfaces of the different solutions are comparable except for ULPFF1 due to its large ULP latch. Regarding the power dissipation in sleep mode, the FF we propose present substantial advantage when compared with LFDFF. ULLPFF2 presents particularly low static leakage due to the fact that low-V_{th} transistors are biased with negative V_{gs} and that there are only two current branches between V_{DD} and V_{SS}.

In MTCMOS circuits, dissipated power can be expressed as:

$$P = \beta . P_{active} + (1 - \beta) . P_{sleep}$$

where $\beta = t_{active}/(t_{active} + t_{sleep})$ represents the portion of time spent by the circuit in active mode, P_{active} is the total power dissipated in active mode (including dynamic and static components), and P_{sleep} is the static power dissipated during sleep mode. However this does not take into account the extra power needed to switch the large sleep transistors. Based on these definitions, Fig. 8 gives the power-delay product (PDP) as a function of β (PDP was divided by β for better legibility of the curve). The curve for a standard dynamic FF (not MTCMOS) is represented as well allowing the β

Table 1. Performances of the different FF structures simulated on 0.12μm PD SOI.

	Active mode power diss. [μW]	Minimum D-Q delay [ps]	PDP [mW.ps]	Active Area [μm²]	Sleep mode power diss. [pW]
ULPFF1	1.29	175.7	0.227	5.55	29.5
ULPFF1b	1.21	161.6	0.195	4.64	30.4
ULPFF2	1.02	208.6	0.213	4.68	21.6
LFDFF	1.38	169.3	0.233	4.84	94.5

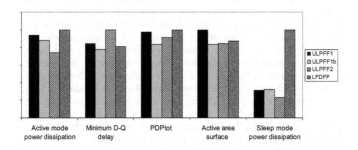

Figure 7. Performances comparison of the different FF structures (simulations on 0.12μm PD SOI).

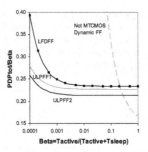

Figure 8. Simulated Power Delay Product (PDP) divided by β as function of β for not MTCMOS standard Dynamic FF, ULPFF, and LFDFF (0.12μm PD SOI).

values to be determined below which MTCMOS schemes become interesting. ULPFF2 clearly gives the best performance. For a β factor of 0.1, its PDP is 91% of that of LFDFF, and for β=0.0001 it represents only 66% of that of LFDFF.

5.   CONCLUSION

To solve the increase of static power dissipation in low V_{DD} / low V_{th} circuits, we proposed a new diode architecture which features a leakage current lying orders of magnitude below standard CMOS diodes. Moreover, a negative resistance effect in its reverse bias region allows to realize memory cells with high temperature functionality and ultra-low static power dissipation by biasing the transistors with negative V_{gs} in very weak inversion at stable states. Based on this principle, we proposed two flip-flop structures, able to store data during sleep mode of MTCMOS circuits, while maintaining ultra-low static power dissipation. In comparison with an existing flip-flop with similar features, they present better Power-Delay products, with sleep power dissipation divided by a factor ~3. We consider that they could be particularly attractive for applications such as wireless sensor networks where individual nodes have to stay in standby for long periods (while gathering data), being only active (for data processing and transmission) occasionally.   Moreover, due to the high temperature functionality of the ULP latch[7], they may present a good alternative to reduce the leakage currents in high temperature environment.

APPENDIX: SLEEP TRANSISTORS SIZING

In all simulated MTCMOS circuits, sleep transistors were sized following the same procedure[11]. The delay of a single gate (τ_d) in the absence of a sleep transistor can be expressed as:

$$\tau_d \infty \frac{C_L V_{DD}}{\left(V_{DD} - V_{thL}\right)^\alpha} \tag{A-1}$$

where C_L is the load capacitance at the gate output, V_{thL} is the low threshold voltage, α is the velocity saturation index for modeling short channel effects[12].

In the presence of a sleep transistor, the delay of a single gate can be expressed as:

$$\tau_d{}^{sleep} \infty \frac{C_L V_{DD}}{\left(V_{DD} - V_X - V_{thL}\right)^\alpha} \tag{A-2}$$

where V_X is the potential of the virtual ground. Assuming $\alpha=1$ and the circuit could tolerate a degradation $\delta = 1 - (\tau_d / \tau_d{}^{sleep})$ in performance, the size of the sleep transistor can be expressed as [11]:

$$\left(W\!\!\Big/\!L\right)_{sleep} = \frac{I_{st}}{\delta\mu C_{ox}(V_{DD} - V_{thL})(V_{dd} - V_{thH})} \tag{A-3}$$

and

$$V_X = \delta\left(V_{DD} - V_{thL}\right) \tag{A-4}$$

where I_{st} is the switching current in the low-V_{th} module and V_{thH} is the high V_{th} of the sleep transistors. In our case the low-V_{th} module is an inverter. The maximum switching current can thus be approximated by:

$$I_{st} \cong \mu C_{ox} \left(W\!\!\Big/\!L\right)_{inv} \frac{\left(V_{DD} - V_{thL} - V_X\right)^2}{2} \tag{A-5}$$

where $(W/L)_{inv}$ is the size ratio of the inverter MOSFETs. We then obtain

$$\left(W\!\!\Big/\!L\right)_{sleep} = \frac{(1-\delta)^2}{2\delta} \frac{(V_{DD} - V_{thL})}{(V_{DD} - V_{thH})} \left(W\!\!\Big/\!L\right)_{inv} \tag{A-6}$$

REFERENCES

1. R. Min, M. Bhardwaj, S.-H. Cho, N. Ickes, E. Shih, A. Sinha, A. Wang, and A. Chandrakasan, Energy-Centric Enabling Technologies For Wireless Sensor Networks, *IEEE Wireless Communications* **9**(4), 28-39, (August 2002).
2. J.T. Kao and A. Chandrakasan, Dual-threshold voltage techniques for low-power digital circuits, *IEEE J. Solid-State Circuits* **35**(7), 1009-1018 (July 2000).
3. L. Wei, Z. Chen, K. Roy, M. C. Johnson, Y. Ye and V.K. De, Design and optimization of dual-threshold circuits for low-voltage low-power applications, *IEEE Trans. on VLSI Systems* **7**(1), 16-24 (March 1999).
4. S. Sirichotiyakul, T. Edwards, C. Oh, J. Zuo, A. Dharchoudhury, R. Panda and D. Blaauw, Standby Power Minimization through Simultaneous Threshold Voltage Selection and Circuit Sizing, *Proc. of the 36ᵗʰ Design Automation Conference*, 436-441, New Orleans (1999).
5. S. Mutoh, T. Douseki, Y. Matsuya, T. Aoki, S. Shigematsu and J. Yamada, 1-V Power Supply High-Speed Digital Circuit Technology with Multithreshold-Voltage CMOS, *IEEE J. Solid-State Circuits* **30**(8), 847-854 (August 1995).
6. D. Levacq, C. Liber, V. Dessard and D. Flandre, Composite ULP diode fabrication, modeling and applications in multi-V$_{th}$ FD SOI CMOS technology, *Solid State Electronics* **48**(6), 1017-1025 (2004).
7. D. Levacq, V. Dessard and D. Flandre, A Novel CMOS Memory Cell Architecture for Ultra-Low Power Applications Operating up to 280°C, *203ʳᵈ meeting of the Electrochemical Society*, ECS Vol. 5, 249-254, Paris (2003).
8. J. Kao and A. Chandrakasan, MTCMOS Sequential Circuits, *IEEE ESSCIRC 2001*, pp. 332-335 (2001).
9. Y. Suzuki, K. Odagawa and T. Abe, Clocked CMOS calculator circuitry, *IEEE J. Solid-State Circuits* **8**(6), 462-469 (1973).
10. V. Stojanovic and V. G. Oklobdzija, Comparative Analysis of Master-Slave Latches and Flip-Flops for High-Performance and Low-Power Systems, *IEEE J. Solid-State Circuits* **34**(4), 536-548, (April 1999).
11. M. Anis, S. Areibi and M. Elmasry, Design and Optimization of Multi-Threshold CMOS (MTCMOS) Circuits, *IEEE Trans. Computer-Aided Design of Integrated Circuits and Syst.* **22**(10), 1324-1342 (October 2003).
12. T. Sakurai and R. Newton, Alpha-Power Law MOSFET Model and Applications to CMOS Inverter Delay and Other Formulas, *IEEE J. Solid-State Circuits* **25**(2), 584-594 (April 1990).

SILICON-ON-INSULATOR CIRCUITS FOR APPLICATION AT HIGH TEMPERATURES

V. Nakov[1], D. Nuernbergk[1], S.G.M. Richter[1], S. Bormann[1], V. Schulze[1] and S.B. Richter[2]

[1]*Institute for Microelectronic and Mechatronic Systems gGmbH, Ilmenau, Germany,* [2]*X-FAB Semiconductor Foundries AG*

Abstract: The capability of SOI circuits to expand the operating temperature range of integrated circuits up to 250°C is demonstrated using a specially suited SOI CMOS technology (XI10 by X-FAB Semiconductor Foundries AG). Basic analog and mixed-signal structures (bandgap, operational amplifier, DAC and ADC) are used for this purpose as an example. Difficulties and challenges in design and measurement are discussed.

Key words: SOI; CMOS; high temperature; analog circuits; automotive.

1.   INTRUDUCTION

One of the outstanding features of SOI devices in comparison to standard CMOS devices is the possibility of operating at higher temperatures (> 150°C) due to reduced substrate leakage currents. Here we present some analog and mixed-signal circuits that were proven to work successfully under such conditions.

2.   TECHNOLOGY XI10 – 1.0 µm CMOS SOI

The basis of this work is the XI10 technology of X-FAB, a semiconductor foundry situated in Erfurt, Germany. A detailed process description can be obtained respectively ordered from the manufacturer's web site[1].

Here is a brief summary of key technology and design kit properties:

145

D. Flandre et al. (eds.), Science and Technology of Semiconductor-On-Insulator Structures and Devices Operating in a Harsh Environment, 145-154.
© 2005 *Kluwer Academic Publishers. Printed in the Netherlands.*

- partially depleted SOI technology
- smallest feature size 1 μm
- 3 metal layers with high temperature option (tungsten) up to 225°C operating temperature
- 5V N- and PMOS transistors in SOI specific layout variations (s. Fig. 2)
- DMOS high voltage capability up to 120 V breakdown voltage
- 13 mask layers, 14 mask layers with linear capacitor, thin gate oxide (tunnel) option
- high resistive poly resistor
- high sensitivity Hall plate
- supported simulators: HSpice, Spectre, SmartSpice; SPICE-Models (bsim3pd, bsim3v3)
- DRC, LVS available (Diva, DRACULA)
- Digital and IO cell libraries available
- Analog IP core cells available

The technology is quite similar to standard CMOS technologies, with only a few additions and thus with comparable processing costs. It is well suited for operating at high temperatures – e.g. the room temperature transistor threshold voltages are accordingly high (about 1.5 V for n-type transistors) to take into account the decrease at higher temperatures, reaching a value of about 0.9 V at 225°C, s. Fig.1.

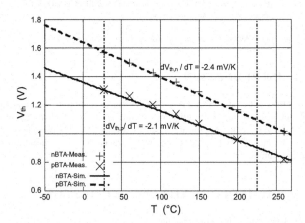

Figure 1. Threshold voltages as a function of temperature

Fig.2 shows a cross section of a n-type MOS and a power transistor and the available layout types: body-tied A-type (BTA), with body tied to source, preferred for digital applications and body-tied H-type (BTH), with body to be connected separately, recommended for analog applications.

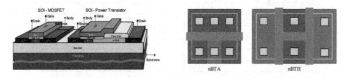

Figure 2. Transistor cross section (n-type) and MOS transistor types.

Fig. 3 and Fig. 4 show typical output characteristics of n- and p-type devices at room temperature and at 200°C. In Fig. 5 one can see the sub-threshold behavior and the comparably small leakage currents ($\leq$ 1 nA) which allow the operation of this devices at high temperatures.

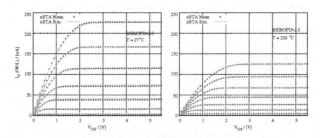

Figure 3. Output characteristics nBTA.

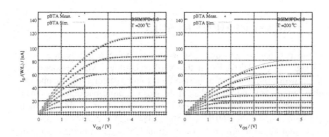

Figure 4. Output characteristics pBTA.

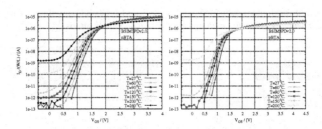

Figure 5. Transfer characteristics nBTA, pBTA.

Digital cells are easily adapted from comparable standard CMOS technologies as schematic but require completely new layout work. They exhibit low diode leakage currents, determined by one transistor since series connected, no leakage to the substrate and are proved to work without problems up to 250°C. It should be pointed out, that due to the body contacts (see Fig. 2) needed there is less or no area saving in the layout as commonly stated for SOI processes. It is currently investigated whether floating body devices can be used for digital cells. The usually undesirable Kink-Effect[2] with higher currents should have a positive effect here, increasing switching speed. Initial measurements on test structures have shown promising results.

Analog devices are on the other hand a challenge and require extra work, new concepts have to be developed and investigated. The recommended primitive transistors for analog application, BTH - see Fig. 2 - is highly symmetric therefore improving noticeably the matching properties. This contribution focus on some standard analog devices to demonstrate their ability to work at high temperatures in the XI10 SOI technology.

3.   MEASUREMENT RESULTS

In this part some basic analog structures are presented, developed within the Design Kit (CADENCE) for the XI10 technology. Measurements were made on wafer by means of a thermo-chuck and a special high temperature capable probe-card and on packaged devices (ceramic packages) by means of a thermo-stream. The later allows the investigation of devices in the temperature range from -60°C to 250°C. The measurements on packaged devices require furthermore special suited boards for the evaluation electronics or one has to separate the device from the measurement board.

3.1　Simple bandgap

The first structure investigated is a simple bandgap topology, see schematic on Fig. 6. Different diode (D1/D2) area ratios were used to achieve an optimal temperature behavior. The corresponding measurement in the temperature range from 10°C to 210°C for the best area ratio is also shown. The voltage drift is about 4.5 mV from 10 to 140°C and 2.5 mV from 140 to 210 °C.

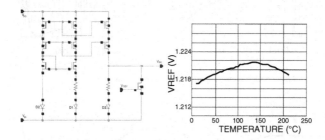

Figure 6. Simple bandgap - schematic and bandgap voltage vs. temperature.

3.2　Operational Amplifier

The next device is a folded cascode operation amplifier with start-up and stand-by subcircuits. Fig.7 shows the schematic of the structure.

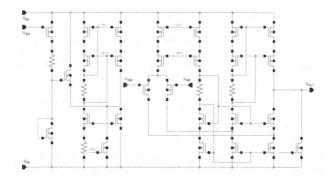

Figure 7. Operational amplifier – schematic.

Fig.8 and Fig.9 show gain and phase responses at 40°C to 210°C. The gain bandwidth for the device is about 1 MHz, which is a typical value for the technology. As mentioned before, the technology is optimised for high temperature applications but not for high speed ones.

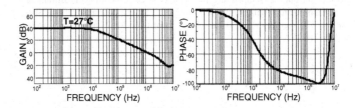

Figure 8. Operational amplifier - gain and phase responses - T=27°C.

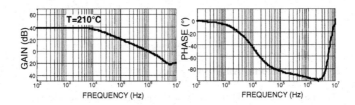

Figure 9. Operational amplifier - gain and phase responses - T=210°C.

In Fig.10 the transient behavior of the amplifier at different temperatures is shown. The input signal is a square wave with 1 V peak-to-peak amplitude. The slew rate varies from 0.5 V/μs at -40°C up to 0.7 V/μs at 225°C.

The temperature drift of the offset voltage is shown in Fig. 11. It has reasonable value of 8.4 μV/K comparable to other CMOS technologies.

3.3   10 Bit ADC/DAC

The last device, a mixed-signal design, is a 10 bit analog-to-digital-converter (ADC) with its corresponding 10 bit digital-to-analog-converter (DAC), available as a separate device.

The working principle is illustrated as a block diagram in Fig. 12: it is based on a linear resistor divider (2 chains for low and high resolution)

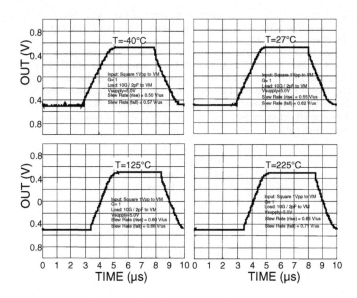

Figure 10. Operational amplifier - transient response.

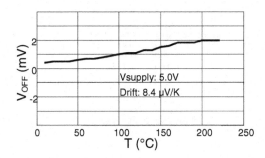

Figure 11. Temperature drift of offset

driven by a successive-approximation-register (SAR) logic. The device operates at 1-2 MHz external clock frequency (f_{clock}) achieving sampling rates of up to 200kS/s (R_{sample}).

Fig. 13 shows the integral (INL) and differential nonlinearity (DNL) error function of the DAC at four different temperatures: -40°C, 27°C, 125°C and 210°C. Fig. 14 contains the corresponding dependencies for the ADC. Some improvement of the properties with higher temperature can be observed which is caused by the decreasing threshold voltages (s. Fig.1). They strongly influence the switching speed of CMOS switches used in the linear resistor chains.

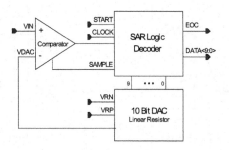

Figure 12. Block diagram ADC10.

Some typical device parameters are summarized in Table. 1.

Table 1. 10 Bit ADC Parameter

Parameter	Value
Temperature Range	-40°C – 225°C
Resolution	10 Bit
INL	±1.0 LSB
DNL	±0.7 LSB
P	4 mW
f_{clock}	1-2MHz
R_{sample}	90-180 kS/s
V_{RP}-V_{RN}	5V
Physical Size	1.0×0.8 mm

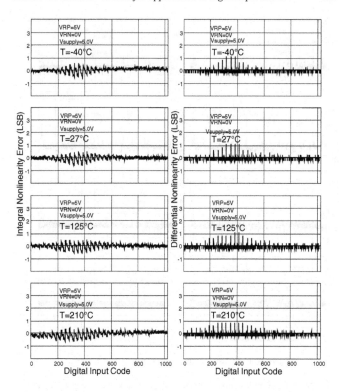

Figure 13. 10 bit DAC - integral and differential nonlinearity. Fclock=1MHz, VRN=0V; VRP=5V; Vsuppl=5V.

4. CONCLUSIONS

In this contribution we have presented some basic analog devices developed within a 'customer available' design kit for a 1 μm SOI partially depleted technology. The devices exhibit the characteristics as designed (simulated) and are shown to work well over the wide temperature range from -40°C to 225°C, which make them attractive especially for high temperature applications, e.g. for automotive or aerospace applications.

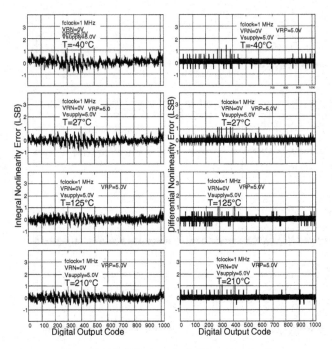

Figure 14. 10 bit ADC – integral and differential nonlinearity.

Furthermore complex mixed signal ICs, not treated here - e.g. a magnetic gear-tooth sensor[3] - were developed and demonstrate the capability of easy making high temperature electronics.

Finally we can state that the technology XI10 together with a design kit developed offer a good basis for designer/customer to create their own integrated circuits for high temperature applications.

REFERENCES

1. http://www.xfab.com
2. J.-P. Colinge, Silicon-on-Insulator Technology: Materials to VLSI. Boston, MA: Kluwer, 1991.
3. D.M.Nuernbergk, V.Nakov et al., "Hitzefest – Design von Mixed-Signal-Hochtemperatur-ICs", in *Design&Elektronik*, Heft 3, März 2002

HIGH-VOLTAGE SOI DEVICES FOR AUTOMOTIVE APPLICATIONS

Jörgen Olsson
The Ångström Laboratory, Uppsala University, Sweden

Abstract: There is a strong trend to increase the level of integration for automotive electronics in modern cars. SOI-technology offers the advantage of integrating both high-voltage devices and low-voltage circuitry on the same chip. This chapter describes both academic research and industrial SOI-processes in the field of high-voltage SOI devices for automotive applications.

Key words: SOI, high-voltage devices, SOI-process, automotive electronics

1. INTRODUCTION

Today's modern cars have an enormous amount of electrical wiring and electronics circuits for a wide variety of applications. Figure 1 illustrates a few of the different areas where automotive electronics are used. Inside the car there are several systems for comfort and convenience, such as ACC (Automatic Climate Control), audio and power seats. Several other applications provide safety and active chassis control, such as ABS (Anti-lock braking system) and TRC (Traction Control). Also in the engine room there are several functions such as the EFI (Electric Fuel Injection), cruise control and cooling fans. These numerous ECUs (Electronic Control Units) in the modern automobile involve different types of devices: low voltage devices for digital computation, analog and digital devices for interfacing and communications, power devices for e.g. motor controls, and also a large number of different sensors. Common to most of these devices is that they have to operate in a very harsh environment, involving high temperatures and EMI (Electromagnetic Interference).

D. Flandre et al. (eds.), Science and Technology of Semiconductor-On-Insulator Structures and Devices Operating in a Harsh Environment, 155-166.

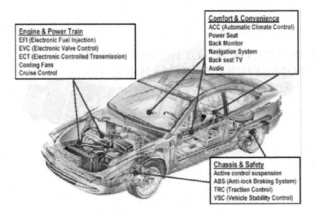

Figure 1. Different application areas in a modern car.

High-voltage semiconductor devices in automobiles are almost exclusively used as switching devices. Traditionally, switching function has been accomplished using electromechanical switches. However, nowadays solid-state devices have replaced the electromechanical switches for many applications, since semiconductor switches have many advantages, such as better reliability.

High-voltage devices with voltage rating of about 35 V is common in the 12 V automotive system. For the emerging 42 V battery automotive system a voltage rating of 120 V is sufficient for most applications. Higher voltages, up to around 500 V, are necessary for some specific semiconductor switches, such as the IGBTs, which are used in the ignition circuits.

As in many other fields of microelectronics there is a strong trend to increase the level of integration also for automotive electronics. Integration will increase functionality and reliability of the systems, and also lower the cost. When high-voltage or power devices are to be integrated with low-voltage control circuitry it often comes to the solution of using multi-chip modules, such as chip-on-chip or chip-by-chip modules[1]. The reason for this is that it is very difficult (or sometimes impossible) to do on-chip integration of high and low voltage devices without using complex and expensive manufacturing processes. It is also simply a matter of cost; a too integrated solution could suffer from lower yield and from higher additional manufacturing costs, and thereby a higher total cost.

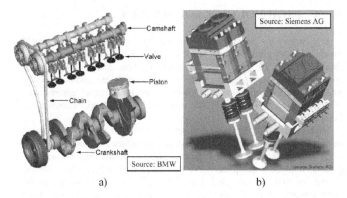

Figure 2. Conventional camshaft engine a), and units for electromagnetic valve control b).

Car manufactures attach great importance to reliability and cost[2], especially for the emerging "X-by-wire" technologies, where X stands for e.g. steer or break. This concept is based on the idea of adding functionality and circumventing constrains of mechanical systems[3]. The cable connections between gas pedal and throttle, as well as between clutch and gears, will be replaced by electronic modules offering significantly greater flexibility[4]. A complete system-on-chip (SoC) concept would provide the ultimate integrated solution for these applications. The SoC concept combines signal-processing parts and power actuators, leading to improved reliability, reduction of EMI and space, and weight reduction[5]. A higher number of functions on a chip is superior to printed circuit boards because electrical connections on chips are more reliable than connections on circuit boards. Failure rates of material interfaces, such as solder joints and packaging adhesives, dominate the total failure rate of electronic components in automobiles[6].

An example of a challenging application for the future automobile is to replace the engine camshaft with an EVC (Electromagnetic Valve Control) unit. Conventionally, the position of the valves is determined by the position of the pistons, see figure 2a, and is independent of the actual engine power. By instead having separate control of the individual valves, as provided by the EVC unit in figure 2b, the valves can be opened and closed independent of the piston position. The valve timing can be then be optimized with respect to fuel economy, emission, or power generation, at all engine speeds. A SoC solution for this particular application would require integration of high-voltage devices with high power handling capability. Furthermore, in this case the electrical module is placed in a very harsh environment.

2.   HIGH-VOLTAGE SOI DEVICES

SOI technology is well suited for automotive electronics and would enable easier integration of high and low voltage devices on the same silicon chip. SOI with its possibility for total isolation, also provides better protection from voltage and current spikes, which are common in the automobile power supply, and also enables operation at higher temperatures. SOI circuits and devices should therefore have better EMC (Electromagnetic Compatibility).

High-voltage SOI devices can either be laterally or vertically integrated on the silicon. A lateral approach is rather straightforward, since it has very much in common with lateral bulk implementation, and does not add any significant process complexity. Vertical implementation of high-voltage devices on SOI, on the other hand, requires some modifications to the manufacturing process, compared to its vertical bulk counterpart. Both vertical and lateral high-voltage devices will be described in following sections.

2.1   Vertical high-voltage SOI process: an academic research project

In order to combine both high-voltage devices and low voltage circuitry on the same piece of silicon, one approach is to integrate vertical high-voltage transistors on thick SOI-layers, isolated from the low-voltage devices by deep trenches, see figures 3. With this approach the low voltage devices, such as CMOS, are very similar to ordinary bulk devices. Thus, those devices will not require any special processing modification. It is however worth noting that CMOS devices on thick SOI do not have the performance advantages of the thin-film SOI fully-depleted or partly-depleted CMOS devices.

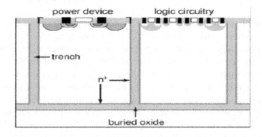

Figure 3. Schematic cross-section of high-voltage devices on thick SOI with trench isolation.

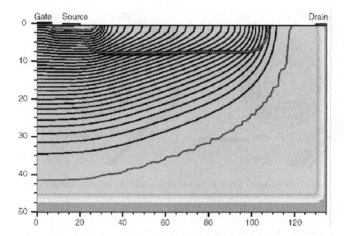

Figure 4. Simulated potential distribution in a 500 V vertical DMOS transistor.

In this thick-SOI type of technology the maximum voltage rating for the high-voltage devices determines the thickness and the doping level of the SOI-layer. It will also determine the required thickness of the buried oxide (BOX) and the dimensions of the trench isolation, i.e. the thickness of the liner oxide. There are also other parameters, which are important to the BOX and trench design, which will be discussed later.

A challenge with this vertical approach is to realize a low-ohmic conduction path to the drain contact at the surface, which for vertical bulk devices normally is at the backside of the wafer. This is accomplished by utilizing buried implanted layers just above the BOX and by dopant diffusion along the trench side-walls. Furthermore, since the high-voltage drain side is at the surface, a lateral high-voltage design is also needed.

This vertical SOI technology has been used to realize both high-voltage (500 V) DMOS transistors on 50 μm SOI-layer[7] and 600 V bipolar devices with a 60 μm layer thickness[8]. Figure 4 shows the example of the 500 V vertical DMOS transistor with simulation results of the electrical potential distribution within the device. The potential is evenly distributed both vertically and laterally, due to a correct charge balance between the n-type substrate and the p-type RESURF (REduced SURface Field) extension along the surface.

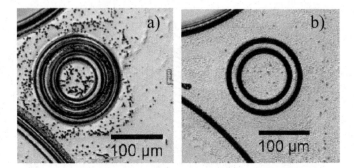

Figure 5. A large number of defects on the surface generated after high-temperature steps a). A significantly reduced number of defects due to back-end trench formation b).

The RESURF layer is formed early in the process flow with a p-type ion-implantation followed by a high-temperature drive-in step for several hours. It was observed that this step gave rise to defect generations in the silicon layer. Figure 5a shows the surface of the wafer of the final processing, and defects are clearly seen after decorations etch. It was found that the defects originated both from the bottom and top trench corners due to high stress in these regions, resulting in slip line dislocations. An alternative process has been found to be successful in minimizing the defect generation. In a modified process the trenches are instead formed at the back-end of the process, and thereby the high-temperature anneal is avoided[9]. Figure 5b shows the corresponding surface with significantly reduced number of slip dislocations.

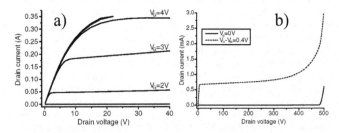

Figure 6. IV-characteristics for a DMOS with a specific on-resistance of 170 mΩcm^2 a). Breakdown characteristics for the 480 V DMOS transistor b). (W_G=30 mm)

Figure 6 shows the measured IV- and the breakdown-characteristics for a 480 V vertical DMOS in a thick-layer SOI-process optimized for suppression of defects originating from the trenches[9]. The measured specific on-resistance is slightly higher than the simulated value for this multi-finger DMOS layout.

2.2 Commercial high-voltage SOI processes

There is a large number of semiconductor manufacturers producing high-voltage devices for automotive electronic application. There are significantly less who offer SOI-processes with high-voltage capability suited for the automotive industry.

Philips Semiconductors has developed several generations of a thin-film SOI BCD process intended for automotive applications[10]. This type of process uses lateral high-voltage devices. Figure 7 shows a cross-section of an LDMOS transistor in this two-metal layer LOCOS isolated SOI-process. The process offers several types of devices, such as 5 V NMOS and PMOS, medium voltage (12-25 V) DMOS and high-voltage (100 V) complementary DMOS transistors, as well as passive devices. It is therefore easy to integrate power, analog and digital functions on the same chip. The latch-up free substrate with improved EMC performance and its high temperature operation makes it especially well suited for automotive applications.

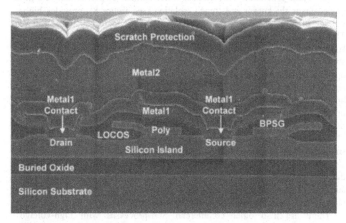

Figure 7. SEM cross-section of first generation thin-film SOI advanced BCD process, showing a lateral DMOS transistor. (With permission from Philips Semiconductors)

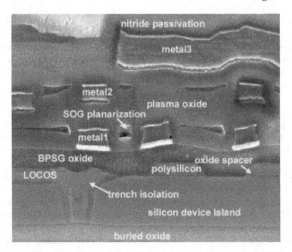

Figure 8. SEM image of Philips Semiconductors' third generation thin film SOI-process with three metal layers and trench isolation. (With permission from Philips Semiconductors)

The latest generation[11], which uses wafer-bonded SOI-substrates (1.5 μm device layer thickness) and trench isolation (see figure 8), provides high-voltage LDMOS transistors with breakdown voltage up to 120 V and 180 V. That voltage rating is sufficient for almost all 42 V automotive applications. Higher voltage, up to around 500 V, is necessary for the semiconductor switches, normally IGBTs, which are used in the ignition circuits. By optimizing a lateral thin-film SOI technology, e.g. by using a graded drift region doping profile, LDMOS and LIGBT can reach breakdown voltages up to around 1000 V. Extremely competitive performance regarding specific on-resistance versus breakdown voltage has been reported for the thin-film SOI LDMOS transistors, such as 0.1 Ωmm^2 for a 120 V device.

The feature size of 0.6 μm in combination with trench isolation results in small die sizes compared to standard BCD technologies. It has also shown about 20 dB better EMC performance and no problems with voltage spikes due to switching of inductive loads. Typical integrated high-voltage automotive applications for this SOI-process are in-vehicle networking chips, sensor systems and also 'under-hood' systems.

There are also other commercial SOI-processes with high-voltage capability suited for automotive electronics. Toshiba has SOI-processes[12-14] where lateral DMOS and IGBTs are integrated on the same silicon as low power CMOS circuits. Voltage rating of up to 400 V for the IGBTs is available. A 1.0 µm SOI process is available from X-FAB, which is compatible with the 42 V automotive battery system. X-FAB also offers a 500 V SOI-process on thick SOI with deep trench isolation[15]. The process has vertical integration of the high-voltage DMOS transistors and is similar to the process described in section 2.1.

3. DISCUSSION

There are several issues that could be mentioned for high-voltage SOI-devices. Some will be discussed in the following.

3.1 Thermal issues

As is well know, self-heating has been an issue for SOI-devices for a long time. It is clear that by adding a buried oxide one also adds to the total thermal resistance, and thereby the risk for device self-heating increases. The use of trench isolation also prevents lateral heat spreading in the device layer, which also adds to the total effective thermal resistance. However, this is normally not a critical issue, since the typical effect of the buried oxide (in steady-state) is negligible to the thermal resistance of the package.

For power transients, parameters such as the SOI film thickness, the BOX thickness and the pulse length will strongly affect the self-heating of the device. The difference in self-heating between SOI and bulk devices have been shown to be greatest for very short power transients, because the initial temperature rise in SOI devices is more rapid[16]. For decreasing film thickness and longer pulses the temperature difference will increase. This is due to the fact that heat spreading in the film will be limited by the BOX and the heat capacity of the film is reduced[17,18].

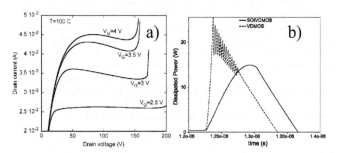

Figure 9. Steady-state self-heating a), and transient power dissipation b) for a 500 V DMOS.

Figure 9a shows measured steady-state self-heating of a 500 V DMOS on thick SOI[19] at the ambient temperature of 100 °C. The characteristics agree fairly well with predictions taking into account the thermal resistance of the BOX. For transients the same transistor shows a rather surprisingly behavior. Simulations of unclamped inductive switching show less power dissipation for the SOI-devices compared to its bulk counter-part, see figure 9b. In this case the BOX is 2 µm and deep trench isolation is used, which results in a significant increase of the thermal resistance. However, when switching the device the inductive load forces current, which charges the BOX capacitance instead of dumping the stored energy over the device. This results in that the maximum current does not coincide with the highest voltage over the device. Consequently the dissipated power in the device will be reduced[20].

3.2   SOI-device layout considerations

One of the most important figures for high-voltage switches is the specific on-resistance versus breakdown voltage. For automotive electronics it is very important to have low saturation voltage in order to minimize power dissipation. This is due to the limited amount of power available, typically around 2.5 kW in modern cars. The thin-film SOI lateral high-voltage transistors have shown very low specific on-resistance. One of the reasons for this is that it is easy to also use the substrate as a back gate, in addition to top surface field plates, to effectively deplete the device layer in off-state. Consequently, a very effective RESURF condition can be obtained. It has also been shown that a three metal layer process is more effective than a double metal process to reduce the on-resistance, due to a more optimized interconnect pattern[11].

For vertical high-voltage SOI devices, as described in section 2.1, additional effects have to be considered. Figure 10a shows the top-view of a multi-fingered DMOS transistor with several mm effective gate width[7]. The gates are in a dense square mesh pattern in order to increase the on-state current density. However, since the current travels both vertically and laterally from the surface drain contact, both the traditional vertical DMOS resistance and the lateral resistance have to be taken into account. In one extreme, the lateral resistance is assumed to be zero, meaning that the lateral n+ layer above the BOX and the n+ along the trench sides have no resistance. In this case the transistor on-resistance is identical to that of an ordinary vertical bulk DMOS transistor. On the other hand, when there is a resistance in the n+ layers the situation is drastically changed.

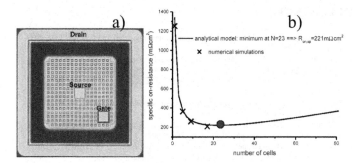

Figure 10. Top view of high-voltage DMOS with multi-fingered gate structure a). Lowest specific on-resistance occurs at an optimum number of cells (gate fingers) b).

The result is that gate fingers at different distance from the drain contact also will experience different lateral resistance. This effect has been treated in 2D and been modeled analytically and verified with numerical device simulations[21]. Figure 10b shows the specific on-resistance as a function of the number of gate fingers. It is clear that it exists an optimum number of cells, in this case 23, which results in the lowest specific on-resistance. Consequently, for high current rating several transistors with an optimum number of cells have to be connected in parallel.

4. SUMMARY

Automotive electronics is a fast growing field of applications. The trend is to integrate more and more functions in order to increase functionality, improve reliability and to reduce the cost. Integration of high-voltage power devices and low voltage control circuitry on one chip is one way to accomplish this. A complete system-on-chip (SoC) concept would provide the ultimate integrated solution for many automotive applications. However, it is important to realize that SoC is only attractive if the overall cost is indeed lower and it is proven to be robust.

SOI-technology does offer easy integration of high- and low-voltage devices. Today, both thin and thick film SOI processes are available, which offer both high-voltage DMOS and IGBT devices on the same chip as CMOS. SOI also has advantages of excellent device isolation, high temperature operation and less sensitivity to EMI. Consequently, SOI is a key technology for complete silicon SoC for automotive applications.

REFERENCES

1. A. Graf, Smart Power Switches for Automobile and Industrial Applications, *VDE ETG Conference*, (2001).
2. J. C. Erskine et al., High Temperature Automotive Electronics: An Overview, *proc. 3rd Intl. High-Temperature Electronics Conference* 1, XIII–21–XIII–31, (1996).
3. J. G. Kassakian and D. J. Perreault, The Future of Electronics in Automobiles, *Proc. ISPSD*, 15–19, (2001).
4. M. Lenz, Systems on Silicon for the System Automobile, *Components*, no. 1, 17–20, (Siemens AG, 1999).
5. C. Contiero, A. Andreini, and P. Galbiati, Roadmap Differentiation and Emerging Trends in BCD Technology, *Proc. ESSDERC*, 275–282, (2002).
6. M. A. Tamor, High-Temperature Electronics for Automobiles, in *High-Temperature Electronics*, edited by R. Kirschman, (IEEE Press, 1998), pp. 153–160.
7. U. Heinle and J. Olsson, Integration of high voltage devices on thick SOI substrates for automotive applications, *Solid-State Electronics*, vol. 45(4) 629-632, (2001).
8. L. Clavelier, et al., 600V Bipolar Power Devices on Thick SOI, *proc. ESSDERC*, 395-398, (2001).
9. U. Heinle, K. Pinardi, and J. Olsson, Vertical high voltage devices on thick SOI with back-end trench formation, *proc. ESSDERC*, 295-298, (2002).
10. J. A. van der Pol, et al., A-BCD: An Economic 100 V RESURF Silicon-On-Insulator BCD Technology for Consumer and Automotive Applications, *proc. ISPSD*, 327-330, (2000)
11. A. W. Ludikhuize, et al., Extended (180V) Voltage in 0.6µm Thin-Layer-SOI A-BCD3 Technology on 1µm BOX for Display, Automotive & Consumer Applications, *proc. ISPSD*, 77-80, (2002)
12. H. Funaki, et al., High voltage BiCDMOS technology on bonded 2 µm SOI integrating vertical npn pnp, 60 V-LDMOS and MPU, capable of 200°C operation, *IEDM Tech. Digest*, 967-970, (1995).
13. A. Nakagawa, Recent advances in high voltage SOI technology for motor control and automotive applications, *proc. IEEE BCTM*, 69-72, (1996).
14. A. Nakagawa, Single chip power integration high voltage SOI and low voltage BCD, *ETG-Fachberichte (Energietechnische Gesellschaft im VDE)*, no. 81, 8-15, (2000).
15. X-FAB, Erfurt (June 3, 2004); http://www.xfab.com.
16. E. Arnold, H. Pein, and S. P. Herko, Comparison of Self-Heating Effects In Bulk-Silicon and SOI High-Voltage Devices, *IEDM Tech. Digest*, 813-816, (1994).
17. H. Neubrand, R. Constapel, R. Boot, M. Füllmann, and A. Boose, Thermal Behaviour of Lateral Power Devices on SOI Substrates, *proc. ISPSD*, 123-127, (1994).
18. J. Olsson, Self-Heating Effects in SOI Bipolar Transistors, *Microelectronic Engineering*, 56(3-4), 339-352, (2001).
19. K. Pinardi, U. Heinle, S. Bengtsson, J. Olsson, and J.-P. Colinge, High-Power SOI Vertical DMOS Transistors with Lateral Drain Contacts: Process Development, Characterization and Modeling, *IEEE Trans. Electron Dev.*, 51(5), 790-796 (2004).
20. K. Pinardi, U. Heinle, S. Bengtsson, J. Olsson, and J.-P. Colinge, Unclamped inductive switching behaviour of high power SOI vertical DMOS transistors with lateral drain contacts, *Solid-State Electroncis*, 46(12), 2105-2110, (2002).
21. U. Heinle and J. Olsson, Analysis of the Specific On-Resistance of Vertical High-Voltage DMOSFETs on SOI, *IEEE Trans. Electron Dev.*, 50(5), 1416-1419, (2003).

HEAT GENERATION ANALYSIS IN SOI LDMOS POWER TRANSISTORS

J. Roig, D. Flores, J. Urresti, S. Hidalgo and J. Rebollo
Centre Nacional de Microelectrònica (CNM-CSIC), Campus UAB, Bellaterra 08193 Barcelona, Catalonia

Abstract: An overview of the heat generation phenomena in SOI LDMOS transistors, mainly due to the Joule effect, is provided in this work. The distribution of the heat generation along the SOI LDMOS cross-section depends on the technological and geometrical parameters and the applied bias. Reported data and results extracted from simulation, theory and experiment are used to give physical insight into the heat generation mechanisms. The analysis of the heat generation is of utmost importance to derive the 3D dynamic temperature distribution at short time operation. An accurate temperature prediction at the source, drain and channel regions is desirable for improved electro-thermal models and for the study of the electromigration in interconnects. Moreover, information on temperature peaks is crucial to understand the failure mechanisms in power LDMOS transistors.

Key words: LDMOS; SOI; self-heating; thermal management; hot spot; heat generation; power dissipation.

1.   INTRODUCTION

Power Integrated Circuits (PICs) are the monolithic integration of power devices with their control circuitry. Nowadays, PICs are a specific case of the Smart Power IC concept, with the integration of low and high power stages in the same substrate. In the 1990s, Smart Power technology was mainly based on Junction Isolated (JI) Bulk Silicon substrates, taking benefits of the MOS gated power devices and their high input impedance, fast switching and CMOS-compatible drive voltages[1]. However, the JI technology suffers from high cross-talk and leakage currents (specially critical at high temperature) and latch-up of parasitic thyristor structures. In

167

D. Flandre et al. (eds.), Science and Technology of Semiconductor-On-Insulator Structures and Devices Operating in a Harsh Environment, 167-178.
© 2005 *Kluwer Academic Publishers. Printed in the Netherlands.*

addition, a large Silicon area is consumed for the efficient isolation between power and signal parts.

The recent evolution of SOI-CMOS technology due to the improvement of SOI substrate quality, and their commercial availability, has favoured the development of SOI Smart Power ICs with full dielectric isolation between low power circuitry and output power devices[2]. Furthermore, the operating temperature and the integration density of SOI technologies are much higher than that of the equivalent Bulk Silicon technologies. Today, SOI-PICs with different circuit topologies, such as half-bridge, full-bridge and forward converters, chopper circuits and inverters are available[3]. These PICs are basically addressed to control electronics lighting and small size motors, flat panel displays, high precision power supplies, IC regulators, telecommunications, portable electronics and automotive electronics. Although SOI technologies are advantageous in terms of integration density, isolation, operating frequency and reliability in aggressive environments, there is still a lot of work to do to improve the voltage capability of SOI lateral power devices and the thermal management.

2.   SOI LDMOS THERMAL MANAGEMENT

Concerning the thermal management of power devices, the ambient temperature (T_{amb}) and the self-heating (S-H) cause the temperature increment (ΔT) inside the device active area. In comparison with Bulk Silicon power devices, ΔT due to S-H in SOI power devices is significantly higher due to the enormous buried oxide (BOX) thermal resistivity (two orders of magnitude higher than the Silicon value). Indeed, the BOX layer represents an important thermal barrier for the heat flowing to the handle Silicon substrate, thus usually exhibiting similar or even higher influence than the ambient temperature in ΔT.

Several techniques are commonly used in order to perform electro-thermal studies on power semiconductor devices[4], evaluating ΔT from T_{amb} and S-H. Among them, the electro-thermal circuit simulation exhibits a good trade-off between accuracy and computational cost. In steady and dynamic state conditions, electrical network models usually consider the electro-thermal coupling by means of an embedded thermal network or sub-circuit. Some authors use the embedded thermal network with a temperature multi-node[5], thus evaluating the temperature distribution in a complete thermal mesh. Other authors regard a unique temperature node which is evaluated through analytical models[4]. Since in both cases one or several electric current sources must be defined to emulate the device heat source, the dissipated power or the heat generation distribution is required to be known.

Subsequently, a heat generation analysis (coming from S-H) has to be carried out in order to obtain accurate 3D temperature mapping, especially for strong non-uniform power dissipation and short time operation. Up to now, very few works include a detailed study of the heat generation when a power device is thermally analysed. In this sense, quantitative analysis have been applied to VDMOS[6], BJT[7] and ESD diodes[8], while a qualitative description, lacking of physical insight, has been devoted to SOI LDMOS transistors[9,10].

An important increment of temperature due to S-H takes place inside the SOI LDMOS active area, thus degrading the static and dynamic electrical characteristics of the power switch and altering the performance of the neighbouring low power circuitry. Hence, S-H effects are critical under prolonged high-temperature operation; i.e., long pulse overloads with high power losses[11,12] and DC operation[10]. Besides the electro-thermal modelling improvement, accurate prediction of the temperature in the source, drain and channel regions is desirable for the study of the electromigration in interconnects[13]. Moreover, information on temperature peaks is crucial to understand the failure mechanisms in power LDMOS transistors.

In accordance with the remarkable role of the heat generation for thermal management purposes, this paper aims to review works devoted to the heat generation analysis in SOI power transistors.

3. FROM THICK TO THIN SOI: A DEGRADATION OF THE THERMAL PERFORMANCE

Once high and low power stages were integrated together in a common die, the evolution of the power devices design was inevitably drawn for the shrinkage of the low power device. Changes in substrate architecture and processing, addressed to the low power stage improvement, demanded a proper optimisation of the LDMOS structure to achieve a good electrical performance. As a result, the appearance of Thin Film Silicon-on-Insulator (TFSOI) substrates limited the thickness of the LDMOS active silicon layer (t_{SOI}) to submicron values for Smart Power ICs[2]. The inherent low load capacitance of TFSOI technologies and the reduction of the power supply have lead to the increase of the operating frequency and the reduction of the power consumption. Hence, TFSOI-CMOS circuits are used to integrate high-speed microprocessors, high speed and low-power LSI (mobile communications) and RF circuits.

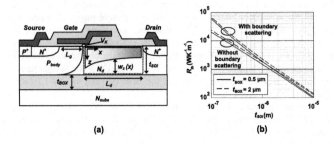

(a)   **(b)**

Figure 1. (a) Schematic cross section of a SOI LDMOS transistor (b) R_{th} vs. t_{SOI} for a punctual heat source on the top of the substrate.

The electrical isolation inherent to TFSOI technology provides numerous advantages in lateral power devices: easy lateral dielectric isolation, latch-up free, low leakage current, suppression of parasitic capacitances and immunity to radiation. In spite of these advantages, new drawbacks related to LDMOS transistor performance arise from TFSOI substrates electrical isolation: limitation of current and voltage capability and strong increase of the thermal resistance. As a matter of fact, a logarithmic increment of the LDMOS thermal resistance (R_{th}) with the t_{SOI} decrement (Fig. 1.a) has been experimentally demonstrated[10] down to 1μm t_{SOI} values. Recently, a simple expression to describe such a tendency in SOI power devices has been reported[14]. The R_{th} dependence with t_{SOI} for a surface punctual heat source is given by :

$$R_{th} = \frac{1}{2\pi\kappa_{si}t_{SOI}}\left[1+\log\left(\frac{m}{t_{SOI}}\right)\right]$$

(1)

where m is the healing length, $m=(t_{SOI}t_{BOX}\kappa_{Si}/\kappa_{Ox})^{1/2}$ and κ is the thermal conductivity of a given material.

The R_{th} logarithmic evolution with t_{SOI} is depicted in Figure 1.b for 0.5 and 2μm t_{BOX} values, corresponding to those used in power applications. Moreover, the dramatic increment of R_{th} at low t_{SOI} values is found to be more pronounced for SOI layers in the submicron range due to the phonon-boundary scattering impact[15]. This effect is introduced in (1) by considering the κ_{Si} dependent on t_{SOI} as expressed in:

$$\kappa_{Si} = \frac{F}{F / \kappa_{Si\_Bulk} + 3/8 t_{SOI}} \tag{2}$$

where F is a temperature factor, evaluated at 300K, and κ_{Si_Bulk} is the thermal conductivity for the Bulk Silicon (κ_{Si_Bulk}=150 WK^{-1}m^{-1}). The relevance of the boundary scattering for submicron t_{SOI} values can be inferred from Fig. 1.b. Most of the existing thermal models for SOI LDMOS transistors use a κ_{Ox} value of 1.5 WK^{-1}m^{-1} and neglect the heat flowing through the interconnects[16]. However, some authors have measured slightly different κ_{Ox} values, depending on the SOI wafer manufacturer[17].

Additionally to the increase of R_{th}, the distribution of the heat generation in SOI LDMOS transistors varies when t_{SOI} is reduced, as it is extensively explained in section 5.

Among the different thermal time constants related to each layer, the one corresponding to the BOX is the most relevant. As a consequence, the thermal impedance is expected not to be significantly modified by a reduction in the value of t_{SOI}, unless dealing with operation times below 10^{-8} sec[18]. At this range the heat generation analysis based on the Joule effect is not suitable since the thermal non-equilibrium effects play an important role[19].

4.   HEAT GENERATION IN THICK FILM SOI LDMOS TRANSISTORS

In order to provide a qualitative description of the heat generation, a reported thick SOI LDMOS transistor[20] has been thermally analysed with the aid of electro-thermal simulations. The selected set of parameters defined in Fig. 1.a to provide a 170 V device is L_d=20µm, L_g=2.5µm, t_{SOI}=4µm, t_{BOX}=1µm and N_d=5·10^{15}cm^{-3}. The described LDMOS transistor is placed on top of a 500µm Silicon substrate and represents a basic cell cross-section of a multifinger power device. Hence, adiabatic walls are defined close to drain and source sides, thus emulating the neighbouring fingers. In addition, the front and back sides are considered adiabatically and isothermally, respectively.

According with 2D conduction steady-state simulations, the location of the maximum temperature in the LDMOS structure does not depend on the gate bias. In fact, a temperature peak is always found at the channel region close to the drift; where the current path is narrow (Fig. 2.a).

The predominant punctual heat source at the end of the channel region can be observed in Fig. 2.a. As the dissipated power is defined as

$P = I_{ch} \cdot V_k$ the q_e peak in the channel region is lower than the corresponding value in the case of a higher V_g due to the V_k fall. In spite of this effect, the punctual heat source still predominates.

5.   HEAT GENERATION IN THIN SOI LDMOS TRANSISTORS

The use of thin SOI layers yields a more complex distribution of the generated heat, as demonstrated by numerical simulations on commercial 120V TFSOI LDMOS transistors[21] with identical boundary conditions as in Section 4. In this case, the geometrical parameters of the optimised LDMOS structure are L_d=10µm, L_g=1.2µm, t_{SOI}=0.2µm, t_{BOX}=1µm and N_d=6·10^{16}cm^{-3}.

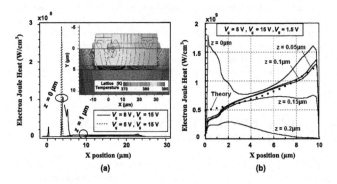

Figure 2. (a) Heat generation at different depth in a Thick Film SOI transistor at V_g=5 and 8V and V_d=15V together with the temperature distribution (inset).   (b) Heat generation at different depth in Thin Film SOI transistor at V_g=5V and V_d=15V.

According to the 2D conduction steady-state simulations, the temperature distribution in the LDMOS structure strongly depends on the V_g value. The temperature mapping inside the LDMOS structure is plotted in Fig. 3 at V_g=5V and 3V and V_d=15V, keeping the source and the substrate grounded. The maximum temperature at V_g=3V is located in the channel region, as in the thick SOI case. On the contrast, the maximum temperature increment at V_g=5V takes place near the drift/drain junction, as it has been reported in other works[22-24].

The different temperature mapping at low and high gate bias can be explained according to the power dissipated in the conduction regime. In the case of power LDMOS transistors, the dissipated power is mainly due to Joule heat generated by the electron current (q_e) flowing through the channel and drift regions. At a low gate bias, the heat generation is mainly located in the channel pinch-off point, thus presenting a similar q_e profile to that shown in Fig. 2.a. At higher gate voltages, the total generated heat is mainly due to the contribution of the drift region. Apart from the V_k fall, the power dissipated in the drift region increases with the gate bias due to the higher device current level, flowing through a section one order of magnitude thinner than in the thick SOI case. Hence, the current density (J_n) and the heat generation in the drift region is drastically increased.

5.1   Uniformly doped drift region

In the case of SOI LDMOS transistors with a uniform doped drift region q_e increases from body to drain at any gate bias, in accordance with the vertical depletion from the BOX layer. The vertical electric field at the BOX/drift interface (E_z) increases from body to drain, accounting for the increasing distribution of the generated heat. As a matter of fact, the current is confined near the surface of the drift region because of the vertical depletion action. Therefore, the effective cross-section of the current flow is shrunk at the drain side since the vertical depletion region increases from body to drain (Fig. 1.a).

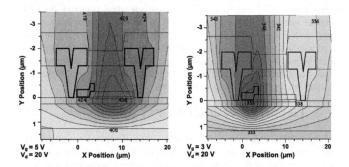

Figure 3. Temperature mapping inside the TFSOI LDMOS transistor, with a uniformly doped drift region, at Vg=5V and 3V. Vd=20V in both cases.

A $q_e(x)$ expression for the heat generation into the LDMOS drift region can be derived, evaluating the physical magnitudes related with the Joule heat equation ($q_e(x) = J_n(x) \cdot E_x(x)$) by means of semiconductor physics principles. Hence, the $q_e(x)$ distribution[21] is finally expressed as:

$$q_e(x) = \frac{J_n(x)^2}{qN_d\mu_d - \frac{J_n(x)}{E_c}} \tag{3}$$

This expression has to be evaluated in a heat source with a thickness defined by $t_{SOI} - w_z(x)$, where $w_z(x)$ is the vertical depletion boundary.

$$w_z(x) = \frac{\varepsilon_{si}}{\varepsilon_{ox}} \cdot t_{BOX} \left[\sqrt{1 + \frac{2\varepsilon_{ox}^2 \varepsilon_0 V(x)}{q\varepsilon_{si} N_d t_{BOX}^2}} - 1 \right] \tag{4}$$

A linear approach of (3) is evaluated in a parallelepiped heat source[25] to easily obtain a dynamic mapping of the temperature. The good match between theoretical and simulation results is shown in Fig. 2.b for the TFSOI LDMOS structure.

Additional simulations with the LDMOS transistor operating in linear, saturation or quasi-saturation regimes have shown that the power density always increases from body to drain, not depending on the conduction regime. The power density profile depends on the amount of injected current into the drift region and on the vertical voltage drop in the BOX layer. Subsequently, the V_g and V_d values are important when analysing the heat generation increment from body to drain. However, the predominance of the channel or drift region heat generation is mainly controlled by the gate bias. The geometrical and technological parameters of the LDMOS structure, as well as the gate bias, have to be taken into account in order to determine the relevance of the drift region dissipated power contribution. Hence, higher q_e is reckoned in the drift region at large L_d and short t_{SOI} values.

5.2   Linearly doped drift region

TFSOI LDMOS transistors with a voltage capability in the range of 600V require the implementation of the Variation of Lateral Doping (VLD) techniques in the drift region. Experimental determinations of the heat generated in SOI LDMOS structures, which include VLD doping profiles, have been reported[9] previously whilst a theoretical model has been recently published[26] giving physical insight into the thermal behaviour. Unlike the case of uniform doped drift LDMOS transistors, the maximum temperature point is always found close to the body region at any applied bias. A thermal model to predict the temperature mapping at short-circuit operation is proposed[26], thus providing a physically based explanation of the temperature peak location.

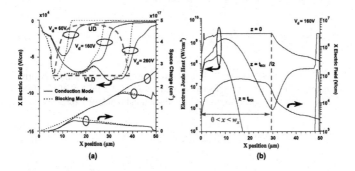

Figure 4. (a) E_x and the space charge distribution along the drift region. (b) E_x and q_e distribution along the drift region.

Thin Film SOI LDMOS transistors with VLD have a linear doping profile in the drift region, increasing from body to drain. An accurate optimisation of the VLD profile, described by $N(x) = N_0 + S_N \cdot x$, is required to improve the R_{on}/V_{br} trade-off (Fig. 4.a). Therefore, the space charge distribution along the drift region is reckoned to differ from the uniform doping case in both conduction and blocking modes. As a matter of fact, the space charge distribution in the conduction mode (Fig. 4.a), above the vertical depletion boundary, confirms the space charge linear increase with x, from $x=0$ up to a certain x value, depending on the applied drain bias. It has also been proved by numerical simulation that the extension of the space charge region is almost identical to the extension of the lateral depletion length (w_x) in the blocking mode. Additionally, the E_x profile shows similar constant value at $0 < x < w_x$ in the conduction and blocking modes, except for small V_d values. If V_d is small enough compared to V_{br}, the space charge and E_x profiles are misshaped due to the field plate action.

The parallelism between blocking and conduction modes leads to some considerations about the heat source shape and dimensions, as well as the power density distribution.

The heat source volume is a rectangular parallelepiped whose dimensions depend on the device biasing conditions and technological/geometrical parameters. As inferred from Fig. 4.b, the heat source length in the x direction is modulated by V_d at a given device geometry. In fact, there is a gradual spreading of the linear space charge region, limited by $0 < x < w_x$, with increasing V_d. As stated before, the E_x profile along the non-depleted region exhibits a constant value for $0 < x < w_x$ and zero for $x > w_x$, as shown in Fig. 4.b. The simplified w_x expression is, finally:

$$w_x = \frac{\varepsilon_{si} V_d}{q S_N} \cdot \left(\frac{t_{SOI}^2}{2} + \frac{\varepsilon_{si}}{\varepsilon_{ox}} \cdot t_{BOX} \cdot t_{SOI} \right)^{-1} \tag{5}$$

The vertical confinement of the current in the drift region is clearly observed in Fig. 4.b. This effect is caused by E_z at the BOX/SOI interface and is modelled by means of the vertical depletion boundary, expressed as:

$$w_z = \frac{\varepsilon_{si}}{\varepsilon_{ox}} \cdot t_{BOX} \left[\sqrt{1 + \frac{2\varepsilon_{ox}^2 \varepsilon_0 V_{br}}{q \varepsilon_{si} L_d S_N t_{BOX}^2}} - 1 \right] \tag{6}$$

and neglecting V_k and N_0. Thus, a rectangular parallelepiped heat source with dimensions $w_x \cdot (t_{SOI} - w_z) \cdot b$ is placed on the surface of the wafer.

The power density inside the heat source is constant in the VLD case, thus leading to the use of the original Joy and Schlig formulation to evaluate the heat conduction[27]. This behaviour is attributed to the constant E_x value ($E_x = V_d / w_x$), extended along the linear space charge region, and the constant current density in the lateral direction ($J_n = I_{ch} / (t_{SOI} - w_z)$).

6. CONCLUSIONS

The thermal performance of LDMOS power transistors integrated on thick and thin SOI substrates is analysed in this paper. The thermal resistance in thin film SOI substrates increases with the reduction of the active Silicon layer thickness, exhibiting a logarithmic dependence. In the case of thick SOI LDMOS transistors, the heat generation is mainly located in the channel region, while in the case of thin SOI LDMOS transistors the maximum heat generation can be placed in the channel or inside the drift region, depending on the biasing conditions. Two heat generation analytical models, suitable for the case where the heat generation in the drift region predominates, are reported in this paper. The models can be applied to thin film LDMOS transistors with uniform or VLD drift region doping profiles. In the case of a uniform drift region, a non-uniform heat distribution is obtained inside the drift region. On the contrast, the heat distribution is almost uniform when the VLD technique is used.

ACKNOWLEDGEMENTS

This work was supported by the Comisión Interministerial de Ciencia y Tecnología (CICYT) (Ref. TIC2002-02564) and by the project "Advanced Techniques for High Temperature System-on-Chip (ATHIS)", contract GRD1-2001-40466, funded by the Growth Programme (EU).

REFERENCES

1. B. Murari, Reliability of Smart Power devices, *Microelectronics Reliability*, **37**(11), 1735-1742 (1997).
2. J. A. van der Pol, A.W. Ludikhuize, H. G. A. Huizing, B. van Velzen, R. J. E. Hueting, J.F. Mom, G. van Lijnschoten, G. J. J. Hessels, E.F. Hooghoudt, R. van Huizen, M. J. Swanenberg, J. H. H. A. Egbers, F. van den Elshout, J. J. Koning, H. Schligtenhorst, J. Soeteman, A-BCD: An economic 100 V RESURF silicon-on-insulator BCD technology for consumer and automotive applications, in: *Proc. ISPSD'00*, 2000, pp. 327-330.
3. F. Udrea, D. Gardner, K. Sheng, A. Popescu, H. T. Lim and W. I. Milne, SOI Power Devices, *Electronics & Communication Engineering Journal*, 22-40 (2000).
4. V. d'Alessandro and N. Rinaldi, A critical review of thermal models for electro-thermal simulation, *Solid-State Electronics* **46**(4), 487-496 (2002).
5. L. Codecasa, D. D'amore and P. Maffezoni, Compact modeling of electrical devices for electrothermal analysis, *IEEE Trans. Circuits and Systems* **50**(4), 465-476 (2003).
6. L. Zhu, K. Vafai and L. Xu, Device temperature and heat generation in power Metal-Oxide Semiconductor Field Effect Transistor, *J. of Thermophysics and Heat Transfer* **13**(2), 185-194 (1999).
7. L. Zhu, K. Vafai and L. Xu, Modeling of non-uniform heat dissipation and prediction of hot spots in power transistors, *Int. J. of Heat and Mass Transfer* **41**(15), 2399-2407 (1998).
8. Y. Wang, P. Juliano, S. Joshi and E. Rosenbaum, Electrothermal modeling of ESD diodes in Bulk-Si and SOI technologies, *Microelectronics Reliability* **41**(11), 1781-1787 (2001).
9. Y. K. Leung, S. C. Kuehne, V. S. K. Huang, C. T. Nguyen, A. K. Paul, J. D. Plummer and S.S. Wong, Spatial temperature profiles due to non-uniform self-heating in LDMOS's in thin SOI, *IEEE El. Dev. Lett.* **18**(1), 13-15 (1997).
10. Y. K. Leung, Y. Suzuki, K. E. Goodson and S.S. Wong, Self-heating effect in lateral DMOS on SOI. in: *Proc. ISPSD '91*, 1991, pp. 27-30.
11. E. Arnold, H. Pein and S. P. Herko. Comparison of self-heating effects in bulk-silicon and SOI high-voltage devices, in: *Proc. IEDM'94*, 1994, pp. 813-816.
12. H. Neubrand, R. Constapel, R. Boot, M. Fullman and A. Boose, Thermal behaviour of lateral devices on SOI substrates, in: *Proc. ISPSD '94*, 1994, pp. 123-127.
13. S. Rzepka, K. Banerjee, E. Meusel and C. Hu, Characterization of self-heating in advanced VLSI interconnect lines based on thermal finite element simulation, *IEEE Trans Compon. Pack.* **21**(3), 406-411 (1998).
14. A. Pacelli, P. Palestri and M. Mastrapasqua, Compact modeling of thermal resistance in bipolar transistors on bulk and SOI substrates, *IEEE Trans. El. Dev.* **49**(6), 1027-1033 (2002).
15. M. Ashegui, B. Behkam, K. Yazdani, R. Joshi and K. E. Goodson, Thermal conductivity model for thin Silicon-on-Insulator layers at high temperatures, in: *Proc. Int. SOI Conf. 2002*, 2002, pp. 51-52.

178   *J. Roig, D. Flores, J. Urresti, S. Hidalgo, J. Rebollo*

16. M. Berger, Z. Chai, Estimation of heat transfer in SOI-MOSFET's, *IEEE Trans. El. Dev.* **38**(4), 871-875 (1991).
17. B. M. Tenbroek, R. J. Bunyan, G. Whiting, W. Redman-White, M. J. Uren, K. M. Brunson, M. S. L. Lee and F. Edwards, Measurement of buried oxide thermal conductivity for accurate electrothermal simulation of SOI devices, *IEEE Trans. El. Dev.* **46**(1), 251-253 (1999).
18. M. Berger, G. Burbach, Thermal time constants in SOI-MOSFETs, in: *Proc. Int. SOI Conf. '91*, 1991, pp. 24-25.
19. A. Raman, D. G. Walker and T. S. Fisher, Simulation of nonequilibrium thermal effects in power LDMOS transistors, *Solid-State Electronics* **47**(8), 1265-1273 (2003).
20. D. M. Garner, F. Udrea, H. T. Lim, G. Ensell, A. E. Popescu, K. Sheng and W. I. Milne, Silicon on Insulator Power Integrated Circuits, in: *Proc. ISPS'00*, 2000, pp. 123-129.
21. J. Roig, D. Flores, J. Urresti, S. Hidalgo, J. Rebollo., Modeling of non-uniform heat generation in Thin-Film SOI LDMOS Transistors, in: *Proc. ISPS'04*, 2004 (to appear).
22. P. Perugupalli, Y. Xu and K. Shenai, Measurement of thermal and packaging limitations in LDMOSFETs for RFIC applications, in: *Proc. IMTC/98*, vol.1, 1998, pp. 160-164.
23. J. M. Park, R. Klima and S. Selberherr, High-voltage lateral trench gate SOI-LDMOSFETs, *Microelectronics Journal* **35**(3), 299-304 (2004).
24. J. Roig, D. Flores, S. Hidalgo, M. Vellvehi, J. Rebollo and J. Millán, Study of novel techniques for reducing self-heating effects in SOI power LDMOS, *Solid-State Electronics* **46**(12), 2123-2133 (2002).
25. J. Roig, D. Flores, S. Hidalgo, M. Vellvehi, I. Cortes and J. Rebollo, A linear heat generation thermal model for LDMOS basic cell self-heating analysis in transient state, in: *Proc. THERMINIC*, 2003, pp. 139-142.
26. J. Roig, D. Flores, X. Jordà, J. Urresti, M. Vellvehi, J. Rebollo, An analytical model to predict the short-circuit thermal failure in SOI LDMOS with Linear Doping Profile, in: *Proc. MIEL'04*, 2004 (to appear).
27. R. C. Joy and E. S. Schlig, Thermal properties of very fast transistors, *IEEE Trans. El. Dev.* **17**(8), 586-594 (1970).

NOVEL SOI MOSFET STRUCTURE FOR OPERATION OVER A WIDE RANGE OF TEMPERATURES

Uritsky V. Ya.
JSC "Svetlana", Saint-Petersburg Electrotechnical University, Saint-Petersburg, Russia

Abstract: There are many advantages of using ultrathin fully depleted silicon on insulator (SOI) wafers in advanced integrated circuit technologies. This includes high-speed and low power circuit operation as well as reduced sensitivity to radiation effects that results from ideal isolation. These properties will make them especially useful for ULSI applications as the device dimensions are scaled to nanometer range. Perspective design of SOI MOSFET with improved characteristics and reliability in wide temperature range (4-600 K) could be realized in nanoscale.

Key words: Drain (source) resistance; gate insulator; fully depleted silicon layer; sectional channel and gate; SOI MOSFET; ultra low temperatures.

1. SECTIONAL INVERSION CHANNEL MOSFET

Nowadays there is a transition from traditional bulk MOSFET[1], formed in Si substrate surface or thick Si layer on insulating substrate, to MOSFET with the inversion channel induced on thin silicon layer and fully depleted by electric field. This depletion is caused not only by traditional upper gate voltage, but also by voltage of the back silicon gate, which is separated from the thin silicon layer by a buried insulating layer – silicon oxide, nitride or oxinitride or insulator with high electrical permeability. The Smart-cut technology[1] is often used to produce silicon on insulator structures.

MOSFET downsizing is impossible without essential evolution of its design. Analysis of above tendencies allowed focusing firstly on the design based on inversion channel with heterogeneous doping concentration (sectional channel design) and heavily doped (degenerated) source and drain regions (Fig. 1)[2].

D. Flandre et al. (eds.), Science and Technology of Semiconductor-On-Insulator Structures and Devices Operating in a Harsh Environment, 179-184.
© 2005 *Kluwer Academic Publishers. Printed in the Netherlands.*

Sectional channel design was intended initially for cryogenic operation and was used in ultra-low temperature ICs[2,3]. It is strategically important to heavily dope source and drain regions including the regions overlapped by the gate edge to decrease resistance of these regions. Firstly a sectional channel was formed in thick silicon layer on non-doped semiinsulating polysilicon substrate insulated by thick thermal oxide and consisted of alternating regions with two different doping levels and correspondingly with two different local gate threshold voltage low (V_t) and high (V_{tsi}) values (Fig. 1, a). Constant source current along the whole channel, excluding channel cutoff region where substrate and gate currents may appear at high drain voltage, provides the maximum of electric field longitudinal component in sectional induced channel in regions with high V_t and V_{tsi} and the minimum one in regions with low V_t and V_{tsi} respectively. It also allows the decrease of the longitudinal electric field in the channel near the drain without using an ordinary design with lightly doped drain and source. Electric field reduction in the sectional channel leads to the substantial reduction of hot carriers generation, thus lowering multiplication factor M (Fig. 2) and also both substrate current and substrate floating potential compared to those in the ordinary MOSFET design. It is preferable to realize this MOSFET in thin a fully depleted silicon film on insulator technology.

It is necessary to mention that longitudinal size of alternating doped regions must not exceed ≈20-50 nm to prevent heating of carriers simultaneously providing abnormally high carrier transport in inversion channel by periodical transfer of carriers from high longitudinal electric field regions to low electric field ones. This produces not only supplementary channel resistance decrease, but also considerable increase of high-speed response and transition frequency[4].

The bulk sectional n-channel MOSFETs were compared to the bulk conventional n-MOSFETs. Source and drain regions of both MOSFET's

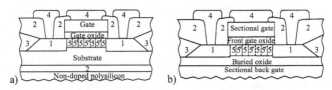

Figure 1. Schematic cross-sectional view of (a) bulk sectional MOSFET and (b) sectional SOI MOSFET on thin fully depleted layer: 1 – heavily doped regions of source and drain, 2 – thick thermal oxide, 3 – field oxide layer, 4 – gate, source and drain metal contacts, 5 and 5' – sectional channel areas with higher and lower local doping concentration.

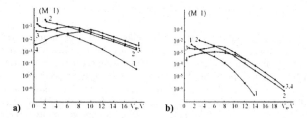

Figure 2. (M–1) via gate voltage V_g at drain voltage V_d = 15 V of (a) conventional n-channel MOSFET and (b) transistor with sectional channel at different temperatures: 298 (1), 77 (2), 20 (3) and 4.2 K (4).

were heavily doped right up to the border of their overlapping by the gate edge. This provides good linearity to the I_d–V_d characteristics at the cryogenic temperatures 4.2K and at low V_d.

Sectional channel MOSFETs were formed by additional local B$^+$ implantation to obtain more highly doped local regions of the multisection channel. The substrate current I_{sub}, the substrate potential U_{sub} and the V_t shift (as a result of opening the substrate contact) were found to be signify- cantly lower for the sectional channel than for the conventional device. As the substrate floating potential U_{sub} (at the grounded or opened substrate contact) reached the turn-on voltage of the source junction, the kink and hysteresis effects appeared. In the multisection channel this occurred at a significantly higher drain voltage than in the conventional device[5].

This design is also well suitable for nano-MOSFETs that are formed on SOI nanostructures with fully depleted silicon layer. This particular layer is blocked by buried layer of silicon dioxide fabricated by Unibond technology based on the Smart-cut process[1]. From the sides it is also blocked by insulator fabricated using an advanced LOCOS process[6].

2.   ROLE OF LIGHTLY DOPED SOURCE AND DRAIN IN MOSFET CHARACTERISTICS

The smaller the channel length the more significant role in the resistance of source and drain regions will play. In case of transistors with lightly doped drain and source it is the resistance of the regions that dominates in strong inversion channel case, when the surface electrostatic potential $| V_s | \geq | 2\varphi_b | + 5kT/e$.

Thus the current in case of a MOSFET with lightly doped source and drain current has the following expression:

$$I_d = \mu_{eff} C_i W L^{-1} (V'_g - V'_{tsi} - V'_d/2) V'_d,$$

where V'_g, V'_d and V'_{tsi} are effective values of gate, drain and strong inversion threshold voltages, respectively, μ_{eff} is the effective carrier mobility in inversion channel, C_i is the gate insulator capacity per unit area, W and L are the inversion channel width and length, respectively. In addition $V'_d = V_d - I_d (R_s + R_d)$, where R_s and R_d are the distributed source and drain resistances, and $V'_g = V_g - V_{s-sub} = V_g - I_d R_s$ and $V'_{tsi} = V_{tsi} + (2e\varepsilon_s \varepsilon_o N)^{1/2} C_i^{-1} [(V_{ssi} + I_d R_s)^{1/2} - (V_{ssi})^{1/2}]$, where N is the doping impurity surface concentration in channel and V_{s-sub} is the surface electrostatical source potential in the gate edge overlap region.

Carrier transport in the inversion channel is determined from (2) not only by V_d, but, what is essential, by a noticeably smaller voltage V'_d with significant and latent rise of strong inversion threshold voltage V'_{tsi} and drain current increase.Consequently, the distributed source and drain resistance will result in a noticeable source surface potential in the gate overlap region, and, as consequence, in an increase of strong inversion threshold voltage V_{tsi} and essential drain voltage drop not only in channel, but also in the lightly doped source and drain regions.

Additionally, even at room temperature, lightly doped peripheral areas at the source may produce a source potential. In this experiment we used γ-irradiation of p-channel MOSFETs with lightly doped source and drain regions with the no gate overlap. It was used as effective and sufficiently precise control of the resistance of the source and drain lightly doped regions. In this case not only does I_d-V_g characteristics a drift at large negative gate voltages takes place, but simultaneous characteristics deformation along channel current scale is observed, which is seen in their shift in lower currents region (Fig. 3.).

This phenomenon is determined by the following: generation of ionized donor centers charge in the gate oxide layer leads to threshold voltage

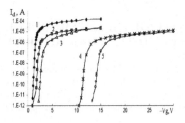

Figure 3. I_d-V_g characteristics of p-MOSFETs with lightly doped ordinary design at V_d=-0.3V before (1) and after γ-irradiation with dose 10^6 rad. V_g = -5 V (2), 0 V (3), +5 V (4), +10 V(5).

increase and simultaneously charge of the above-mentioned centers, but already localized in the oxide layer, grown on the lightly doped regions of source and drain near the gate edge, causes depletion of this regions. Resistance growth in these lightly doped regions of source (drain), non-covered by gate edge, in turn results in the appearance of a source potential, causing transistor drain (source) current decrease (Fig. 3). Presence of parasitic resistance of drain and source also significantly influences output characteristics, resulting in prolonged increase of drain current especially in short-channel structures. As channel length increases the above-mentioned phenomena decrease noticeably.

3.   SECTIONAL CHANNEL ADVANTAGES

From the above considerations we can infer that only sectional channel with heavily doped drain and source will provide a monotonous decrease of current at gate voltage increasing over a wide temperature range.

Drain (source) current in strong inversion in sectional channel transistor and strongly doped drain and source regions at low values of V_d is expressed by the following equation:

$$I_d = \mu_{eff} C_i W L^{-1} (V_g - V_{tsi} - V^\bullet_d/2) V^\bullet_d ,$$

where $V^\bullet_d$ is the drain voltage drop in the homogeneously doped section with length L_i with defined local values of doping impurity concentration and threshold voltage of strong inversion V_{tsi} . In case of a drain voltage producing no channel cut off region the following equalities take place $\Sigma L_i = L$ and $\Sigma V^\bullet_d = V_d$. Local drain voltage drop $V^\bullet_d$ on more heavily doped channel sections is larger than on lightly doped sections. At high values of V_d the following equation takes place: $\Sigma V^\bullet_d = V_d - \Delta V_d$ and $\Sigma L_i = L - \Delta L$, where ΔV_d is the drain voltage drop in channel cutoff region with length ΔL. Noticeable increase of voltage drop $\Sigma V^\bullet_d$ directly in the sectional channel results in a decrease of drain voltage drop in the cut off channels.

The above-mentioned trend of MOSFET downsizing and operating voltage decrease allows one to consider several alternatives or supplementary variants of sectional channel realization. Thus, in case of thin gate insulator there is an effective design of sectional gate including alternating regions with different work functions. This gate may be realized using a poly-Si layer with alternating local regions heavily doped by donors or acceptors (Fig. 1, b) providing variation of work function ≈1 eV. In case of forming transistor in thin fully depleted semiconductor layer on a thin insulator (oxide) potential variation may be produced by alternating regions with different work function in surface layer of back gate. Double sectional gate can also be considered.

As mentioned above, all types of MOSFETs with sectional channel may be formed in SOI-structure by joining and bonding oxided silicon wafers and implanting one of them with hydrogen[1], which causes during subsequent anneal splitting of thin layer from donor wafer saving monolith bond through buried oxide layer with base silicon wafer, which will be substrate (back gate) of the MOSFET. In such SOI structures the silicon layer is separated from underneath by buried insulator layer (oxide) and from sides by insulators of local oxidation[6], thus resulting in abrupt decrease of drain and source regions (Fig. 1b) and the elimination of wells in case of complementary devices. This gives leakage currents several orders less than in conventional silicon MOSFETs formed on thick silicon substrate and thus allows significant increasing maximal operating temperature up to ≈ 600 K[7]. We can also expect extra rapid ballistic transfer in sectional channel in wide temperature range[4].

The junction area of source and drain is very small in SOI nano-MOSFETs. As result its leakage current at high temperature is several orders smaller than in ordinary MOSFETs. Besides the source and drain junction area must be reduced in order to decrease junction leakage currents and to increase maximal operating temperature. Thus the SOI MOSFET with sectional channel presents interest for both ultra-low and high temperature applications (up to 600 K) due to reduced junction leakage current[7,8].

REFERENCES

1. M. Bruel, A new silicon on insulator material technology, *Electronic letters.* **31**(11), 1201–1205 (1995).
2. V. Lepilin, I. Mamichev, S. Prokofiev and V. Uritsky, MIS-transistor, A1 № 1355061 (1986). Official bulletin *"Izobreteniya"*. (6), 209 (1994) (Rus.).
3. V. Lepilin, I. Mamichev, S. Prokofiev and V. Uritsky, MIS-transistor, A1 № 1507145 (1986). Official bulletin *"Izobreteniya"*. (3), 208 (1994) (Rus.).
4. V.A. Gergel' and V. G. Mokerov, Significant improvement of the transistor transconductance and speed by using a graded channel, *Russian Microelectronics.* **30** (4), 286–288 (2001).
5. E. Simoen, B. Diericks and C. Claeys, Analytical model for the kink effect in n-MOST's operating at liquid-helium temperatures, *Solid State Electronics.* **33**(4), 445–454 (1990).
6. A. H. Johnston, Radiation effects in advanced microelectronics technologies, *IEEE Transactions on nuclear science.* **45** (3), 1353 (1998).
7. J. P. Colinge, Trends in silicon on insulator technology, *Microelectronic Engineering.* **19**, 795–802 (1992).
8. G. K. Celler and S. Cristoloveanu, Frontiers of silicon-on-insulator, *J. Applied Physics.* **93**(9), 4955–4978 (2003).

MOSFETS SCALING DOWN: ADVANTAGES AND DISADVANTAGES FOR HIGH TEMPERATURE APPLICATIONS

V. Kilchytska, L. Vancaillie, *K. de Meyer and D. Flandre
Microelectronics Laboratory, Universite Catholique de Leuvain, Louvain-la-Neuve, Belgium
IMEC, Leuven, Belgium

Abstract: With technology advances into deep submicron era, new physical phenomena appear and the relative importance of existing phenomena for high-temperature behaviour can change. This paper is focused on the influence of scaling down technology, particularly the decrease in gate oxide thickness and the increase in doping levels on the high-temperature characteristics of SOI and bulk MOSFETs. By examining different device properties, major evolutions in high-temperature behaviour with regards to previous device generations have been identified.

Key words: SOI MOSFETs, high-temperature, device scaling, leakage currents, gate currents, gate-induced kink effect, threshold voltage shift

1.   INTRODUCTION

One of the main markets for silicon-on-insulator (SOI) devices is high-temperature (HT) applications. In the last decade, the technology advances to deep submicron to improve device performance, lower power consumption, etc., has resulted in the appearance of new phenomena or significant changes to existing ones. The question to be addressed is "how these advances in technology will affect the high-temperature behaviour of the devices?" This paper outlines results, obtained by measurements and simulations, and analyses the consequences of scaling down technology on the HT behaviour of SOI and bulk MOSFETs, paying particular attention to the influence of two main factors: 1) increase in doping levels and 2) decrease in gate oxide thickness.

D. Flandre et al. (eds.), Science and Technology of Semiconductor-On-Insulator Structures and Devices Operating in a Harsh Environment, 185-190.

2.   DEVICES

The partially-depleted (PD) and fully-depleted (FD) SOI transistors were fabricated in IMEC on UNIBOND SOI wafers with a buried oxide thickness of 400 nm and 200 nm, respectively. The final Si film thickness of the FD and PD devices are 30 and 100 nm, respectively. A post-annealed NO gate oxide of 2.5 nm is grown, followed by a 150 nm poly silicon deposition. High dose tilted HALO implants have been used to optimise the short-channel device performance[1]. For comparison, the bulk devices followed a process similar to PD SOI. The devices under investigation have a variation in gate length from 0.12 to 10 μm and a width of 10 μm. The measurements were performed in the temperature range from 20 °C upto 300 °C.

3.   THRESHOLD VOLTAGE

Figure 1a presents the experimental variation of the threshold voltage, V_T, with temperature for FD, PD and bulk devices. It is seen that a V_T shift as low as 0.5 mV/°C can be obtained in modern FD devices. More significantly, the V_T shift for modern PD (and bulk) devices decreases even faster, than for FD and approaches the values of 0.7-0.9 mV/°C, which is much lower than the typical values reported for the PD and bulk devices and is similar to the values reported for FD devices from previous technology generations[2,3].

In describing the changes in temperature dependence of threshold voltage two different cases must be distinguished: partially-depleted (or bulk) and fully-depleted devices. In the case of FD devices, the variation of the threshold voltage with temperature can be expressed as:

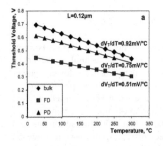 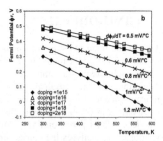

Figure 1. (a) Experimental T-dependences of threshold voltage in bulk, PD SOI and FD SOI MOSFETs. (b) Fermi potential variation with temperature for different doping levels.

$$\partial V_T / \partial T = \partial \phi_F / \partial T \qquad (1a)$$

where ϕ_F is a Fermi potential. The shift of V_T with temperature increase is independent of the gate oxide thickness, depending only on the doping levels. As shown in Figure 1b, higher doping level results in a weaker decrease of Fermi potential with temperature, and hence it can be expected that the variation of the threshold voltage with temperature for FD devices to be slightly smaller in deep submicron technologies than in previous technology generations, which is supported by experimental data (Fig. 1a).

In the case of PD (or bulk) devices, the variation of threshold voltage with temperature depends not only on the variation of Fermi potential, but also on the variation of depletion width:

$$\frac{\partial V_T}{\partial T} = \frac{\partial \phi_F}{\partial T} + \frac{1}{C_{ox}} \cdot \frac{\partial Q_{depl}}{\partial T} \qquad (1b)$$

where Q_{depl} is the depletion charge and C_{ox} is the gate oxide capacitance.

Therefore, in PD (or bulk) devices both changes in doping and oxide thickness influence the change of V_T with temperature. In this case the dependence on doping level in the 1st term, related to the Fermi potential, and in the 2nd term, related to the depletion charge, play in opposite direction. It can be shown that the higher the doping level the stronger the variation of depletion charge, Q_{depl}, with temperature. Hence, for PD devices, the doping dependence of V_T shift with temperature depends on which term will prevail. This is directly related to the gate oxide thickness, t_{gox}. For previous technology generations, the term related to the depletion charge was dominant and, therefore, V_T shift vs temperature was greater for more highly doped devices (e.g. for t_{gox}=30 nm in Fig.2a). Decrease in gate oxide

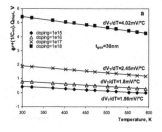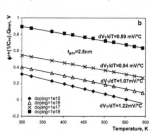

Figure 2. ϕ_F+$(1/C_{ox})$×Q_{depl} variation with temperature for different doping levels and gate oxide thickness of 30 nm (a) and 2.5 nm (b).

thickness reduces the influence of the "depletion charge" term and hence, similarly to the case of FD devices, with doping level increase, V_T variation with temperature becomes weaker (Fig.2b for t_{gox}=2.5nm). Another important conclusion is that with oxide thinning $\partial V_T/\partial T$ reduces for PD devices, due to the diminishing effect of the 2nd term in eq.(1b) and so the difference between FD and PD devices becomes smaller and smaller, as proved by experimental results (Fig. 1a).

4.   LEAKAGE CURRENTS

In scaling down technology, particularly the thinning of the Si film and the increase in the channel doping level, lower values of p-n junction leakage currents can be expected, if it is done without lifetime degradation. Unfortunately, it is not the case, because with gate oxide thinning, other sources of leakage currents such as gate currents and gate-induced drain leakage (GIDL) become more important. Figure 3a shows the temperature dependence of physical leakage current (measured at negative gate voltage, when current reaches minimum value) for different devices. For all devices T-dependence keeps proportionality to n_i, i.e. generation processes are dominant in leakage current, even for PD devices, in contrast to previous technology generations, for which diffusion current dominated leakage currents with temperature increases greater than 150-200°C[3,4]. One explanation is the different doping dependence of diffusion and generation currents. Indeed, with doping concentration increase, the diffusion current decreases more strongly ($\sim 1/N_A$) than generation current ($\sim 1/\sqrt{N_A}$) and so the transition from generation to diffusion mechanism appears at higher temperatures, as shown in Figure 3b (which presents a qualitative estimation

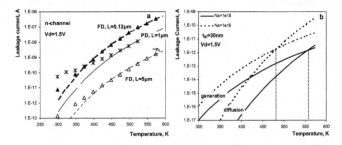

Figure 3. (a) Experimental T-dependence of leakage currents and (b) Simulated T-dependences of generation and diffusion currents for different doping levels .

of doping level influence on the different components of leakage current).

Increased leakage currents at "low" temperature are caused mainly by GIDL, which is especially strong in PD devices due to floating body effects[5]. Fortunately, these currents are relatively T-independent, and so less influent the high-temperature leakage currents. While these currents can serve as an additional reason for weaker temperature dependence of leakage currents.

5. GATE CURRENT AND GATE-INDUCED KINK-EFFECT

When scaling the oxide thickness down to 3nm and less, current through gate oxide is strongly increased and dominated by direct tunnelling[6,7]. It exhibits stronger temperature dependence, especially at low gate bias (Fig. 4a), comparing to nearly temperature independent Fowler-Nordheim tunnelling, which was dominant in previous technology generations. Estimated activation energy of these currents (insert in Fig. 4a) is smaller than half of Si bandgap. One may assume at first, that at low V_g, it is band-to-trap, and then at higher V_g, band-to-band tunnelling could be a limiting factor[6]. One of the new effects, so-called gate-induced[8] or linear kink effect[9], is directly related to the increased currents through the gate oxide, which cause body charging. It appears as a 2nd peak in the transconductance vs V_g curve. With temperature increases, the gate-induced kink is progressively attenuated and eventually disappears, as shown in Figure 4b. In fact, due to the exponential temperature dependence of generation-recombination processes, the contribution from the source and drain junctions increases more rapidly than gate currents with temperature increase. This results in ever smaller changes in the body potential, caused by the gate current, up to total offset.

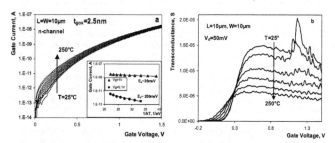

Figure 4.(a) Gate current variation with temperature. (b) Temperature evolution of transconductance.

6. CONCLUSIONS

The consequences of scaling down technology on high-temperature characteristics of MOSFETs have been analysed based on experimental and simulation results. The analyses performed give an explanation to the decrease of threshold voltage variation with temperature in modern devices and convergence of the values for FD and PD devices by the gate oxide thinning and doping level increase. At the same time the increase in the doping level is one of the factors, which is responsible for the weakening of the high-temperature dependence of the leakage currents. It was also shown, that without regards to the gate currents increase with temperature, the gate induced kink effect disappears at elevated temperatures.

ACKNOWLEDGEMENTS

This research has been funded by the European Union through IST-1999-12342 "SPRING" project.

REFERENCES

1. H. van Meer, and K. De Meyer, A 2-D Analytical Threshold Voltage Model for Fully-Depleted SOI MOSFETs With Halos or Pockets", *IEEE Trans. on El. Dev.* **48** (10), 2292-2302 (2001).
2. J.-P. Colinge, Silicon-on-insulator technology: materials to VLSI, 2nd edition (Kluwer Academic Publishers, Dordrecht, 1997).
3. D. Flandre, A. Terao, P. Francis, B. Gentinne, and J.-P. Colinge, Demonstration of the potential of accumulation-mode MOS transistors on SOI substrates for high-temperature operation (150-300°C), *IEEE Electron. Dev. Lett.* **14** (1), 10-12 (1993).
4. T.E. Rudenko, V.S. Lysenko, V.I. Kilchytska, and A.N. Rudenko, in: *Perspectives, Science and Technologies for Novel Silicon on Insulator Devices*, edited by P.L.F. Hemment, V.S. Lysenko, and A.N. Nazarov (Kluwer Academic Publishers, Dordrecht, 2000), pp. 281-293.
5. Najeeb-un-din, et al., Analysis of floating body effects in thin film conventional and single pocket SOI MOSFETs using the GIDL current technique, *IEEE Electron. Dev. Lett.*, **25** (4), 209-211 (2002).
6. C.-H. Lin, B.-C. Hsu, M. H. Lee, and C.W. Liu, A comprehensive study of inversion current in MOS tunnelling diodes, *IEEE Trnas. On El. Dev.* **48** (9), 2125-2130 (2001).
7. Y.T. Hou, M.F. Li, Y. Jin, and W.H. Lai, Direct tunnelling hole currents through ultrathin gate oxides in metal-oxide-semiconductor devices, *J. of Appl. Phys.* **91** (1), 258-264 (2002).
8. J. Pretet, et al., in: Proceedings of the 32nd European Solid-State Device Research Conference, edited by G. Baccarani, E. Gnani, and M. Rudan (University of Bologna, Firenze, 2002), pp. 515-518.
9. A. Mercha, et al., in: *Silicon-on-Insulator Technology and Devices XI*, edited by S. Cristoloveaunu (The Electrochemical Society, Inc., Pennington, 2003), pp. 319-325.

TEMPERATURE DEPENDENCE OF RF LOSSES IN HIGH-RESISTIVITY SOI SUBSTRATES

D. Lederer and J.-P. Raskin

Microwave Laboratory of the Université catholique de Louvain, Louvain-la-Neuve, Belgium

Abstract: This paper analyzes RF substrate losses in High-Resistivity (HR) SOI and oxidized HR bulk silicon wafers. Through experimental and simulation data, it is shown that when sufficiently high trap densities are introduced at the buried SiO_2/Si interface, HR silicon substrates are virtually lossless up to approximately 100°C, remain acceptable for high temperature RF applications up to 120~150°C, depending on the oxide thickness, but show no significant improvement compared to standard resistivity substrates above 200°C.

Key words: CPW, High temperature, RF losses, interface traps, oxide charges

1.   INTRODUCTION

High-Resistivity (HR) Si substrates with resistivity values higher than 3 kΩ.cm are suitable for High-Frequency applications due to their negligible ohmic losses[1,2]. However, oxide-passivated HR Si substrates usually suffer from resistivity degradation near the SiO_2/Si interface due to the presence of fixed charges (Q_{ox}) in the oxide[3]. These charges attract free carriers near the substrate surface, forming an accumulation or inversion layer. Coplanar structures made on such substrates are very sensitive to the presence of this layer, which can also be formed underneath metallic lines upon the application of a DC bias[4]. In all cases the presence of free carriers underneath the oxide leads to substantial RF loss increases. As shown in[4] and [5], an efficient way to (partially or entirely) remove those carriers is to introduce a large density of traps (D_{it}) at the SiO_2/Si interface. This can be achieved by adjusting the oxide deposition parameters[4] or by introducing an additional trap-rich layer between the oxide and the Silicon substrate. In [5] and [6], a polySilicon layer was used to form an Oxide-Polysilicon-Silicon (OPS) substrate. In both cases, the RF structures designed on the OPS

191

D. Flandre et al. (eds.), Science and Technology of Semiconductor-On-Insulator Structures and Devices Operating in a Harsh Environment, 191-196.

substrate clearly exhibited the lowest substrate losses. In this work we investigate for the first time the suitability of HR SOI and HR OPS substrates for high temperature RF applications. The proposed analysis combines the effects of high temperature (T) and interface trap density on resistivity degradation in HR SOI and HR OPS substrates. The temperature range of interest is from 20 to 200°C.

2.   STRUCTURE AND MODEL

The RF structures used in this work to investigate the suitability of HR SOI wafers for high temperature applications are CoPlanar Waveguides (CPW) designed for 50 Ω-characteristic impedance (Figure 1). The CPW lines were analyzed on three types of substrates: SOI with etched silicon film, oxide-passivated bulk Si and oxide-polysilicon-silicon substrates. Two substrate doping levels were considered: $N_a = 5 \times 10^{12}$ /cm^3 and $N_a = 6.5 \times 10^{13}$ /cm^3, respectively corresponding to nominal values of substrate resistivity (ρ_{DC}) of 3 kΩ.cm and 20 Ω.cm. For all simulations the backside of the substrate was grounded. For the simulations performed with interface traps, a constant trap distribution was considered throughout the Si bandgap (as in[4]). The AC simulations were performed from 1 to 40 GHz with 1 GHz step at 20, 70, 100, 150 and 200°C. The total simulated high frequency conductance (resp. capacitance) between the central and the lateral electrodes is noted G_{tot} (resp. C_{tot}).

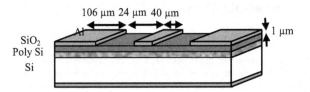

106 µm 24 µm 40 µm

SiO$_2$
Poly Si
Si

1 µm

Figure 1. Cross-section of RF structure under study

The quality factor considered in this work to characterize the substrates is their *effective resistivity* (ρ_{eff}), which is the resistivity actually 'seen' by the coplanar structures at RF and which accounts for surface variation of Si resistivity created by accumulation, inversion and/or depletion but also for the presence of the insulating oxide layer. As shown in[7], ρ_{eff} is given by:

$$\rho_{eff} = \frac{K}{2G_{tot}} \sqrt{\frac{C_{tot}}{C_{eff}}} \tag{1}$$

where K [m] is a geometry factor related to the shunt capacitance of the device. This equation is simply obtained by considering an *equivalent, unpassivated* Si ($\varepsilon_{equ} = 11.7$) substrate with an homogeneous (i.e., without space charge effect) value of resistivity (i.e., ρ_{eff}) such that RF losses are identical on that substrate and the analyzed wafer. Its lineic capacitance is noted C_{eff}.

3. RESULTS AND DISCUSSION

3.1 Simulations

For frequencies higher than 1 GHz the total losses (α) in a CPW transmission line can be approximated by:

$$\alpha \approx \frac{1}{2} \left[R\sqrt{\frac{C}{L}} + G\sqrt{\frac{L}{C}} \right] \tag{2}$$

where R, G, C and L are the elements of the line distributed circuit [7]. The first and the second terms in the sum represent conductor and substrate ohmic losses, respectively. Figure 2 shows simulated data of α *vs.* frequency for different values of ρ_{eff} at 20 °C and includes both conductor *and* ohmic losses[2]. The data clearly indicate that HR Si substrates are quasi-lossless if

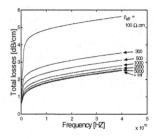 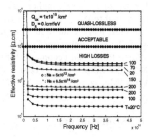

Figure 2. Total losses vs frequency for varying ρ_{eff} values

Figure 3. ρ_{eff} vs frequency for different T and N_a values.

$\rho_{eff} > 3$ kΩ.cm; their ohmic losses are small (< 0.5 dB/cm) when 1 kΩ.cm $<$ $\rho_{eff} < 3$ kΩ.cm but they suffer from significant ohmic losses (> 0.5 dB/cm) and are thus inadequate for RF applications if $\rho_{eff} < 1$ kΩ.cm.

Figure 3 presents the extracted value of ρ_{eff} with respect to f for the HR ($\rho_{DC} = 3$ kΩ.cm) and the standard resistivity ($\rho_{DC} = 20$ kΩ.cm) SOI substrates with a buried oxide thickness of 145 nm, a low Q_{ox} density (1×10^{10}/cm^2) and no D_{it}. It is interesting to notice that for all temperature points the value of ρ_{eff} is much smaller than 3 kΩ.cm for the HR SOI substrate. This is related to the inversion layer formed at the substrate surface and induced by Q_{ox}, reducing ρ_{eff} by more than one order of magnitude and rendering the HR substrate unsuitable for RF applications.

For temperatures below 100 °C the data indicate that ρ_{eff} is a slowly increasing function of T, which is explained by a temperature-related mobility reduction inside the inversion layer.

However, above 100 °C ρ_{eff} drastically decreases and drops down to 80 Ω.cm at 200 °C. This is a direct consequence of the intrinsic carrier concentration (n_i) increase in the substrate. The figure also shows that the standard resistivity SOI substrate exhibits an increase of ρ_{eff} with T, which is also due to mobility reduction of free carriers at higher temperatures. However, another interesting feature highlighted by these simulations is that the value of ρ_{eff} becomes independent on the doping level at around 200°C (Figure 4, two bottom curves). This occurs because as T is increased the Fermi energy level of the substrate is shifted towards midgap, reducing the free carrier concentration inside the substrate to a value close to n_i, and hence, independent on substrate doping level[8].

As expected from previous results made at room temperature[4,5] ρ_{eff} can be largely improved for HR SOI substrates when higher trap densities are present at the SiO$_2$/Si interface (Figure 4). The improvement is the most obvious in the lower temperature range and reaches a saturation value close to 3 kΩ.cm for D_{it} levels higher than a few 10^{10} /cm^2/eV. This increase is directly related to the absorption of free carriers from the SiO$_2$/Si interface. The data show that the substrate remains quasi-lossless up to 100°C for high values of D_{it}. For temperature points higher than 120°C the HR substrates exhibit high losses and become therefore inappropriate for RF applications. At 200°C, the value of ρ_{eff} becomes as low as ~60 Ω.cm (regardless of D_{it} density), a value comparable to that of lower quality standard resistivity SOI wafers.

3.2   Experiments

In this work three distinct wafers were investigated up to 200 °C to confirm the trends unveiled by the simulation results (Table 1). To increase the trap density at the SiO_2/Si interface an additional 20 nm-thick layer of polysilicon was LPCVD-deposited on top of the HR Si substrate for one of the wafers. This layer was then passivated with a 3 μm-thick PECVD oxide to form the OPS substrate. On each measured wafer CPW lines were patterned on the top oxide after substrate processing. The lines S-parameters were obtained with a Anritsu VNA from 40 MHz to 40 GHz. After using a classical TLR de-embedding method and a characteristic impedance extraction method developed in[9], their lineic shunt parameters (C_{eff} and G_{eff}) were computed with the theory of quasi-TEM modes.

Table 1. Technological details of the physical wafers investigated in this work

Wafer	Process (Main steps)	Oxide	T_{ox}	Q_{ox} [/cm^2]	D_{it} [/eV/cm^2]
HR SOI	top Si etching	thermal (BOX)	145 nm	~1×10^{10}	Negligible
STD BULK	PECVD oxide	PECVD	1 μm	>1×10^{11}	//
OPS	LPCVD PolySi (20nm) PECVD oxide	PECVD	3 μm	>1×10^{11}	~1×10^{12}

The effective resistivity of each wafer was then extracted and plotted in Figure 5 with respect to T. Additional simulation data have also been included with technological parameters (t_{ox}, Q_{ox} and D_{it}) corresponding to those of the wafers and closely agree with the experimental results. For the simulations of the PECVD-oxide wafers (resp. thermally oxidized HR SOI wafer), a high (resp. low) value of Q_{ox} was considered: 20×10^{10}/cm^2 (resp. 1×10^{10}/cm^2). The OPS substrate was simulated without traps and with a trap density of 1×10^{12} /ev/cm^2, according to[10]. It can be seen that when no traps are present at the SiO_2/Si interface, the value of ρ_{eff} is extremely low (~100 Ω.cm) despite the large oxide thickness of the OPS structure (3 μm).

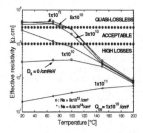

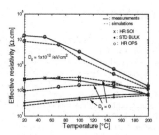

Figure 4. ρ_{eff} vs T for a varying D_{it} density and a fixed value of Q_{ox} (1×10^{10}/cm^2).

Figure 5. ρ_{eff} vs T for the three measured substrates and their corresponding simulated data

This is clearly due to the high Q_{ox} density present in PECVD oxides[7]. However, when a high trap density is present at the SiO_2/Si interface, the effective resistivity climbs up to about 10 kΩ.cm at room temperature. Similarly to results plotted in Figure 3, ρ_{eff} decreases at higher temperatures but remains larger than 3 kΩ.cm up to 100°C. This seems to indicate that when trap densities are high enough to completely absorb the inversion layer induced by fixed oxide charges, HR oxidized substrates remain quasi-lossless up to around 100°C for t_{ox} values in the range of a few microns. Their loss level remains however acceptable up to 120°C (Figure 3) or 150°C (Figure 5), depending on the oxide thickness. At 200°C, no significant improvement is obtained by using HR wafers.

4.   CONCLUSION

Atlas simulations have shown that High-Resistivity (HR) SOI substrates and HR oxidized substrates with an initial resistivity of 3 kΩ.cm are unsuitable as such for RF applications, even at room temperature, if fixed charges are present in the oxide. However, when the trap density at the SiO_2/Si interface is such that free carriers can be totally absorbed the HR substrates become and remain quasi-lossless up to around 100°C. They still remain acceptable for RF applications ($\alpha_{sub} < 0.5dB/cm$) up to 120~150°C (depending on the oxide thickness), above which point they start suffering from significant RF losses and become therefore inadequate for RF applications. At 200°C, the data show that the quality of HR wafers is comparable to that of low-cost wafers with standard resistivity.

REFERENCES

1. A. Reyes et al., IEEE MTT-S Int. Microwave Symp. Dig. 1996, pp. 87-90.
2. D. Lederer et al., IEEE 2003 Int. SOI Conf.,Newport Beach, CA, pp. 50-51.
3. A. Reyes et al., IEEE Trans. Microwave Theory Techn., vol 43 (9), 1995, pp. 2016-2022.
4. D. Lederer and J.P. Raskin, Journ. Sol. St. Electron., vol 47, 2003, pp. 1927-1936.
5. Gamble et al. IEEE Microwave and Guided Wave Lett.,vol 9 (10) ,1999, pp. 395-397.
6. H.-T. Lue and T.-Y. Tseng, J. Appl. Phys, vol 91(8), 2002, pp. 5275-5282.
7. D. Lederer and J.-P. Raskin, submitted to IEEE EDL, April 2004.
8. Y. Tsividis, McGraw-Hill Co., Inc., New York, 1987.
9. M. Dehan, PhD Thesis, Université catholique de Louvain, Belgium, 2003
10. S. Hirae *et al.*, J. Appl. Phys., vol. 51 (2), 1980, pp. 1043-1047.

REVIEW OF RADIATION EFFECTS IN SINGLE AND MULTIPLE-GATE SOI MOSFETS

Sorin Cristoloveanu
Institute of Microelectronics, Electromagnetism and Photonics (CNRS, INPG, UJF)
ENSERG, BP 257, 38016 Grenoble Cedex 1, France

Abstract: The radiation effects in SOI structures are discussed in the context of the device miniaturization. We review radiation data accumulated in the recent years and point out new directions of research. The properties of standard and innovative SOI wafers with buried alumina are presented. The scaling of partially depleted SOI MOSFETs results in new mechanisms, such as gate-induced floating body, which combine with more classical effects. On the other hand, fully depleted transistors with ultra thin body are prone to giant coupling effects. The operation principles of transistors with double, triple or quadruple gates are briefly addressed. It is demonstrated that the volume conduction, enabled in multiple-gate MOSFETs, exhibits superior resistance to radiations.

Key words: SOI, radiation, MOSFETs, device physics, multiple-gates

1. INTRODUCTION

Under its primitive form SOS, SOI technology has been invented for radiation-hard applications.[1,2] The context has radically changed. SOI technology is no longer restricted to the niche of rad-devices and now covers the mainstream microelectronics. However, the commercial integrated circuits become more and more sensitive to natural radiations, which means that the related degradation mechanisms become of general interest. SOI is being adopted by major companies because there is no other clear possibility to continue the CMOS scaling. The 10-nm MOSFET barrier can be broken with SOI only, using advanced concepts for multiple-gate transistor architecture.[3-5]

197

D. Flandre et al. (eds.), Science and Technology of Semiconductor-On-Insulator Structures and Devices Operating in a Harsh Environment, 197-214.
© 2005 *Kluwer Academic Publishers. Printed in the Netherlands.*

In this exciting context, we discuss the impact of radiation effects from various angles. Section 2 is dedicated to the innovative SOI materials. The electrical properties and the radiation-induced degradation of state-of-the-art SOI wafers are monitored with the pseudo-MOSFET technique. A different avenue in SOI material science is to replace the buried oxide with a different dielectric that offers improved thermal conductivity and attenuated self-heating. Section 3 describes the radiation effects in partially depleted SOI MOSFETs. We review typical radiation-related implications and focus on a new mechanism, the gate-induced floating-body effect (GIFBE). Fully depleted SOI MOSFETs are governed by the coupling between the front and back channels. In section 4, we show that under irradiation, the trapped charge in the BOX acts as a built-in back-gate voltage, which modifies the threshold voltage, subthreshold swing, transconductance and series resistance.

Multiple-gate MOSFETs are discussed in section 5. Double-Gate and Gate-All-Around transistors offer excellent scalability as well as improved immunity to radiation effects. The 4-gate transistor is a fascinating device which exhibits striking features including resistance to radiations.

2.   SOI MATERIALS

The signature of the radiation effects depends on the device structure as well as on the intrinsic nature of SOI materials.[1,2] The market of device-grade SOI wafers is currently shared by Unibond and SIMOX. The trend is to achieve thinner films (< 100 nm) and BOX (< 150 nm).

2.1   SIMOX

SIMOX is synthesized by deep implantation of high doses of oxygen and high temperature annealing, which results in a BOX with special microstructure. Typical defects are interface traps, hole and electron traps in the BOX, which are activated by exposure to radiation.[6,7] The quality of the BOX and interfaces has dramatically been improved and the defects in the film have been erased, except the threading dislocations.

2.2   Unibond

Unibond is a very flexible technology which provides 300 mm wafers with unlimited combinations of BOX and film thickness. The BOX and film-BOX interface have thermal-oxide quality. The bonding interface is more defective and sensitive to radiation effects. However, the bonding interface has a minor impact on the device performance because it is located underneath the BOX. As far as the front-gate interface is concerned, the density of interface states that may be induced by the released hydrogen is not significant up to very high total doses (about 1 Mrad).[8]

2.3   Wafer-Level Radiation Effects

The large variety of SOI structures makes the radiation-induced defects differ according to the conditions of synthesis. The pseudo-MOS transistor (Ψ-MOSFET) is a simple and efficient method to monitor the generation of BOX, film and interface defects during radiation.[9] SOI is a natural upside-down MOS structure, where the Si substrate acts as a gate terminal and can be biased to induce a conduction channel (inversion or accumulation) at the interface. The BOX plays the role of a gate oxide and the Si film represents the transistor body. Low-pressure probes are placed on the film and form source and drain point contacts (Fig. 1).

Very pure MOSFET-like characteristics are produced and used for parameter extraction: mobility of electrons and holes, threshold and the flat-band voltage, density of traps at the film-BOX interface and fixed charges in the BOX, carrier lifetime.[9] Figure 1 shows the remarkable shift in the Ψ-MOSFET curves during radiation-induced positive charge trapping in the BOX.[10]

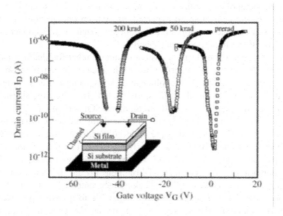

Figure 1. Configuration of the pseudo-MOS transistor and typical $I_D(V_G)$ characteristics in a thin SOI film, before and after several radiation steps.[10]

2.4   Novel SOI Materials

The recent demand for thinner buried oxides ($t_{box} < 50$ nm) is motivated by two key arguments: reduction of the short-channel effects and improvement of the thermal dissipation. As the film becomes thinner, the heat path through the source/drain regions is squeezed which increases the thermal resistance and makes the transistor body temperature rise. Self-heating is responsible for performance degradation in SOI MOSFETs: mobility and threshold voltage lowering, increase in leakage current and subthreshold swing, etc.[1,2] Self-heating in SOI transistors is primarily due to the poor thermal conductivity of the BOX which acts as a barrier for heat dissipation. It has recently been proposed to modify the generic SOI structure by replacing the standard SiO_2 with buried alumina or other dielectric, able to offer improved thermal conductivity.[11,12]

The thermal advantage of buried Al_2O_3 over buried SiO_2 was evaluated by comparing the total thermal conductance (derived from an equivalent thermal circuit) and the self-heating of corresponding MOSFETs.[12] It was concluded that the BOX/substrate combination dominates the thermal behavior in SOI MOSFETs. The superiority of buried alumina in advanced MOSFETs is illustrated in Figure 2a. Even for very thin BOX, the advantage of buried Al_2O_3 is maintained.

A reduction in lattice temperature by 40 K enables a gain of more than 20% in mobility ($\mu \sim T^{-1.5}$), governed by acoustic phonon scattering). This improvement applies simultaneously to electrons and holes and compares favorably to the gain in speed expected either from one generation CMOS scaling or from the use of strained silicon. What is the impact of an alternative BOX on the electrical properties?

The change in the dielectric constant (3 times higher for alumina than for SiO_2) impedes on the 2-D distributions of the electric potential and field in the transistor, hence on the coupling and short-channel effects. Extensive simulations show that the classical short-channel effects (charge sharing and DIBL) are marginally degraded in alumina SOI MOSFETs. The main concern in sub-100 nm SOI MOSFETs is the drain-induced virtual substrate biasing (DIVSB).[13,14] Increasing the drain voltage allows the fringing fields to penetrate into the BOX and underlying substrate. A depletion region develops underneath the BOX and reduces its effective capacitance. Even more importantly, the potential distribution in the BOX is modified, resulting in a net increase of the back-surface potential at the film-BOX interface. The fringing field effect is comparable to a virtual positive bias (DIVSB) applied to the substrate which drives the back interface from depletion to weak inversion.[13] Due to interface coupling, the front-channel threshold voltage is lowered, the subthreshold swing is degraded, and a parasitic back-channel can be activated. This degradation is more acute for alumina BOX but can be cured by improving the device architecture.[12]

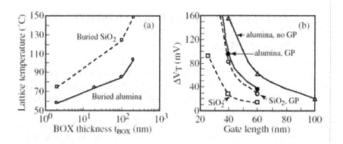

Figure 2. (a) Simulated body temperature versus BOX thickness in 70 nm long, 10 nm thick SOI MOSFETs, and (b) threshold voltage roll-off, induced by DIBL and DIVSB, versus channel length in various SOI MOSFETs (SiO2 and Al2O3 BOX, ground-plane and no GP, after Oshima et al[12]

The ground plane (GP) structure has been proposed as a solution to cancel the penetration of the fringing fields into the substrate.[14] A GP is a conductive layer, located underneath the BOX and synthesized by ion implantation through the BOX or by shallow surface diffusion prior to wafer bonding. The GP suppresses the depletion region in the Si substrate underneath the BOX. Without GP, the typical DIVSB effects increase exponentially in MOSFETs shorter than 50 nm beyond the acceptable limits for threshold voltage lowering (100 mV/V) and subthreshold swing degradation (100 mV/decade). The slight disadvantage of alumina as compared to SiO_2 BOX is practically erased in GP MOSFETs. The conclusion is that no trade-off between the thermal and electrical performance of SOI MOSFETs with buried alumina is necessary, even for extremely short device.[12]

In a more general context, novel SOI wafers are needed for the fabrication of advanced nano-structures (quantum dots, wires, single-electron circuits, tunneling devices, etc). New materials, such as stacked layers for 3-D circuits, light transmission and optical devices, are being explored. The combination of strained layers (Si, SiGe) or compound semiconductors (III-V, II-VI) on various substrates (SiC, diamond, glass, air, flexible polymers, etc) opens exciting fields in nano- and micro-electronics. All these emerging materials, still labeled as SOI for *Semiconductor On Insulator*, can be synthesized by adapting the Smart-Cut process.

A refreshing and challenging field of research will be the analysis of radiation effects in these new SOI wafers.

3.   PARTIALLY DEPLETED SOI MOSFETS

3.1   Classical Effects

In partially-depleted SOI MOSFETs, a neutral region subsists and gives rise to *floating-body* effects. The charging of the body by majority carriers, generated by impact ionization, enables the *kink effect, hysteresis, latch, parasitic bipolar* action and transient currents.

The radiation effects, critical for space and defense applications, become more and more severe in commercial devices operated at the ground level. As the ULSI technology evolves, the charge stored in elementary nodes decreases down to a few femto-coulomb, and can be easily upset by the radiation deposited charge.[6] Transient damage or *soft errors*, like Single Event Upset (SEU), originate from the parasitic transient current. SOI MOSFETs exhibit excellent tolerance (one order of magnitude better than for bulk transistors) for two basic reasons: (i) full isolation from the substrate and from neighboring devices (no latch triggered by parasitic thyristor structure), and (ii) minimum transistor volume which limits the amount of generated charge.

However, the parasitic bipolar action is effective when the body is left floating. After an ionizing particle strikes, the majority carriers tend to accumulate in the body. They contribute to forward bias the body-source junction and to activate the parasitic bipolar, which amplifies the charge collection.[15,16] As devices are scaled down, competing effects (smaller collecting body volume but higher amplification factor) occur and lead to an overall decrease of the device cross section.

The bipolar amplification can even lead to snap-back.[17] Additional carriers generated by impact ionization maintain the body-source junction forward biased, so that the transistor can not return instantly to its initial off-state. The adjunction of body ties allows to reduce the bipolar amplification in both fully and partially depleted devices.[6]

Permanent or *cumulated dose* damage is due to the charged trapped in the insulators, mainly lateral dielectrics and buried oxides. This causes a permanent shift of device parameters such as threshold voltage, off and on currents, etc.

In SIMOX and UNIBOND buried oxides, a positive charge trapping occurs being partially compensated by electron trapping. The contributions of both species has been demonstrated by back-channel measurements.[7]

Figure 3 reveals several distinct mechanisms according to the bias applied to the back gate during exposure to radiations. For positive bias, radiation-induced electron-hole pairs are separated by the field and the holes are driven towards the upper interface of the BOX. A net positive charge is build-up, yielding a very large decrease in the back-channel threshold voltage V_{T2}.

For zero bias, most of the electron-hole pairs recombine instantly, hence V_{T2} shift is attenuated. But, for negative bias, the pairs again separate and the holes are attracted to the bottom BOX interface. The positive charge effect still dominates for low doses, whereas beyond 100 krad a rebound is observed. The large increase in V_{T2} accounts for electron trapping near the film-BOX interface, which modifies the surface potential as well as the series resistance of the transistor.[7]

In n-channel PD MOSFETs devices, positive charge trapping in the BOX induces the induces a gradual negative shift of the back threshold voltage which facilitates the back-channel activation (Figure 4).[18] Under normal bias conditions (back gate grounded), a leakage current is observed due the parasitic conduction of the back-gate transistor. This 'leakage' current cannot be controlled by the front gate and prevents the main transistor from switching off.

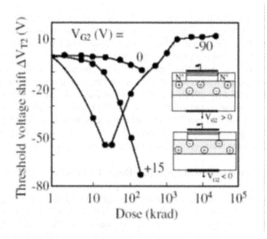

Figure 3. Radiation-induced shift of the back-channel threshold voltage in a SIMOX MOSFET for various back-gate bias during radiation.[7]

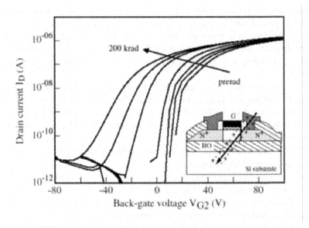

Figure 4. Drain current versus back gate bias in a partially depleted n-channel SOI MOSFET, at low drain bias and after several doses of radiation. The transistor was biased with $V_{G1} = 1V$ and $V_{G2} = 40V$ during radiation.[18]

Below the critical dose, the neutral region can in principle protect the front channel from the detrimental impact of radiation-induced damage in the BOX. This is not totally true, because second order mechanisms arise. For example, a positive charge in the BOX lowers the series resistance of LDD regions, increasing the impact ionization rate and the body charging. As a consequence, hysteresis and latch can be activated during radiation. The transient and history effects are also modified as soon as the generation-recombination rate in the volume or at the BOX and sidewall interfaces is increased. Besides the bipolar transistor and back channel, a lateral parasitic transistor is activated by charge trapping in the lateral isolation (LOCOS, STI) and induces a leakage current on the transistor edges.

3.2   Gate-Induced Floating-Body Effects

As the gate oxide is scaled down, the gate tunneling current becomes increasingly relevant. Since the body of the SOI MOSFETs can be charged by the tunneling current, gate-induced floating-body effects (GIFBE) are enabled at low drain voltage,[19,20] being therefore different from the conventional floating-body effects induced by impact ionization. The body potential increases and leads to a second peak in the transconductance curves (Figure 5).

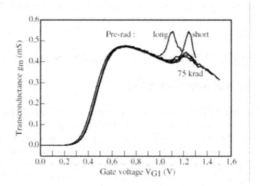

Figure 5. Typical GIFBE signature in SOI MOSFETs: second peak in transconductance curves before and after radiation (V_D = 0.1 V, W = 10 mm, L = 0.6 mm, after Jun et al[22]).

The body potential is governed by the balance of body charging (via tunneling) and body discharging (via junction leakage and/or carrier recombination). The amplitude of GIFBE decreases in shorter and narrower MOSFETs, and depends on the scanning speed and direction: the longer the measurement time is, the lower the gate voltage which triggers GIFBE. A hysteresis is observed due to the difference between gradual body charging (for increasing gate bias) and body discharging (for decreasing gate voltage).[19-21]

Recent measurements show that GIFBE is strongly attenuated after radiation (Fig. 5).[22] This is explained by the degradation in carrier lifetime and the increase in leakage current. Both mechanisms contribute to a reduction in the charge stored in the body.

Since the tunneling current enables a faster recovery of the steady-state charge, the transient overshoot and undershoot currents are dramatically modified by GIFBE. According to the initial state of the body (charged or discharged), pulsing the gate from high to low voltage can give rise to an overshoot or, respectively, undershoot. This qualitative change in the transient nature obviously impacts on the propagation delays and the so-called 'history effects' in digital circuits. In an inverter chain, GIFBE are responsible for slowing down the first switch and accelerating the second switch. The steady-state is reached more rapidly when GIFBE are turned on, leading to faster history effects.[20]

Other implications of GIFBE include the modification of the low-frequency noise spectra. A GIFBE excess Lorentz noise is superimposed on the conventional $1/f$ noise, increasing the noise figure-of-merit by 1-2 orders of magnitude.[23]

4.   FULLY DEPLETED SOI MOSFETS

4.1   Typical Radiation Effects

Fully depleted (FD) devices are characterized by the coupling between the front and the back gate transistor. Carriers flowing at one interface are influenced by the presence of defects at the opposite interface. This *defect coupling* is responsible for an apparent degradation of the front-channel characteristics (threshold voltage, transconductance and subthreshold swing), which is actually due to radiation-induced damage in the BOX.

When irradiating FD devices, the trapped charge in the BOX acts as a build-in positive back-gate voltage. The front-channel threshold voltage shifts in proportion to the back-channel threshold voltage shift induced by charge trapping into the BOX. Even more importantly, the build-up of a fixed charge in the BOX results in a modification of the back surface potential (accumulation, depletion or inversion), changing the mode of operation of the front channel. The transconductance may also be affected by the modulation of the series resistance.

A thinner BOX does not significantly improve the radiation hardness of fully depleted devices.[24] Indeed, the coupling coefficient is enhanced and the front transistor becomes more sensitive to a smaller charge trapped in the BOX. In FD MOSFETs, GIFBE modifies the potential at the back interface (film-BOX): the front-channel threshold voltage is indirectly lowered by interface coupling effect.[25] The second peak in transconductance is amplified when driving the back interface towards accumulation (by substrate biasing or radiation effects).

4.2   Ultra-Thin Body Effects

While most of the commercial SOI circuits are currently fabricated with PD technology, the future is definitely imposing FD devices. They have a striking advantage for complying with the ITRS roadmap constraints, because their scaling does not rely on a heavier doping (as for bulk-Si or PD transistors). Indeed, the thinning of the Si film is a more efficient strategy for miniaturization:[3,4] short-channel effects are well controlled if the film thickness is about 25-35% of the channel length. The threshold voltage can be adjusted by the use of gate materials with appropriate work-function;

since the film can be left *undoped* in order to preserve a high carrier mobility.

An interesting question is to define the actual boundary between PD and FD transistors. It is admitted that the transition depends on the balance between the film thickness, t_{Si}, and the gate-induced depletion depth w_G: $t_{Si} <$ w_G for FD.

This implies that the depletion region is homogeneous along the channel and abrupt in the vertical direction. In short-channel MOSFETs, these hypotheses do not hold. Figure 6 shows that the condition for full depletion involves not only the body thickness and doping but also the channel length.[26] The reason is that the proximity of source and drain junctions lowers the apparent doping seen from the gate, which makes full depletion easier. The curve has been computed with an analytical model for this 2-D coupling effect, also called *length-doping transformation*.[26] Above the curve, the transistor is PD whereas below it operates in FD mode. For a given doping, a transistor can behave as PD or FD according to its length. As a consequence, the nature of radiation effects can be fundamentally different in short (FD) and long (PD) transistors. A simple solution to this puzzle is to avoid any doping at all by focusing on FD transistors.

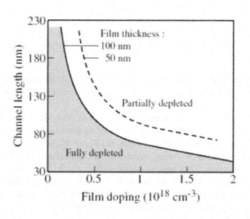

Figure 6. Transition between partial depletion and full depletion in SOI MOSFETs as a function of channel doping and length (after Allibert *et al*[26]).

We now consider the case of ultra-thin FD films, where the interface coupling effects are strongly amplified. The threshold voltage of the front channel (respectively back channel) decreases with increasing the back-gate voltage (respectively the front-gate voltage).[27] When drawing both characteristics $V_{T1}(V_{G2})$ and $V_{G1}(V_{T2})$ on the same graph, the two curves are in general different (Figure 7).

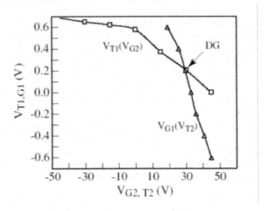

Figure 7. Front (back) channel threshold voltage versus back (front) gate bias in a SOI MOSFET (L = 10μm, W = 10μm, 47-nm-thick film). The intercept point DG shows the correct bias for pseudo double-gate operation (after Pretet *et al*[20]).

The intercept point DG represents the *unique* couple of front and back gate voltages which enables the *simultaneous* activation of the two channels.[20]

In sub 10-nm-thick films, the two curves tend to coincide.[20] This means that an arbitrary back-gate bias V_{G2} is promoted as threshold voltage V_{T2} as soon as the front gate is biased at threshold ($V_{G1} = V_{T1}$). In other words, when one channel reaches strong inversion, the opposite channel is also dragged into inversion. This *super-coupling* effect results from the vertical profile of the body potential which is rather flat between the two gates. It is worth noting that the notion of front and back channels needs to be replaced by the concept of *volume inversion* across the whole film.[28]

In films thinner than 10 nm, quantum confinement becomes relevant.[29,30] The sub-band splitting results in an increase of the threshold voltage for thinner films. The carrier mobility is affected by competing effects such as enhanced phonon scattering and lowered effective mass. Since the centroid of the inversion charge is moved towards the volume of the film, surface roughness scattering is in principle reduced.[29] Monte Carlo simulations suggest that the mobility is maximum in 3-5 nm thick films.[31] However, early experimental results show that the mobility is degraded in thinner films. This may be due to a degradation of the crystal quality during thinning by sacrificial oxidation. Our recent measurements point out a difference between long channels (constant mobility) and short channels, where the apparent mobility degradation is the consequence of the very large effect of series resistances.

5.   MULTIPLE-GATE SOI MOSFETS

Innovative transistors with double gate, triple gate and quadruple gate are currently being explored for enhanced performance and functionality.

Double-gate (DG) including Gate-All-Around (GAA) MOSFETs are ideal devices in terms of electrostatic control and ultimate scaling, far below 10 nm channel length: numerical simulations show that even 2-nm-long DG MOSFETs can be envisaged.[32] In an extremely thin body, volume inversion and quantum confinement make it possible to reach outstanding transconductance, higher than twice the value in single gate MOSFETs.[29,33] The key difficulty is to figure out a practical technology. Planar process is suitable but a critical issue is the self-alignment of the two gates.

The DG technology can be greatly simplified if some degree of gate misalignment is tolerated. This implies making the bottom gate longer, whereas the channel length is still defined by the source/drain implantation through the shorter top gate. Numerical simulations show that such non-ideal DG MOSFETs are competitive: the subthreshold swing is minimum (60 mV/decade) and, surprisingly, the transconductance and drive current are higher than for the ideal DG transistor with symmetrical self-aligned gates.[34] The reason for the transconductance improvement is the double action of the longer gate: as the back inversion channel is formed, the source/drain and extension regions are *accumulated*. The latter field-effect-junction mechanism contributes to the dynamic lowering of the series resistance.

GAA transistors have been shown to exhibit excellent tolerance to radiations.[2] This is because the charge generated and trapped in the thin gate oxide that envelops the body is very limited. The problem of radiation-induced BOX charge is here totally erased.

The same arguments apply to DG MOSFETs. Figure 8a compares simulation results in single-gate (SG) and DG MOSFETs with symmetrical gate stack and thin front and back oxides.[35] The radiation effect is summarized by assuming a given amount ($\sim 10^{12}$ cm^{-2}) of interface traps and fixed charge in the two oxides. The carrier mobility decreases with film thickness, whereas the threshold voltage increases.

The good news is that the amount of degradation is less acute for DG operation than for SG operation (with back gate grounded). This difference is explained by the charge centroid which is located either in the middle of the film (DG mode) or next to the front interface where the oxide defects are more effective (SG mode). It is concluded that volume inversion allows the current to become less sensitive to radiation damage. Recent experimental data confirms the tremendous advantage of DG devices (Fig. 8b).[18] In this case, the DG mode was emulated in regular FD MOSFETs with thick BOX by simultaneously biasing the two gates.

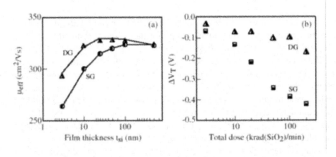

Figure 8. Comparison of single-gate (SG) and double-gate (DG) characteristics after radiation. (a) Simulated effective mobility versus film thickness (after Cirba *et al*[35]). (b) Experimental threshold voltage shift versus dose (after Jun *et al*[18]).

FinFETs are attractive non-planar DG transistors because the process flow is easier to implement. In a FinFET (Fig. 5a), the gate covers three sides of the body but the top channel is deactivated by using a thicker dielectric.[36] The current is controlled by the two lateral sections of the gate and flows along the body sidewalls. The FinFET performance is very promising, however there are two scaling issues: (i) the control of the crystal quality on the sidewalls and (ii) the thinning of the transistor body (inter-gate distance) in order to control the short-channel effects.

A FinFET with a thin dielectric on all the three sides of the fin exhibits three channels controlled by a single gate. This device, abusively named triple-gate MOSFET, exhibits excellent performance. The magnitude of the current can be adjusted by increasing the fin width, in other words the contribution of the top surface channel. This advantage is debatable because volume inversion and scaling capability are lowered. In FinFETs, several channels are simultaneously activated and interacting. Their discrimination is possible using appropriate biasing of the back gate and variable fin widths.

It is found that the carrier mobility is substantially degraded on the fin sidewalls as compared to the top or even bottom channel.[37]

Note that FinFETs are not immune to radiation effects. Charge trapping in the BOX has a two-fold action: (i) the back channel can be turned on and (ii) the modification of the back interface (fin-BOX) potential affects the lateral channels by coupling. Moreover, the fin corners are responsible for a parasitic current and need to be investigated under radiation.

The G^4-FET is a genuine four-gate transistor, operated in accumulation-mode.[38] It has the structure of an inversion-mode SOI MOSFET with two body contacts, except that the current is perpendicular. The majority carriers flow between the body contacts which here play the role of source and drain (Fig. 9a). The G^4-FET features the usual front and back MOS gates (controlling the accumulation and vertical depletion regions), plus the two lateral junctions that govern the effective width of the body (by horizontal depletion regions). The conduction path is modulated by mixed MOS-JFET effects: from a tiny quantum wire, surrounded by depletion regions, to strongly accumulated front and/or back interface channels. Different models apply to surface accumulation or pure volume conduction mechanisms. Each gate has the capability of switching the transistor on and off.[39] The G^4-FET exhibits high current and transconductance, low subthreshold swing and, in volume conduction mode, low noise and intrinsic radiation hardness. The independent action of the four gates opens exciting perspectives for novel applications: mixed-signal circuits, nanoelectronic devices (quantum wires), 4-level logic schemes, etc.

In volume conduction mode, when the channel is double shielded from the interfaces by inversion charges and depletion regions (Fig. 9b), the radiation tests have indicated an outstanding intrinsic immunity.

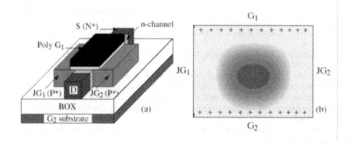

Figure 9. (a) Configuration of the four-gate transistor and (b) majority carrier distribution in the channel cross-section for G^4-FET operation in volume conduction mode.

6.   CONCLUSIONS

SOI wafers and devices play an increasing, vital role in the CMOS technology, where SOI brings enhanced performance and capabilities. The horizon of SOI is even brighter because CMOS can hardly be scaled down without SOI structures. The evolution of the MOS transistor will continue by shrinking the dimensions of the body, gate oxide and BOX, and by reconsidering the nature of the various layers, in order to infuse new functionality and performance. In parallel, the transistor architecture will evolve by adopting multiple gate configurations. These trends open the space for new device-physics mechanisms and circuit topologies. In addition, multiple-gate MOSFETs exhibit exceptional potential for limiting the radiation effects. However, extensive work is needed before fully understanding the radiation behavior of the new SOI materials and devices.

ACKNOWLEDGMENTS

Most of this work has been performed at the Center for Projects in Advanced Microelectronics (CPMA), operated by CNRS, LETI and universities. Special thanks to J. Pretet, F. Allibert, R. Schrimpf, C. Cirba, B. Jun, M. Cassé, A. Ohata, T. Poiroux, K. Oshima, H. Iwai, S. Deleonibus, B. Dufrene, N. Bresson, K. Akarvardar, F. Daugé, A. Vandooren, J-H. Lee, and C. Mazuré.

REFERENCES

1. S. Cristoloveanu and S.S. Li, Characterization of Silicon-On-Insulator Materials and Devices, Kluwer, Boston, 1995.
2. J-P. Colinge, Silicon-On-Insulator Technology: Materails to VLSI, Kluwer, Boston, 3rd ed., 2004.
3. D.J. Frank, R.H. Dennard, E. Nowak, P.M. Solomon, Y. Taur, H.S.P. Wong, *IEEE Proc.*, **89**(3), 259-288 (2001).
4. G. K. Celler and S. Cristoloveanu, *J. Appl. Phys.* **93**, 4955-4978 (2003).
5. S. Cristoloveanu, in H.R. Huff, L. Fabry, S. Kishino (Eds.), *Semiconductor Silicon 2002*, Electrochem. Soc. Proc., vol. PV-2002-2, Pennington, USA, 2002, p.328.
6. S. Cristoloveanu and V. Ferlet-Cavrois, *Int. J. High Speed Electronics and Systems* in press (2004).
7. T. Ouisse, S. Cristoloveanu, and G. Borel, *IEEE Electron Device Letts.*, vol. EDL-12, no. 6, pp. 312-314, (1991).
8. D.M. Fleetwood, *Microelectron. Reliab.*, vol. 42, pp. 1397-1403 (2002).
9. S. Cristoloveanu, D. Munteanu, and M. Liu, *IEEE Trans. Electron Devices*, vol. 47, no. 5, pp. 1018-1027 (2000).
10. B. Jun et al, Proc. *NSREC'03*, 2003.

11. S. Bengtsson, M. Bergh, M. Choumas, C. Olesen, K.O. Jeppson, *Jpn. J. Appl. Phys.*, **35**, 4175 (1996).
12. K. Oshima, S. Cristoloveanu, B. Guillaumot, H. Iwai and S. Deleonibus, *Solid-St. Electron* **48** 907-917 (2004).
13. S. Cristoloveanu, T. Ernst, D. Munteanu, T. Ouisse, *Int. J. High Speed Electronics and Syst.*, **10**(1), 217-230 (2000).
14. T. Ernst, C. Tinella, and S. Cristoloveanu, *Solid-State Electron.* **46**(3), 373-378 (2002).
15. O. Musseau, J. L. Leray, V. Ferlet-Cavrois, Y. M. Conc, and B. Giffard, *IEEE Trans. Nucl. Sci.* **41**(3), 607-612 (1994).
16. V. Ferlet-Cavrois, G. Gasiot, C. Marcandella, C. D'hose, O. Flament, O. Faynot, J. de Pontcharra, and C. Raynaud, *IEEE Trans. Nucl. Sci.* **49**(6), 2948-2956 (2002).
17. P. E. Dodd, M. R. Shaneyfelt, J. R. Schwank, G. L. Hash, B. L. Draper,and P. S. Winokur, *IEEE Trans. Nucl. Sci.* **47**(6), 2165-2174 (2000).
18. B. Jun et al, Proc. *NSREC'04*, in press, 2004.
19. J. Pretet, T. Matsumoto, T. Poiroux, S. Cristoloveanu, R. Gwoziecki, C. Raynaud, A. Roveda and H. Brut, in G. Baccarani, E. Gnani and M. Rudan (Eds.), *Proc. ESSDERC'02*, Univ. of Bologna, 2002, pp. 515-518.
20. J. Pretet, A. Ohata, F. Dieudonn\'e, F. Allibert, N. Bresson, T. Matsumoto, T. Poiroux, J. Jomaah and S. Cristoloveanu, in R.E. Sah, M.J. Dean, D. Landheer, K.B. Sundaram, W.D. Brown and D. Misra (Eds.), *Silicon Nitride and Silicon Dioxide Thin Insulating Films VII*, Electrochem. Soc. Proc., vol. PV-2003-02, Pennington, USA, 2003, pp. 476-487.
21. Mercha, J.M. Rafi, E. Simoen, E. Augendre, C. Claeys, *IEEE Trans. Electron Devices*, **50**(7), 1675-1682 (2003).
22. Jun et al, Proc. *NSREC'04*, in press, 2004.
23. F. Dieudonné, S. Haendler, J. Jomaah, F. Balestra, *Solid-State Electron.*, **48**(6), 985-997 (2004).
24. V. Ferlet-Cavrois, O. Musseau, J. L. Leray, J. L. Pelloie, C. Raynaud, *IEEE Trans. Electron Devices*, **44**(6), 965 (1997).
25. M. Cassé, J. Pretet, S. Cristoloveanu, T. Poiroux, C. Raynaud and G. Reimbold, *Solid-State Electron.*, **48**(7), 1243-1247 (2004).
26. F. Allibert, J. Pretet, G. Pananakakis,S. Cristoloveanu, *Appl. Phys. Letts.*, **84**, 1192-1194 (2004).
27. H. K. Lim and J. G. Fossum, *IEEE Trans. Electron Devices*, **30**, 1244-1251 (1983).
28. F. Balestra, S. Cristoloveanu, M. Bénachir, J. Brini, and T. Elewa, *IEEE Electron Device Lett.*, **8**(9), 410-412 (1987).
29. T. Ernst, S. Cristoloveanu, G. Ghibaudo, T. Ouisse, S. Horiguchi, Y. Ono, Y. Takahashi and K. Murase, *IEEE Trans. Electron Devices*, **50**, 830-838 (2003).
30. C. Fiegna, A. Abramo, E. Sangiorgi, in S. Lury, J. Xu, and A. Zaslavsky (Eds), *Future Trends in Microelectronics}*, Wiley (1999), pp. 115-124.
31. F. Gamiz, J. B. Roldan, J. A. Lopez-Villanueva, P. Cartujo-Cassinello, J. E. Carceller and P. Cartujo, in *Silicon-On-Insulator Technology and Devices X*, eds. S. Cristoloveanu *et al*, Electrochem. Soc. Proc., PV-2001-3, Pennington, USA, 2003, pp. 157-168.
32. K.K. Likharev, in J. Greer, A. Korkin and J. Labanowski (Eds), *Nano and Giga Challenges in Microelectronics*, Elsevier, Amsterdam (2003).
33. D. Esseni, M. Mastrapasqua, G.K. Celler, C. Fiegna, E. Sangiorgi, *IEEE Trans. Electron Devices*, **50**(3), 802-808 (2003).
34. F. Allibert, A. Zaslavsky, J. Pretet and S. Cristoloveanu, *Proc. ESSDERC'2001* (2001) p. 267.

35. C.R. Cirba, S. Cristoloveanu, R.D. Schrimpf, L.C. Feldman, D.M. Fleetwood and G.K. Galloway, *Silicon-On-Insulator Technology and Devices X*, eds. S. Cristoloveanu *et al*, Electrochem. Soc. Proc., PV-2001-3, Pennington, USA, 2003, pp.~493-498.
36. D. Hisamoto, W-C. Lee, J. Kedzierski, E. Anderson, H. Takeuchi, K. Asano, T-J. King, J. Bokor, C. Hu, *Technical Digest IEDM'98*, 1032-1034 (1998).
37. F. Daugé, J. Pretet, S. Cristoloveanu, A. Vandooren, L. Mathew, J. Jomaah, B-Y. Nguyen, *Solid-State Electron.*, **48**, 535-542 (2004).
38. B.J. Blalock, S. Cristoloveanu, B. Dufrene, F. Allibert, M.M. Mojarradi, in Y.S. Park, M.S. Shur and W. Tang (Eds), *Frontiers in Electronics – Future Chips*, World Scientific, **26**, 305-314 (2002).
39. K. Akarvardar, B. Dufrene, S. Cristoloveanu, B. Blalock, T. Higashino, M. Mojarradi and E. Kolawa, in J. Franca and P. Freitas (Eds), *Proc. ESSDERC'03*, Lisbon (2003) pp. 127-130.

RADIATION EFFECTS IN SOI: IRRADIATION BY HIGH ENERGY IONS AND ELECTRONS

I.V.Antonova, J.Stano*, O.V.Naumova, V.A.Skuratov*and V.P.Popov
*Institute of Semiconductor Physics, Novosibirsk, Russia, *Joint Institute of Nuclear Research FLNR, 141980, Dubna, Russia*

Abstract: The following radiation effects are considered in silicon- on- insulator structures irradiated with either 2.0 MeV electrons or 245 MeV multi charged Kr ions in the dose range of $10^5 - 10^6$ rad: (1) accumulation of the positive charge in the buried oxide, (2) generation / transformation of traps at the Si/SiO_2 interfaces, (3) introduction of electrically active radiation defects in the top silicon layer and the substrate.

Key words: silicon- on- insulator, irradiation, electrons, high energy ions, charge in oxide, interface traps, radiation defects

1.   INTRODUCTION

Building of devices in the top silicon layer of SOI wafers provides many advantages. The radiation hardness of SOI structures calls our attention in the present study. The typical absorbed doses used for testing of SOI structures are in the range in $10^5 - 10^7$ rad.[1] For cases of electron or γ-ray irradiation these absorbed doses correspond to relatively low fluences and concentrations of the radiation defects in the SOI are negligible. Another situation is observed in the case of high energy ion implantation.[2] High concentrations of defects are formed in silicon along the ion tracks already after relatively low fluences. As a result, ion irradiation gives the unique possibility to study radiation defects formed in the top silicon layer and the substrate of SOI structures for absorption doses of 10^5 rad and higher. The aim of the present study was to examine the effects of irradiation (by electrons or Kr ions) on the formation of radiation defects in the top silicon layer, generation of traps at the Si/SiO_2 interfaces, and accumulation of

D. Flandre et al. (eds.), Science and Technology of Semiconductor-On-Insulator Structures and Devices Operating in a Harsh Environment, 215-220.

charge in the buried oxide of SOI structures fabricated by wafer bonding and hydrogen cutting.

2.   EXPERIMENTAL

SOI structures were fabricated from float zone grown silicon of n-type conductivity. Thickness of the top silicon layer was 0.63 μm, the buried oxide thickness was 0.4 μm. The electron concentrations in the top silicon layer and the substrate were 1×10^{16} cm^{-3} and 2×10^{14} cm^{-3}, respectively. The Si/SiO$_2$ interface between the top silicon layer and the buried oxide in our SOI structures is a bonded interface, and interface between the buried oxide and the substrate was created by thermal oxidation.

The irradiation sources were 2.0 MeV electrons and 245 MeV Kr ions delivered by U-400 FLNR Dubna cyclotron. Main characteristics of irradiations, such as particle type, fluence, incident energy, electronic $(dE/dx)_e$ and nuclear $(dE/dx)_e$ stopping powers, absorbed dose (D), primary defect concentration (N_v) in the SOI layers are listed in Table 1. The fluences were chosen so that the absorbed doses were the same (about 10^5 and 10^6 rad) for ions and electrons, respectively. To calculate data for Kr ions, SRIM-

Table 1. Parameters of ion and electron irradiation for Si.

Particle	Energy, R_p	Fluence, cm^{-2}	dE_n/dx, keV/nm/part.	N_v, cm^{-3}	dE_e/dx, keV/nm/part.	D, rad
Kr	245MeV, 31.4 μm	2.2×10^8 1.4×10^9	2.3×10^{-3}	3.7×10^{14} 2.4×10^{15}	9.5	$1..6 \times 10^5$ 9.2×10^5
e	2 MeV	3×10^{12} 3×10^{13}	9.5×10^{-9}	2.5×10^{13} 2.5×10^{14}	$3,5 \times 10^{-4}$	1×10^5 1×10^6

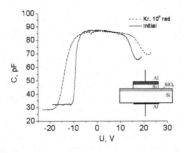

Figure 1. CV characteristics for the SOI structure before and after irradiation with Kr ions with a dose of 1.6×10^5 rad

2000 code was employed and a displacement energy threshold of 13 eV was used.

Experimental methods were capacitance-voltage (CV) measurements, performed at a 1-MHz frequency and charge deep-level transient spectroscopy (Q-DLTS).

3. RESULTS AND DISCUSSION

The charge at the bonded interface, Q_b, at the thermal interface, Q_t, and total charge, $Q = Q_b + Q_t$, were deduced from the flat-band voltages taken from the CV characteristics. The dose dependences of Q_b, Q_t, and Q are given in Table 2. Positive charge accumulated under irradiation with the high-energy ions is lower than that found for electron irradiation.

Table 2. Dose dependences of the positive charge at the bonded interface, Qb, and at the «thermal» interface, Qt, and the total charge in the buried oxide Q = Qt+ Qb

D, rad	2, MeV			Kr 245 MeV		
	Q_b, cm^{-2}	Q_t, cm^{-2}	Q_s, cm^{-2}	Q_b, cm^{-2}	Q_t, cm^{-2}	Q_s, cm^{-2}
initial	5.8×10^{11}	3.7×10^{11}	9.5×10^{11}	5.8×10^{11}	5.8×10^{11}	5.8×10^{11}
10^5	9.8×10^{11}	5.8×10^{11}	1.6×10^{12}	3.5×10^{11}	6.2×10^{11}	9.7×10^{11}
10^6	1.1×10^{12}	1.0×10^{12}	2.1×10^{12}	8.4×10^{11}	8.0×10^{11}	1.6×10^{11}

From the CV data we have chosen for the DLTS measurements a voltage range where the depletion regime in the top silicon layer and accumulation in the substrate (capacitance decrease at positive voltages in Fig.1) occurs. Voltage within this range allows us to recharge the traps at the silicon layer / buried oxide interface and deep levels in the top silicon layer. More details of the method of measurement are described in ref 3. The dose dependence of the DLTS spectra measured before and after electron or ion irradiation for bonded interface is presented in Fig.2. The value of the applied voltages and magnitude of filling pulse are determined for each structure individually. Fig.3 shows the Q-DLTS spectra for the top layer, the substrate, and reference bulk Cz-Si irradiated with 245 MeV Kr ions with dose of 10^6 rad.

Wide peaks in the spectrum of initial and electron irradiated SOI correspond to the interface traps. Peaks, labeled A, B, C1, C2, D are traps introduced in the top silicon layer, the substrate of SOI and bulk Cz-Si. Parameters defining the deep levels observed in irradiated SOI are given in Table 2. Peaks A, B are well known radiation defects and they are found in all irradiated crystals: peak A is connected with complexes of vacancy-oxygen, and interstitial and substitution carbon atoms, peak B corresponds to the divacancy $W^=/W^-$. Contamination - related peaks C1 and F are typically observed in the metal-oxide-silicon (MOS) structure after irradiation.[4,5]

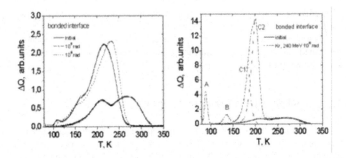

Figure 2. DLTS spectra of SOI structures, measured before, after irradiation for the bonded interfaces. Duration of the filling pulse was. $\tau = 1$ ms.

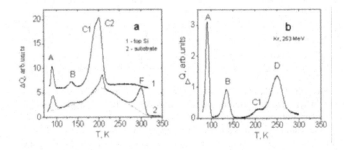

Figure 3. DLTS spectra of SOI structures (a) and bulk Cz-Si (b), irradiated with 245 MeV Kr ions with dose of 10^6 rad

Peaks marked as C2 and D are attributed to centers with very similar energy levels but different cross sections. Center D is the well-known second level of the divacancy (W^-/W^o). We suggested that peak C2 also is associated with the divacancy.

The electron and nuclear stopping powers are very similar in the top silicon layer and the substrate near the SiO_2/Si interface. As it is seen from Table 2, peaks A – C2 have concentrations more than one order of magnitude higher in the top silicon layer then in the substrate and the bulk silicon. It is suggested that this effect is due to different defect and impurity concentrations in the top silicon layer and the substrate (or the initial bulk silicon).

To find the distribution of the interface state density in the forbidden gap from Q-DLTS spectrum, we followed the method described elsewhere.[3] The capture cross-section of interface states was found to be equal to 10^{-18} cm^2 for both interfaces in the initial[3] and Kr irradiated SOI. After electron irradiation the value at the bonded interface was $\sigma = 10^{-19}$ cm^2

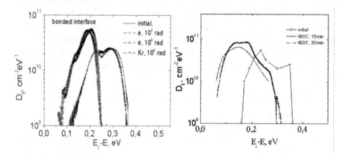

Figure 4 Energy distribution of interface states for the top silicon layer/buried oxide interface and the substrate/ buried oxide interface of the SOI structures before and after irradiation.

Table 3. Parameters of deep levels (level energy, cross section and concentration) in SOI observed after irradiation with Kr ions with dose of 10^6 rad (Fig.3).

Peak	Energy level, Cross section	Interpretation	SOI, Fz-Si		Cz-Si
			Top Si	Substrate	
A	0.16 eV, 2×10^{-14} cm^2	V-O + C_j-C_s	4.7×10^{14}	6.7×10^{13}	7.4×10^{13}
B	0.21 eV, 5×10^{-16} cm^2	Divacancy	2.4×10^{14}	9×10^{12}	3.3×10^{13}
C1	0.28 eV, 3×10^{-16} cm^2	C-X, Cu-X[4]	8×10^{14}	8×10^{12}	1.5×10^{13}
C2	0.42 eV, 3×10^{-13} cm^2	Divacancy?	1.6×10^{15}	4.5×10^{13}	-
D	0.40 eV, 5×10^{-16} cm^2	Divacancy	-	-	5.0×10^{13}
F	0.57 eV, 1×10^{-14} cm^2	Au-X[5] ?	-	8.1×10^{13}	-

The energy spectra of traps at the bonded interface are shown in Fig.4a. The high concentration of deep levels in the DLTS spectra for the bonded interface in SOI irradiated with ions allows us to extract only part of the interface trap spectrum. As seen from Fig.4, the spectrum of traps at the bonded interface is not changed after the Kr irradiation.

An energy spectrum of traps at the bonded interface in the SOI structure is transformed during electron irradiation (Fig.4.): the energy range is shifted towards the conduction band, and the cross section is decreased. It should be noted that the same transformation of the interface traps was found in SOI structures after annealing at a temperature of 430°C in a hydrogen ambient.[3] This allows us to suppose that the trap transformation is due to relaxation of the stressed Si/SiO_2 interface due to different external factors (irradiation or annealing). The mechanism and origin of this transformation of traps are unknown. But hydrogen assistance can be suggested in the both cases. Hydrogen remains in some concentration at the Si/SiO_2 interface and in the buried oxide of SOI structure. During irradiation protons are known to be released from bonds in the buried oxide and have high mobility.

Ion irradiation does not cause a pronounced change in the energy spectrum of the interface traps, most likely due to the large distance between ion track regions (about 270 nm for the maximum dose). The amount of hydrogen, which is released in the buried oxide under ion irradiation, has to be lower in this case

4.   CONCLUSIONS

A comparison of the radiation effects in SOI structures irradiated with electrons and high-energy Kr ions has been made. The fixed positive charge introduced in the buried oxide was found to be higher after electron irradiation. In the dose range of $10^5 - 10^6$ rad, electrons cause transformation of the energy spectrum of interface traps at the bonded interface. Formation of radiation defects was the main effect of high energy ions in SOI. Defect concentration in the top silicon layer is found to be higher than that in the substrate.

REFERENCES

1. R.K.Lawrence, D.E.Ioannou, W.C.Jenkins, S.T.Liu, Gate-diode characterization of the back-channel interface on irradiated SOI wafer, *IEEE Trans. Nucl. Sci.,* **48**, 2140-2145, (2001).
2. P.Mangiagalli, M.Levalois, P.Marie, A comparative study of induced damage after irradiation with swift heavy ions, neutrons and electrons in low doped silicon, *Nucl. Instr. Meth. B*, **146**, 317-322, (1992).
3. I.V. Antonova, O.V. Naumova, V.P. Popov, J. Stano, V.A. Skuratov, Modification of the bonded interface in silicon-on-insulator structures under thermal treatment in hydrogen ambient, *J. Appl. Phys*, **93**, 426-431, (2003).
4. D.Vuillaume, J.C.Bourgoin, Transient spectroscopy of Si-SiO₂ interface states, *Surf. Science* **162**, 680-686, (1985).
5. B.G.Svensson, C.Jagadish, A.Hellen, J.Lalita, Generation of vacancy-type point defects in single collision cascades during swift –ion bombardment of silicon, *Phys. Rev. B* **.55**, 10498-10507, (1997).

RADIATION CHARACTERISTICS OF SHORT P-CHANNEL MOSFETS ON SOI SUBSTRATES

A. Evtukh[1], A. Kizjak[1], V. Litovchenko[1], C. Claeys[2] and E. Simoen[2]
[1]Institute of Semiconductor Physics, 45 Prospekt Nauki, Kiev, 03028, Ukraine; [2]IMEC, Kapeldreef 75, B-301, Leuven, Belgium.

Abstract: The influence of γ-radiation with doses of (10^4-10^6 rad) on the threshold voltage and maximum transconductance of p-channel MOS field effect transistors with channel lengths of 0.5-2.0µm prepared by 0.5µm SOI technology has been investigated. A radiation-induced negative charge is built-up in silicon oxide - silicon structures after irradiation with a dose of 10^4rad for transistors with L>0.9µm. Higher γ-radiation doses caused the building up of positive charge. The negative charge can be thermally annealed at T=315°C and then recovered again after annealing at temperature T=415°C. The positive charge after annealing at T=315°C is significantly reduced. Annealing at T=415°C promotes further reduction of positive charge and the appearance of an effective negative charge for radiation doses of 10^5rad and for lower channel length after a dose of 10^6rad. The maximum tranconductance is increased by 15% by irradiation with a dose of 10^6rad. Subsequent annealing decreases the transconductance in comparison with the initial value (before radiation). The improvement of the MOS transistor parameters after irradiation with low doses (10^4rad) has been observed. To explain the variety of observed radiation effects a model based on radiation induced defect generation and their annihilation during thermal annealing has been proposed. It takes into account boron-oxygen and phosphorus-oxygen defect centers in SiO_2.

Key words: radiation; radiation hardness; MOSFET; SOI; threshold voltage; transconductance; mobility; charge.

1. INTRODUCTION

It is well known that radiation, in particular γ-radiation, adversely affects semiconductor device performance and stability. However, there is much data pointing out the influence of interface and dielectric formation conditions on radiation hardness of the MOS devices.[1] On the other hand, γ-

D. Flandre et al. (eds.), Science and Technology of Semiconductor-On-Insulator Structures and Devices Operating in a Harsh Environment, 221-226.

irradiation with subsequent annealing under controlled conditions promotes an ordering effect.[2-4] The improvement of the electrical properties, reduction of the defect concentration and recombination loses lead to substantial improvement of microelectronic devices and IC parameters. It is very important to investigate the influence of γ-radiation on MOS short p-channel transistors with a thin gate oxide on wafers with buried oxide layer (SOI).

In this work the authors investigate the influence of radiation dose from a γ-source ^{60}Co and low temperature annealing on the threshold voltage (V_{th}) and maximum transconductance ($G_m = dI_d/dV_g$) of p-MOS transistors with short channel length in the range of 0.5-2.0 µm fabricated by 0.5 µm SOI technology.

2.   EXPERIMENT

The LDD p-MOS transistors with LOCOS isolation formed in n-wells have been prepared by a 0.5 µm SOI technology on SIMOX substrates. The schematic cross-section of a typical transistor is shown in Fig.1. The channel length was varied in the range 0.5-2.0µm. For all transistors the channel width was W=20 µm. To obtain the required threshold voltage LDD p-MOS transistors were formed using additional ion implantation with boron into the channel through an SiO_2 layer, 13nm thick, in next regime E=35 keV, $D_b=2\times10^{12}$ cm^{-2} plus E=15 keV, $D_b=0.4\times10^{12}$ cm^{-2} (D_b-boron channel dose). The gate oxide 15 nm thick was grown at T=900°C in dry oxygen. A polycrystalline 0.35µm thick silicon layer was used as the gate electrodes. The MOS transistors were irradiated with γ-rays from a ^{60}Co source with doses of 10^4, 10^5, and 10^6 rad. Thermal annealing of the transistors was at T=315°C and T=415°C in argon.

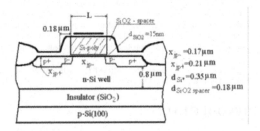

Figure 1. The cross-section of the investigated p-MOS transistors

3.   RESULTS AND DISCUSSION

3.1   INFLUENCE OF THE LOW-DOSE γ-RADIATION

The dependences of threshold voltage on channel length for different irradiation doses are shown in Fig.2 (curves 1). After irradiation with doses of 10^5 and 10^6 rad a negative shift of V_{th}, due to a built up of the total positive charge in the MIS structure is observed. In contrast for the lower irradiation dose 10^4 rad and a channel length more that 0.9 μm a positive shift ΔV_{th} occurs. This is indicative of a building–up of the total effective negative charge. The change of threshold voltage after the low dose (10^4 rad) γ- irradiation depends strongly upon the channel length. But for higher doses (10^5, 10^6 rad) the length dependence of ΔV_{th} is weaker. For the irradiation dose 10^4 rad the increases of ΔV_{th} are in range –(3% - 9%), 10^5 rad – (12% - 15%), and 10^6 rad up to 50%. As can be seen from Fig.3 the maximum transconductance, in general, is increased after γ- irradiation. This increase is a maximum after 10^6 rad. However for specific channel lengths after irradiation with a dose 10^5 rad a decrease of transconductance is observed (see Fig.3).

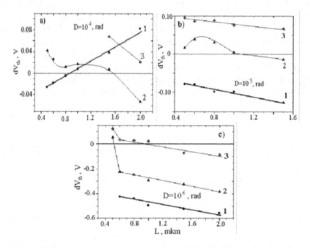

Figure 2. Threshold voltage shift showing the channel length dependences after γ-radiation with different doses: a)- D=10^4rad, b)- D=10^5rad, c)- D=10^6rad (1- after radiation; 2- after radiation and annealing T=315°C; 3- after radiation and annealing T=315°C + T=415°C).

3.2   ANNEALING

Thermal treatment at T=315°C in argon annihilates the positive charge in the "spacer" region and negative charge in p-MOS transistors with L<1 μm for radiation doses of 10^4 and 10^5 rad. For p-channel transistors irradiated with doses of 10^6 rad only part of positive charge is annealed. The transconductance is reduced in comparison with the initial value by 1.5%, 5% and 15% for irradiation doses of 10^4, 10^5, and 10^6 rad, respectively (Fig.3 curve 2). After additional annealing at T=415°C the negative charge dominates in transistors irradiated with doses 10^4, 10^5 rad. The transconductance is no more than 5% lower than the initial value (before irradiation). In the case of the irradiation dose of 10^6 rad the total negative charge is observed only for transistors with L<1 μm, for longer channels a small positive charged is created. The transconductance is reduced by approximately 7%.

There are many experimental results which point out that radiation induced charge and surface traps can be entirely annealed at T=415°C or lower.[1] In work [5], for example, it was shown that positive charge builds up in the oxide of MOS transistors with a polysilicon gate after –radiation from ^{60}Co and annealing at 300°C for 30 minutes. The interface states are annihilated during an anneal at the same temperature for 45 minutes.

In our case the negative charge is annealed at T=315°C. At this temperature the positive charge is annealed only in spacers and partially in SiO_2 under the channel. The full annealing of positive charge is observed only at T=415°C. But for the case of a radiation dose of 10^6 rad and L>1.1μm the positive charge is not annealed fully even at that temperature.

After annealing the μ_p is decreased to approximately the initial values before radiation.

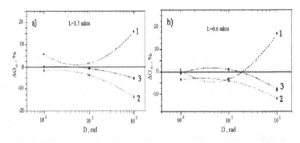

Figure 3. Transconductance dependence upon radiation dose: 1- after radiation; 2- after radiation and annealing at T=315°C; 2- after radiation and annealing at T=315°C+T=415°C (a)- L=1.5μm, b)- L=0.6μm).

3.3 MODEL

According to the model proposed in work [6], (1) positive charge connected with a breakdown of the strained Si-O bonds and capture of holes and (2) negative charge connected with the breakdown of B-O bonds and capture of electrons both build up in the SiO_2 and at the Si-SiO_2 interface during γ- radiation (Fig.4).

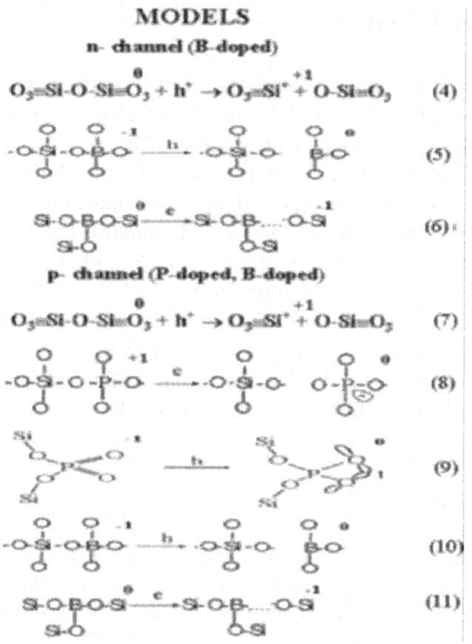

MODELS

n- channel (B-doped)

$$O_3{\equiv}Si\text{-}O\text{-}Si{\equiv}O_3 + h^+ \rightarrow O_3{\equiv}Si^{+1} + O\text{-}Si{\equiv}O_3 \qquad (4)$$

$$(5)$$

$$(6)$$

p- channel (P-doped, B-doped)

$$O_3{\equiv}Si\text{-}O\text{-}Si{\equiv}O_3 + h^+ \rightarrow O_3{\equiv}Si^{+1} + O\text{-}Si{\equiv}O_3 \qquad (7)$$

$$(8)$$

$$(9)$$

$$(10)$$

$$(11)$$

Figure 4. Schematic of the processes in SiO_2 during γ-radiation.

During thermal treatment the annealing of positive and negative charge centers occur. The annealing process of negative charge centers is fastest. As a result, build up of positive charge is observed after annealing at $T=315°C$ for an irradiation dose of 10^4 rad (see Fig.2a). For higher γ- radiation doses the build up of positive charge is significantly higher and the role of the negative charge is negligible. Subsequent treatment at $T=415°C$ promotes additional annealing of the positive charge. As a result, a further shift of V_{th} is observed. These observations highlight the need to design and build MOS transistors using a technology that will constrain the build up of charge during irradiation.

So, the radiation ordering effect is realized after radiation with doses 10^4-10^6 rad for case of L≥1 μm p-channel MOSFETs. But radiation – thermal annealing ordering effect was observed only for L<1 μm p-channel transistors after irradiation with D=10^6 rad and following annealing at T=415°C. The dependences of the threshold voltage shift during γ-irradiation and subsequent annealing is influenced by the "spacer" regions.

4.   CONCLUSION

The diffusion of boron and phosphorus in SiO_2 strongly influences the γ-radiation hardness of p-channel MOS transistors. Both positive and negative charges accumulate in the gate dielectric and at the Si-SiO_2 interface during γ-radiation. These processes complicate the threshold voltage and transconductance behavior during irradiation and annealing. After low dose radiation (10^4rad) the negative charge dominates in p-channel MOS transistors. As a result of annealing at 415°C, the total effective negative charge is built in for cases of radiation doses 10^4, 10^5, and 10^6 rad (for lower channel lengths). In the case of L≥1 μm the transconductance is also increased during radiation with doses in the range 10^4 -10^6 rad. This highlights the effects of radiation and radiation-thermal annealing of p-channel MOSFETs with a thin (d=15 nm) gate oxide.

REFERENCES.

1. T.P. Ma, P.V. Dressendorfer, *Ionizing radiation effects in MOS devices and circuits* (A Wiley - Interscience Publication, John Wiley & Sons, New York, 1989).
2. V.G. Litovchenko, V.Ya. Kiblik, R.O. Litvinov, The influence of radiation -thermal treatments on semiconductor structures characteristics. *Optoelectronics and Semiconductor Technique* **1**, 69-73 (1982) (in Russian).
3. O.Yu. Borkovskaya, N.L. Dmitruk, V.Ya. Kiblik, V.G. Litovchenko, and R.O. Litvinov, Radiation stimulated annealing of structural defects in layer structures based on Si and III-V compounds. *Acta Physica Hungarica* **59 (3-4)**, 327-332 (1986).
4. I.P. Lisovskii, R.O.Litvinov, V.G. Litovchenko, Effect of radiation ordering in Si-SiO_2 structures at ion and plasma-thermal treatments of surface. *Poverchnost. Phys. Chem. Mechan.* **6**, 69-75 (1986) (in Russian).
5. A.G. Sabnis, Characterization of annealing of ^{60}Co gamma-ray damage at Si-SiO_2 interface. *IEEE Trans. Nucl. Sci.* **NS-30**, 4094-4099 (1983).
6. C. Claeys, E. Simoen, V.G. Litovchenko, A. Evtukh, A. Efremov, A. Kizjak, Yu. Rassamakin, Influence of γ- radiation on short channel MOSFETs with thin SiO_2 films. In: *Progress in SOI Structures and Devices Operating at Extreme Conditions*, edited by Balestra F. et al (Kluwer Academic Publ., the Netherlands, 2002), pp.211-220.

TOTAL DOSE BEHAVIOR OF PARTIALLY DEPLETED DELECUT SOI MOSFETS

O.V. Naumova, A.A. Frantzusov, I.V Antonova and V. P. Popov
Institute of Semiconductor Physics, RAS, Novosibirsk, Russia

Abstract: The effect of gamma-irradiation on the electrical characteristics of MOSFETs
fabricated in DeleCut SOI wafers was defined. Properties of gate-oxide and
BOX (buildup of radiation-induced charge in the oxide, interface state density
and the initial concentration of traps in the oxide) were determined for
DeleCut SOI MOSFETs and compared with that for thermal oxide on bulk
silicon, Unibond and SIMOX wafers.

Key words: gamma radiation; interface states; MOSFETs; SOI.

1.   INTRODUCTION

SOI technologies, as it is known, were initially developed and used for radiation-hardened applications. The immunity to latch-up due to complete dielectric isolation of transistors and the small volume of active silicon are the main SOI advantages compared to bulk technologies[1]. These advantages are achieved due to the existence of the relatively thick layer of oxide (buried oxide, BOX), which separates a top silicon layer from the silicon substrate. But, on the other hand, the radiation-induced positive charge built-in in the BOX creates leakage channel in the top silicon layer for the n-channel MOSFET. Thus, the total dose behavior of SOI MOSFETs is determined by the charge accumulation both in gate- and buried-oxide and the initial properties of the BOX are no less important than the properties of the gate-oxide.

Commercial methods of SOI wafers fabrication are frequently based on ion implantation. This technology involves oxygen implantation and high temperature annealing to form SIMOX wafers, alternatively hydrogen implantation is used in the Smart Cut[2] technology to form Unibond wafers.

*D. Flandre et al. (eds.), Science and Technology of Semiconductor-On-Insulator Structures
and Devices Operating in a Harsh Environment, 227-232.*

In the last case the buried oxide, which initially is a thermally grown oxide, is subjected to hydrogen implantation. Properties of these buried oxides are similar to that of thermal SiO_2 after high temperature annealing however some differences are observed[3,4].

In the present report we examine the gamma radiation response of SOI MOSFETs, in which the buried oxide is a thermal oxide but has not been subjected to any implantation during the SOI wafer fabrication[5]. These SOI wafers were called DeleCut SOI. Properties of gate-oxide and BOX were determined for DeleCut SOI MOSFETs and compared with that for bulk silicon, Unibond and SIMOX wafers.

2.   EXPERIMENTAL

SOI wafers were made using a technology which is similar to Smart Cut, that involves hydrogen-induced transfer of silicon layers onto a handle wafer. However there are some differences: (1) buried oxide was not subjected to a hydrogen implantation unlike the Smart Cut process and (2) the interface between the buried oxide and the top silicon layer is a bonded interface. This method of SOI formation is described elsewhere[5].

Partially depleted p-channel transistors have been processed on a 120 nm thick silicon film. The thickness of the buried oxide was either 80 or 350 nm. The MOS technology has an effective gate length of 0.5 µm, a gate width of 30 µm. The thickness of a gate oxide was 18 nm. A N^+-doped polysilicon gate was used, the p-channel transistors have a boron doped body region with a doping level of about 2×10^{17} cm^{-3}. The body-source connection was realized for every transistor. Isolation between SOI MOSFETs was obtained by the formation of mesa structures. No special precautions were been taken to intentionally harden this technology.

The effect of irradiation on the electrical characteristics of the SOI MOSFETs was examined using Cs-137 as the radiation source. The dose rate was of 100 rad/s (Si). Terminals were grounded during irradiation. The temperature during irradiation did not exceed 30°C, the total dose was varied from 10^5 to 10^7 rads.

Measurements of drain current (I_{ds}) versus gate-voltage (V_g) were performed at a drain bias of V_{ds}=0.15 V before and after irradiation. The interface state density and radiation-induced charge accumulation in the gate-oxide and BOX were extracted from the subthreshold swing and the threshold-voltage shift of I_{ds}-V_g characteristics of the top- and back-gate transistors, respectively.

SOI capacitors with aluminum contacts to both the top silicon layer and the substrate were fabricated. The initial charge value in the buried oxide and the filling of traps in the oxide under voltage stress were determined from measurements of the high-frequency capacitance-voltage (CV) characteristics. Stress time was 30 min.

3. RESULTS AND DISCUSSION

Typical series of pre- and post-irradiation subthreshold characteristics for back- and front-gate SOI MOSFETs are shown in Fig.1. One can see the almost parallel shifts in the subthreshold characteristics with dose in the interval of 10^5-10^7 rads for both types of transistors.

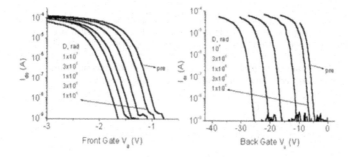

Figure 1. Front- and back-gate subthreshold I_{ds}-V_g characteristics versus dose for p-channel DeleCut SOI MOSFET

Fig.2 shows the dose dependencies of the charge buildup in the BOX and gate-oxide, which was extracted from the threshold voltage shifts, ΔV_{th}, and capacitance of the oxide, C_{ox}, as $\Delta Q_{ox} = \Delta V_{th} C_{ox}$. Dose dependencies of ΔQ_{ox} for Unibond and SIMOX SOI[1,6] are shown in Fig.2 for comparison. One can see that the buildup of charge in the buried oxide with dose for the DeleCut SOI wafer is 1) weakly dependent on the buried oxide thickness (80 nm or 350 nm) and almost the same as for the 18 nm front gate-oxide, and 2) it is less than that for the Unibond and SIMOX wafers.

It is well known, that the main effect of ionizing radiation upon the gate oxide of MOS transistors is the accumulation of positive charge in oxide and the generation of fast states, D_{it}, at the Si/SiO$_2$ interface.

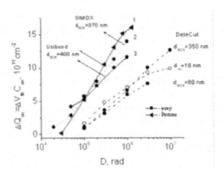

Figure 2. The dose dependencies of the charge accumulation in the gate- (empty symbol) and buried oxides (filled symbol) for different SOI wafers. The data for SIMOX and Unibond BOX were extracted from device characteristics (ΔV_{th} and C_{ox}) taken from: 1- 6; 2,3 - 1

The radiation-induced increase of the density of the interface traps, ΔD_{it}, can be extracted from the increase of subshreshold swing, ΔS, as:

$$\Delta D_{it} = \frac{\Delta S}{\frac{kT}{q}\ln 10}\frac{C_{ox}}{q}$$

The value of ΔD_{it} at doses 1 Mrad and 10 Mrad are shown in the Table1 for different MOSFETs. One can see, that the density of interface traps introduced by the irradiation of the silicon layer/BOX interface in DeleCut SOI is small which is the main reason for the dose dependencies of ΔQ_{ox} in the dose interval of 10^5-10^7 rad (Fig.2). On the other hand, the same result (weak increase in D_{it}) is observed for the three other cases, namely: 1) thermal oxide on silicon irradiated by hydrogen (Unibond wafer), 2) thermal oxide bonded with hydrogen irradiated silicon and 3) thermal oxide grown on the hydrogen irradiated top silicon layer. From these results we conclude that more stable hydrogen-related complexes are formed at the Si/SiO_2 interfaces in DeleCut and Unibond wafers due to the hydrogen implanted in silicon.

As a contribution by the charge on interface states is not essential, we conclude that the dose dependencies of the charge, ΔQ_{ox}, in DeleCut (and Unibond) BOX layers is connected with the introduction of positive charge into the oxide. Filling of preexisting traps or the generation of new oxide traps may be the cause.

Table 1. Subthreshold swing, ΔS, and interface state densities, ΔD_{it}, for different Si/SiO_2 interfaces in MOSFETs

dox, nm	interface	1 Mrad		10 Mrad		Refer enc.
		ΔS mV/dec	ΔD_{it}, 10^{11} $eV^{-1}cm^{-2}$	ΔS mV/dec	ΔD_{it}, 10^{11} $eV^{-1}cm^{-2}$	
18	gate-oxide/top Si(DeleCut)	1	0.2	3	0.59	
25	gate-oxide/bulk Si		1.6-2.4			7
100	gate-oxide/bulk Si		14-20			7
80	BOX/top Si (DeleCut)	33	0.15			
350	BOX/top Si (DeleCut)	15	0.15	186	1.87	
400	BOX/top Si (Unibond)		<10%		<20%	8
510	BOX/top Si (SIMOX)		5.5			9

Table 2. The charge accumulation under voltage stress ($E=5 \times 10^5 V/cm$) in the buried oxide in different SOI materials.

SOI	ΔQ, 10^{11} cm^{-2}	References
SIMOX	1-12	3
	10	10
Unibond	3.5	11
DeleCut	0.9	our data

Accumulation of charge under electrical field stress was investigated for the as prepared BOX in DeleCut wafers to determine the density of pre-existing traps. Table 2 shows data for the density of hole traps in DeleCut wafers. The data for Unibond and SIMOX BOX are included in Table 2 for comparison. One can see, that no more than 10 percent of the charge, accumulated during the irradiation, can be stored on the preexisting-traps of DeleCut material. (Fig.2), however this component is more significant in SIMOX and Unibond wafers.

4. CONCLUSION

The dose dependencies of the interface state densities at gate-oxide/silicon and BOX/silicon interfaces of DeleCut SOI MOSFETS is found to be similar to the values in Unibond SOI. The radiation-induced increase of the interface state density in DeleCut SOI MOSFETS is lower than that for Si/SiO_2 interfaces on bulk silicon and in SIMOX wafers. The residual hydrogen in DeleCut SOI (and Unibond) is considered to be the reason for the more toughened (complicated) structure of the interface states (higher stability of states) compared with that at the thermal oxide interface on bulk silicon. The charge on the preexisting-traps in BOX of DeleCut wafers is no more than 10 percent of the total radiation induced charge for doses up to 10^7 rad. It is assumed that this is one of the reasons for the higher radiation tolerance of DeleCut BOX in comparison to Unibond and SIMOX BOX.

5.   REFERENCES

1. J.R.Schwank, P. E. Dodd, V. Ferlet-Cavrois, R. A. Loemker, P. S. Winokur, D. M. Fleetwood, Paillet, J.-L. Leray, B. L. Draper, S. C. Witczak and L. C. Riewe, Correlation Between Co-60 and X-Ray Radiation-Induced Charge Buildup in Silicon-on-Insulator Buried Oxides, *IEEE Trans.Nucl.Sci.*, **47**(6), 2175-2182 (2000)
2. M. Bruel, Application of hydrogen ion beams to Silicon On Insulator material technology, *Nucl.Instr. and Math. In Phys Res.* **B 108.**, .313-319 (1996)
3. A.G.Reverz and H.L.Hughes, The defect structure of buried oxide layers in SIMOX and BESOI structures, in: *Physical and Technical Problems of SOI Structures and Devices*, edited by J.P.Collinge, (Kluwer Academic Publishers, Netherlands, 1995), pp.133-156
4. A.N.Nazarov, Electrical instability in silicon on insulator structures and devices during voltage and temperature stressing, in *Perspectives, Science and Thechnologies for Novel Silicon on Insulator Devices*, edited by P.L.F.Hemment, (Kluwer Academic Publishers, Boston/Dordrecht/London, 1995) **73**, pp.163-186
5. V. P. Popov, A. I. Antonova, A. A. Frantsuzov, L. N. Safronov, G. N. Feofanov, O. V. Naumova, and D. V. Kilanov, Properties of structures and devices on SOI substrates, *Semiconductors*, **35**(9), 1030–1037 (2001).
6. Y.Li, G. Niu, J. D. Cressler, J. Patel, P. W. Marshall, H. S. Kim, M. S. T. Liu, R. A. Reed and M. J. Palmer, Proton Radiation Effects in 0.35-μm Partially Depleted SOI MOSFETs Fabricated on UNIBOND, *IEEE Trans.Nucl.Sci.*, **49**(6), 2930-2936 (2002)
7. C.M.Dozier, D.B.Brown, R.K.Freitag and J.L.Throckmortont, Use of the subdhreshold bechavior to compare X-ray and Co-60 radiation-induced defects in MOS transistors, *IEEE Trans.Nucl.Sci.*,**NS-33**(6),.1324-1329(1986)
8. V.Ferlet-Cavrois, P.Paillet, O.Musseau, J.L.Leray, O.Faynot, C.Raynaud and J.L.Pelloie, Total dose behavior of partially Depleted SOI Dynamic Threshold Voltage MOS (DTMOS) for Very Low Supply Voltage Applications (0.6-1V), *IEEE Trans.Nucl.Sci.*, **47**(3), 613-619 (2000)
9. S.T.Liu and L.P.Allen, Back channel uniformity of thin SIMOX wafers, *IEEE Trans. Nucl.Sci.*, **38**(6), 1271-1275 (1991)
10. A.N.Nazarov, V.I.Kilchitska, I.P.Bachuk, A.S.Tkachenko, and S.Ashok, RF Plasma Annealing of positive charde generated by Fowler-Nordheim electron imjection in buried oxide in silicon, J.Vac.Sci.Thechnol., **B18**, 1254-1261 (2000)
11. A.Nazarov, V.Kilchytska, and A.Tkachenko, Trapping Annealing of Charge Generated by FN Electron Injection in Buried Oxide of SIMOX and Unibond SOI Structures, Abstracts of 199-th ECS Meeting, **N502** (2001)

RADIATION EFFECT ON ELECTRICAL PROPERTIES OF FULLY-DEPLETED UNIBOND SOI MOSFETS

Y. Houk, A. N. Nazarov, V. I. Turchanikov, V. S. Lysenko, *S. Adriaensen, and *D. Flandre
*Institute of Semiconductor Physics, National Academy of Science of Ukraine, Prospekt Nauky 45, 03028, Kyiv, Ukraine, * DICE, Universite Catholique de Louvain, Louvain-la-Neuve, Belgium*

Abstract: A radiation effect on edgeless FD accumulation mode (AM) p-channel and inversion mode (IM) n-channel MOSFETs, fabricated on UNIBOND SOI wafers, is investigated. The method of second derivative is used to determine the threshold voltages of front and back channels in the MOSFETs from the measurements of front-gate transistors only. Stronger irradiation effect on IM n-MOSFET than that on AM p-MOSFET is revealed. It has been showed, that radiation-induced positive charge in the BOX inverted back interface causes back channel creation in IM n-MOSFET but no such effect in AM p-MOSFET has been observed. It is demonstrated that small-doses have the effect of improving the quality of both interface.

Key words: SOI, FD MOSFET, radiation, small-dose improvement

1. INTRODUCTION

Fully depleted (FD) silicon-on-insulator (SOI) MOSFETs are very attractive as basic elements of CMOS integrated circuits for high-temperature applications.[1] However, in consequence of the charge coupling effect[2] and thick buried oxide (BOX) in SOI wafers such elements have to be less radiation-resistant to total dose. Studies of the radiation effect on FD SOI MOSFETs that are dedicated to high-temperature applications have not been performed yet. Additionally, the extraction of radiation charge generated in the BOX is usually carried out by back-gate SOI MOSFETs.[3, 4] In this paper the double derivation method[5-7] is applied to front-gate

233

D. Flandre et al. (eds.), Science and Technology of Semiconductor-On-Insulator Structures and Devices Operating in a Harsh Environment, 233-239.

characteristics of FD SOI MOSFETs to extract the charges generated in the gate and buried oxides during irradiation.

2.   EXPERIMENTAL

Edgeless IM n- and AM p-MOSFETs with L/W=3/172 μm, fabricated in the same chip, have been measured during these experiments. The devices were produced using UNIBOND SOI material. After device processing the BOX, silicon film and gate oxide thicknesses were 360, 80 and 38 nm, respectively. Doping channel concentration was 5×10^{16} cm^{-3}. For irradiation a ^{60}Co gamma-ray source with a flux of 390 Rad/s has been used. The MOSFETs were shorted during irradiation.

Drain current vs. gate voltage (I_D-V_{G_1}) characteristics as a function of the bias applied to the silicon substrate (V_{G_2}) were measured at room temperature in the linear regime ($V_D = 0.1$V). First and second derivatives of drain current vs. gate voltage were calculated numerically.

3.   METHODOLOGY

The I_D-V_{G_1} characteristics for IM n-MOSFETs and AM p-MOSFETs at room temperature are presented in Figs. 1a and 1a', respectively.

Mobility of the charge carriers in the front channel of the MOSFETs was calculated from the maximum transconductance under depletion condition at the BOX/silicon film interface (see Figs. 1b and 1b'). The mobility of the charge carriers in the back channel was calculated from the plateau observed in the front-channel transconductance (see Fig. 1b) by the following formula:[7]

$$\mu_2 = \mu_1 \frac{g_{plateau}}{g_{max}} \left(1 + \frac{C_{Si}}{C_{ox_1}} + \frac{C_{it_1}}{C_{ox_1}} \right), \tag{1}$$

neglecting C_{it_1} to obtain a minimal estimation of back channel mobility.

Hereinafter $\mu_{1,2}$ are the charge carriers in the front and back channels, $C_{it_{1,2}} = qD_{it_{1,2}}$ are the interface-trap capacitances ($D_{it_{1,2}}$ are the average concentrations of interface traps), C_{Si} is the depleted-film charge, and $C_{ox_{1,2}}$ are the oxides capacitances, $Q_{Si} = -qN_a t_{Si}$ is the depletion charge (t_{Si} is the thickness of Si-film), $\phi_{ms_{1,2}}$ are the metal-semiconductor work functions, $Q_{f_{1,2}}$ are the oxide fixed charges, $Q_{c_{1,2}}$ are the surface channel charges, and $\psi_{s_{1,2}}$ are the surface potentials.

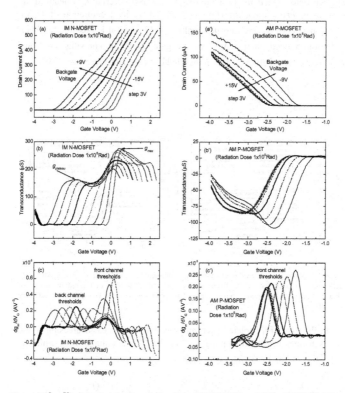

Figure 1. I_D-V_G characteristics for edgeless IM n-MOSFET (a) and AM p-MOSFET (a'), irradiated with total dose of 1×10^6 Rad, IM (b) and AM (b') MOSFETs' transconductance, and derivatives of transconductance for IM (c) and (AM) (c') MOSFETs.

Transconductance derivatives with respect to V_{G_2} for the MOSFETs are shown in Figs. 1c and 1c'. Front-gate threshold voltages of the MOSFETs for front and back channels were determined from the maximums of the transconductance derivative, as depicted by Figs.1c and 1c'.

Dependence of AM p-MOSFET front-gate channel voltages for front and back channels on conditions at the back interface is illustrated in Fig. 2a.[7]

From available dependencies it is possible to extract radiation-induced dynamical change of the fixed charge on the gate oxide and BOX as well as surface state capacitances for both front and back interfaces. Indeed, in the case of **inverted back interface** ($\psi_{s_2} = 2\phi_F$) we have saturation of V_{th_1} vs. V_{G_2} dependence for high positive values of V_{G_2}, and it is possible to obtain radiation-induced charge ΔQ_{f_1} in the gate oxide by knowing the front-channel voltage shift ΔV_{th_1} for an inverted back interface (see Fig. 2b):[7]

$$V_{th_1} = -\frac{Q_{f_1}}{C_{ox_1}} - 2\phi_F \frac{C_{Si}}{C_{ox_1}} - \frac{Q_{Si}}{2C_{ox_1}}. \qquad (2)$$

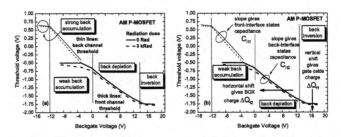

Figure 2. Front-gate threshold voltages for front and back channels of AM p-MOSFET depending on conditions at the back interface (a) and illustration of the radiation-induced charge parameters extraction from the dependencies of the threshold voltages upon back-gate voltages (b).

In the case of a depleted back interface we have linear decreasing of the front-channel threshold voltage V_{th_1} vs. back-gate voltage V_{G_2}, which is described by the following expression:[7]

$$V_{th_1} = -\frac{Q_{f_1}}{C_{ox_1}} - \frac{Q_{Si}}{2C_{ox_1}} - \frac{C_{Si}C_{ox_2}}{C_{ox_1}(C_{ox_2} + C_{Si} + C_{it_2})}(V_{G_2} - [\phi_{ms_2} - \frac{Q_{f_2}}{C_{ox_2}} - \frac{Q_{Si}}{2C_{ox_2}}]). \quad (3)$$

From the slope of this linear dependence we can obtain the capacitance of the back-interface states C_{it_2}. Now, knowing ΔQ_{f_1} and C_{it_2} we can find the radiation-induced charge ΔQ_{f_2} in the BOX from the horizontal shifts ΔV_{G_2} of the V_{th_1} vs. V_{G_2} curves in the same region (see Fig. 2b). Finally, from the slope of the linear dependence of the front-gate threshold voltage for the back channel with the condition of **weak accumulation at the back interface** (see Fig. 2b) we can obtain the capacitance of the front-interface states C_{it_1}:[7]

$$V_{th_2} = -\frac{Q_{f_1}}{C_{ox_1}} - \frac{Q_{Si}}{2C_{ox_1}} - \left(1 + \frac{qC_{it_1} + C_{Si}}{C_{ox_1}}\right)\frac{C_{ox_2}}{C_{Si}}(V_{G_2} - [\phi_{ms_2} - \frac{Q_{f_2}}{C_{ox_2}} - \frac{Q_{Si}}{2C_{ox_2}}]).\,(4)$$

4.   RESULTS AND DISCUSSION

The threshold voltage shift in the front and back transistors at $V_{G_2} = 0V$ as a function of irradiation dose is presented in Fig. 3a, which demonstrates a stronger irradiation effect on the threshold of the front channel for the AM p-MOSFET than for the IM n-MOSFET. Such a phenomenon is possibly associated with a compensative effect of the radiation induced positive oxide charge in the gate oxide and negatively charged surface states on the threshold voltage of the IM n-MOSFET and the overall effect of the positive oxide charge and positively charged surface states on the threshold voltage of the AM p-MOSFET.

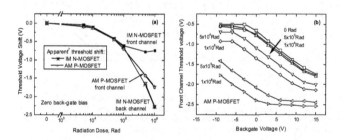

Figure 3. Threshold voltage shifts for IM n- and AM p-MOSFETs (a), and the front channel threshold voltage of AM p-MOSFET vs. backgate voltage for different radiation doses (b).

Additionally, an apparent threshold voltages shift in the MOSFETs, which is defined as the front-gate voltage from which the drain current starts to increase significantly in the linear regime[7], is also presented in Fig. 3a. It should be noted that, in the case of the apparent threshold voltage shift with irradiation, the AM p-MOSFET appears more radiation-resistant than IM n-

MOSFET. This is associated with inversion channel creation at the back interface of the IM n-MOSFET with positive charge generation in the BOX during gamma irradiation. Thus, for the IM n-MOSFET we can observe a good correlation of the shift of apparent threshold voltage with the physical threshold voltage shift for the back channel (see Fig. 3a). That is why it is impossible to use the apparent threshold voltage to determine charges both in gate and buried oxides for IM n-MOSFETs.

Fig. 4a depicts the change of electron and hole mobility in the front and back channels of the MOSFETs. It should be noted that in the dose range from 10^3 to 10^4 Rad an increase of the electron and hole mobility is observed. This effect can be related to a radiation ordering of the gate oxide/silicon film interface as well as the BOX/silicon film interface.

As described in the previous section, the radiation induced charge in the gate oxide, buried oxide and interfaces can be calculated from the front channel threshold voltage V_{th_1} vs. back-gate voltage V_{G_2} characteristics as a function of irradiation dose (see Fig. 3b). Such calculations can be performed most simply for the AM p-MOSFET, for which I_D-V_{G_1} characteristics are weakly affected by a back channel current, and the front channel threshold voltage can be extracted easily.

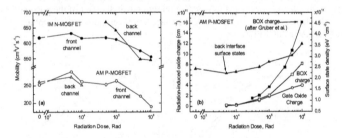

Figure 4. The front and back channel mobilities for IM n- and AM p-MOSFETs (a) and radiation-induced charges in the gate and buried oxides of AM p-MOSFET and surface states density at the BOX/Si film interface (b). Radiation-induced BOX charge in MOSFET made on similar UNIBOND technology is plotted after Gruber et al.[4]

The results of the calculation are presented in Fig. 4b. The important point is smaller radiation induced positive charge in the BOX in our MOSFETs than that presented in Ref. 4 (see Fig. 4b) for similar UNIBOND technology. Fortunately improvements during recent years of the UNIBOND technique have improved the quality of the BOX in such SOI structures. Additionally it is worthy noting that the interface surface state density in the BOX/Si film interface reveals a minimal value in the dose range from 10^3 to 10^4 Rad. This phenomenon attests increase of the electrical quality of the back interface as well as the gate oxide/Si film interface during small dose irradiations.

5. CONCLUSIONS

Thus, it is possible to distinguish unambiguously the front-gate threshold voltages for the front and back channels and to extract the concentration of the radiation induced charges in the FD SOI MOSFETs by carrying out front-gate measurements only. Following small doses of irradiation, an increase of charge carrier mobility both in front and back channels and decrease of surface states density in the BOX/Si film interface are observed, that are due to an increase of the quality of the structural and electrical properties of the gate oxide/film interface as well as the BOX/film interface.

ACKNOWLEDGEMENTS

The authors thank the technical stuff of the UCL Microelectronics Lab for device fabrication. This work has been supported by STCU project No 2332.

REFERENCES

1. J.-P.Colinge, Silicon-on-Insulator Technology: Materials to VLSI, Dordrecht: Kluwer, 2nd edition, 1997.
2. H.K.Lim and J.G.Fossum, Threshold voltage of thin-film silicon-on-insulator (SOI) MOSFETs. *IEEE Trans. Electron Devices*, **39**, 1244-1251 (1983).
3. P.Paillet, J.L.Autran, O.Flament, J.L.Leray, B.Aspar, A.J.Auberton-Herve, X-radiation response of SIMOX buried oxides: influence of the fabrication process. *IEEE Trans. Nucl. Sci.*, **43**, 821-825 (1996).
4. O.Gruber, P.Paillet, J.L.Autran, B.Aspar, A.J.Auberton-Herve, Charge trapping in SIMOX and UNIBOND® oxides. *Microelectron. Eng.*, **36**, 387-390 (1997).
5. H.-S.Wong, M.H.White, T.J.Krutsick, and R.V.Booth, Modeling of transconductance degradation and extraction of threshold votage in thin oxide MOSFET's. *Solid-State Electron.*, **30**, 953-968 (1987).
6. A.Terao, D.Flandre, E.Lora-Tamayo, and F.Van de Wiele, Measurement of threshold voltages of thin film accumulation-mode PMOS/SOI transistors. *IEEE Electron. Dev. Lett.*, **EDL-12**, 682-684 (1991).
7. D. Flandre and A. Terao, Extended theoretical analysis of the steady-state linear behavior of accumulation-mode, long-channel p-MOSFETs on SOI substrates. *Solid-State Electron.*, **35**, 1085-1092 (1992).

LOW COST HIGH TEMPERATURE TEST SYSTEM FOR SOI DEVICES

G.Russell, Y. Li and H.Bahr
School of Electrical, Electronic & Computer EngineeringUniversity of Newcastle upon Tyne, England UK

Abstract: Silicon on Insulator technology has found widespread use in high temperature electronic applications in the automotive, aerospace and oil well logging industries. To ensure the integrity and reliability of these devices in their applications, a major challenge is to be able to test the devices, economically, under similar operating conditions to those they would experience in the field, that is in operating temperatures of around 300C. Furthermore, due to the criticality of the applications it is also necessary to perform stress testing to identify potentially weak devices before they are used in the field. Several companies now produce 'dynamic burn-in' systems, however, these are very expensive. Consequently, there is a need for the development of low cost bench top high temperature test systems which can be used for high temperature testing of prototype devices to be used in 'harsh environments'.

Key words: High temperature electronics; Dynamic Burn-in, IDDQ test, Memory test.

1.   INTRODUCTION

Over several decades, with the continual improvements in semiconductor processing technology, millions of transistors can now be integrated into a single die; consequently microelectronic circuits and systems have insidiously pervaded many aspects of our daily lives, particularly regarding communications and consumer electronics sectors of the market. More recently, however, there have been advances in another sector of the market place, which although equally important has not had such a high profile, namely 'High Temperature Electronics'. High temperature electronics has been developed to operate in 'harsh environments' where the operating conditions exclude the use of conventional semiconductor technologies as

D. Flandre et al. (eds.), Science and Technology of Semiconductor-On-Insulator Structures and Devices Operating in a Harsh Environment, 241-246.
© 2005 *Kluwer Academic Publishers. Printed in the Netherlands.*

employed, for example, in the manufacture of consumer products. Typical 'harsh environment' applications are related to the aerospace, avionics, automotive and well logging industries, where not only are the operational temperatures too high but also the inability to provide cooling would cause conventional electronic systems to fail. However, with the development in materials, processes, encapsulation techniques etc, semiconductor technologies have evolved, such as Silicon on Insulator (SoI), which can operate at elevated temperatures without the need for any external cooling systems (with subsequent reduction in size, weight and cost) and can be placed, to good advantage, in the 'hot zone' close to the point of application.

With the advances in high temperature electronic technology it is estimated by 2005 that the world market for high temperature electronics will approximate to $1B, with ninety percent of the market being driven by aerospace, automotive and well logging applications. However, to ensure that this technology can be used to good advantage it is essential to guarantee the integrity and reliability of these devices in their applications. A major challenge is to be able to test these devices, economically, under similar operating conditions to those they would experience in the field, that is in operating temperatures of around 300C; due to the criticality of the applications it is also necessary to perform stress testing to identify potentially weak devices before they are used in the field. Several companies now produce 'dynamic burn-in' systems however these are very expensive. Consequently, there is a need for the development of low cost bench top high temperature test systems which can be used for high temperature testing of prototype devices to be used in 'harsh environments'.

2.   HIGH TEMPERATURE TEST SYSTEM

The system implemented features a modular architecture which can configured as a pattern generator, a logic analyser or a data acquisition system.
- the pattern generator – creates the test stimuli for the circuit under test.
- the logic analyser – compares actual and expected responses from the circuit under test.
- the data acquisition system – stores and manipulates the test results.

It can also be used to perform "current signature" measurements via an off-chip current monitor', dynamic burn-in and the analysis of high temperature test runs.

Overall the system comprises a PC, interface unit and a high temperature oven. The interface is a mix of software and hardware. The software aspects of the interface have been implemented in LabVIEW©, which is a graphical programming language developed by National Instruments with built-in functionality for data acquisition, instrument control, measurement analysis, and data presentation. Within LabVIEW 'Virtual Instruments' (VIs) are created, which are customised programmable functions used to realise a range of instrumentation functions rather than employing dedicated hardware. This approach has the major advantage of flexibility and can accommodate changes in requirements with relative ease.

3. APPLICATION OF THE TEST SYSTEM

To demonstrate the capabilities of the test system the test vehicle used was a small SoI memory, which will eventually be incorporated in an automotive application.

It has been accepted for some time that standard logic(voltage) testing of digital systems no longer ensures a good fault coverage of all possible faults which can occur in a circuit and that a combination of current and voltage testing is required. Current testing detects those faults, which do not manifest themselves as gate output stuck-at 1 or 0, for example bridging faults and pin holes in gate oxides.

3.1 Current Testing

Current testing methods are generically referred to as IDDQ testing[1]. IDDQ is defined as the current, which exists in a CMOS circuit after all of the switching activity has ceased, due to a change in input stimuli, and all circuit nodes attained a steady state value. Ideally IDDQ should be zero, however in the presence of certain physical defects a current greater than zero will be drawn from the supply. Consequently by monitoring the supply current the presence of a fault can be detected. The technique has been extremely useful in detecting faults, which would affect the long term reliability of the circuit, eg pin holes in gate oxide. However one of the limiting factors of using IDDQ testing is setting the upper test margin for the allowable 'fault free' current, a value which is process dependent; accommodation of this process dependency can result in 'yield loss' and 'escapes'. To overcome this problem several IDDQ variations have been developed, for example ΔIDDQ[2] and Current Signatures[3], however these techniques require multiple current measurements and some post-processing. The current testing technique adopted in this system was developed by Maxwell and O'Neill[4] which requires neither multiple current measurements nor the setting of absolute test margins to be established in order to determine the upper limit of fault free current. Using this approach a realistic upper threshold value can be obtained which is independent of process variations, inter-die variations, and other increases background leakage current due, for example, to rises in temperature.

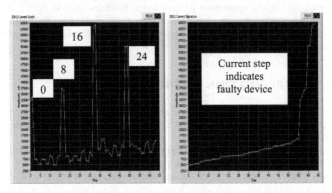

Figure 1. . Current Measurement test Results

3.2   Memory Test

The most efficient method of testing memory devices is to use functional testing techniques which ensure that each memory cell can be read or written to, can store a logic 1 and 0   and also that there is no interaction with surrounding cells. For the purposes of test all possible faults which can occur in a memory are mapped into a small number categories, namely Address Faults (AF), Stuck at Faults (SAF), Transition Faults (TF), Coupling Faults (CF) and Pattern Sensitivity Faults (PSF). As the memory is a regular structure test vectors are generated algorithmically, a considerable number of memory test patterns[5] are available. However, the most widely used algorithm in industry is the March test of which there are several variants. The March Test variant used in this test system is called March C- and is capable of detecting all AF, SAF, TF and CF faults. It is generally considered that tests targeted on PSF faults are very inefficient. Within this test system memory test algorithm used can be readily altered by the simple expediency of changing a LabView VI in the test pattern generation block in the tester interface.

The results obtained from the system when a memory with some defects present was tested is shown in Figures 1 and 2. The IDDQ measurements were made using QD1010 Current Monitor[6] from Q-Star©, the large current spikes in Figure 1 indicating the presence of faults, this is also reflected in the large steps in the 'current signature'. In this instance the results from March Test, shown in Figure 2, also detect the same faults.

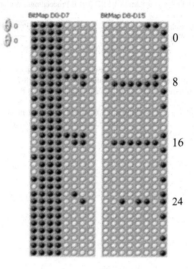

Figure 2. March Test Results

4. CONCLUSIONS

The test scheme outlined above offers a very flexible low cost solution to the challenge of performing functional tests on devices used in high temperature applications. It is also capable of performing 'dynamic burn-in' permitting reliability characterisation to be undertaken in a 'functional test' environment. The use of LabView ensures that the system can be readily modified to suit user needs, for example by modifying the March Test VI, the system can be used for endurance testing of EEPROMs.

ACKNOWLEDGEMENT

This research is part of the EU funded Advanced Techniques for High Temperature System on Chip(ATHIS) Project No. G1RD-CT-2002-00729.

REFERENCES

1. R.Rajsuman,'IDDQ Testing for CMOS VLSI', IEEE Proceedings , Volume 88 Number 4, April 2000, pp 544-566.
2. A.C.Miller, 'IDDQ Testing Deep Submicron Integrated Circuits', Proceedings International Test Conference, 1999, pp 724-729.
3. A.E Gattiker, W.Maly, 'Current Signature Applications', Proceedings International Test Conference, 1998, pp 1168-1177.
4. P.Maxwell, P.O'Neill,' Current Ratio – A Self Scaling Technique for production IDDQ Testing', Proceedings International Test Conference, 1999, pp 738-746.
5. A.K.Sharma, 'Semiconductor memories: Technology, Testing and Reliability', IEEE Press, ISBN 0-7803-1600-4.
6. V.Stopjakova, H.Manhaeve, M.Sidiropulos, ' On –Chip transient Current Monitor for Testing Low Voltage CMOS ICs', Proceedings Design Automation and Test Europe, March 1999, pp 538-542.

CHARACTERIZATION OF CARRIER GENERATION IN THIN-FILM SOI DEVICES BY REVERSE GATED-DIODE TECHNIQUE AND ITS APPLICATION AT HIGH TEMPERATURES

T. E. Rudenko[1], V. I. Kilchytska[1,2], and D. Flandre[2]
[1] Institute of Semiconductor Physics, NASU, Kyiv, Ukraine, [2] *Microelectronics Laboratory, UCL, Louvain-la-Neuve, Belgium*

Abstract: This paper presents a revision of the reverse gated-diode technique for application to thin-film SOI devices. Based on modeling of gate-controlled volume and surface generation components, a reliable approach for extracting generation parameters in thin-film SOI devices from reverse gated-diode measurements is developed and validated for high temperatures. The proposed approach is used for characterizing generation processes and evaluating generation parameters in UNIBOND SOI devices operating at high temperatures (in the temperature range $100 - 300^0$ C).

Key words: Carrier generation lifetime; Surface generation velocity; Thin-film SOI devices; High-temperature operation; UNIBOND.

1.   INTRODUCTION

The carrier generation lifetime and surface generation velocities are two of the most important SOI parameters, being closely related to the performance of both CMOS and bipolar SOI devices. Besides, they can be used to monitor the quality of the SOI material and the device fabrication. As a rule, generation parameters in SOI devices are determined from the transient drain current characteristics of SOI MOSFETs, that requires accurate measurements of the transients and a rather complicated data analysis [1-3]. In this work, we present the reverse gated-diode technique as a simple and reliable measurement tool for extracting generation parameters in fully-depleted (FD) SOI MOS devices. SOI-specific aspects of this

D. Flandre et al. (eds.), Science and Technology of Semiconductor-On-Insulator Structures and Devices Operating in a Harsh Environment, 247-254.

technique are discussed. It is demonstrated that, in contrast to bulk Si devices, in thin-film SOI devices the technique is applicable at high temperatures, allowing the separate characterization of the Si film volume, front and back interfaces.

2.   SIMULATION

For better understanding the generation current behavior in a FD SOI device, we simulated the gate-controlled volume and surface generation components. Simulations were performed using the Shockley-Read-Hall theory on the assumptions of single-level traps with mid-gap energy position and equal capture cross sections for electrons and holes, as well as uniform doping and trap distribution in the device body. For the sake of generality, we used simulations of *reduced* generation components (that is, normalized to their maximum values), thus avoiding the necessity of knowing the concrete values of the carrier lifetime and surfaces generation velocities.

Fig.1 shows the calculated normalized generation current components in the n^+-p-p^+ SOI gated-diode as a function of the front–gate voltage V_{gf} for various negative back-gate voltages V_{gb}. The similar family of curves calculated for positive V_{gb} is not shown here. For front-gate operation, the behavior of the front surface component (Fig.1(a)) is similar to that in a bulk device [4], that is, it arises approximately when $\varphi_{sf} \approx 0$ (where φ_{sf} is the front surface potential), and vanishes at the onset of strong inversion (when $\varphi_{sf} \approx 2\varphi_F + V_R$, with φ_F the Fermi potential and V_R the reverse bias of the p-n junction). However, in a FD SOI device, due to interface coupling rise and fall edges shift linearly along the V_{gf}-axis with V_{gb} when the back interface is depleted and saturate when the back interface is accumulated.

The front-gate voltage dependence of the back surface component for various V_{gb} is shown in Fig.1(b). This dependence can have a rectangle-like shape ($V_{gb} = 0$, -2 V), like the front-surface component in Fig.1(a), or present a single rising edge, followed by a plateau. The back surface-related plateau reaches its maximum value when the back interface is depleted and can be suppressed by biasing the back interface into strong accumulation or inversion ($V_{gb} = $ -20 V).

Fig.1(c) shows the behavior of the generation current in the Si film volume. Under proper conditions ($V_{gb}=0$), the gate voltage dependence of the volume generation component is entirely identical to that of the surface generation in Fig1(a). This is observed when measurement conditions are similar to the double-gate regime. When the front surface is inverted (or accumulated), the volume generation current shows up as a plateau. The height of the plateau strongly changes with the opposite gate voltage and saturates when one of interfaces is in strong accumulation while the other is in strong inversion. Note that the volume generation component sharply decreases with the onset of inversion at the front interface, which is displayed as the fall regions on the reverse current curves. The magnitude of these volume-related fall regions is not constant but strongly depends on V_{gb}. The reason is that in a FD SOI device at the onset of inversion (or accumulation) the free carrier charge is built up not only at the interface but also deeper in the film, resulting in the "switching-off" of generation in the significant part of the Si film. Since in a FD SOI film the carrier distribution depends on both V_{gf} and V_{gb}, the width of the "dead zone" for generation (where n, $p \ll n_i$) also depends on V_{gb}, and at the conditions close to the double-gate regime it occupies the whole film thickness.

The above analysis shows that, for the case of a FD SOI device, the conventional bulk Si description of reverse gated-diode measurements [4] can lead to erroneous results. In contrast to a bulk device, in a FD SOI device

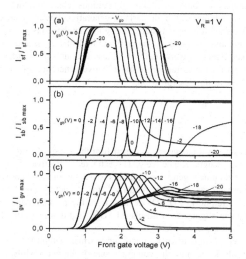

Figure 1. Normalized generation current components in a thin-film n+-p-p+ SOI gated diode calculated as a function of the front gate voltage for various negative back gate voltages for $V_R=1$ V (a) the front surface component; (b) the back surface component; (c) volume component. The Si film, the gate oxide and buried oxide thicknesses are $t_{Si}=85$ nm, $t_{of}=30$ nm, $t_{ob}=380$ nm, respectively. The body doping is $N_A=3 \times 10^{16}$ cm^{-3}, T=300 K.

rise and fall regions on the reverse current curves are not necessarily due to the surface generation at the scanned interface, but they may arise also from the generation in the Si film volume (see Fig.1(c)) or generation at the opposite interface (see Fig.1 (b), V_{gb}=0; -2 V). Furthermore, the current in the saturation range may be associated not only with the volume generation, but it may be due also to generation at the opposite interface (see Fig. 1(b), V_{gb}= -2 ÷ -18 V). Thus the reverse-biased gated diode measurements in a FD SOI device should be interpreted with great care.

3.   PARAMETER EXTRACTION

The performed analysis shows that there are two types of measurement conditions that simplify data interpretation to give straightforward and reliable methods for the extraction of generation parameters. One of them is realized when front- and back-gate biases are adjusted to provide identical surface potentials at both interfaces (conditions close the double-gate regime). Then volume and both surface generation current components exhibit similar gate-voltage dependencies being simultaneously suppressed in accumulation and inversion and a maximum in depletion. In this case, the maximum experimental reverse current I_{Rmax} is the sum of all components and can be used for the extraction of the effective generation lifetime τ_{geff} using the following simple expression:

$$\tau_{geff} = \frac{q \cdot t_{Si} \cdot n_i \cdot W \cdot L}{I_{R\max}} \tag{1}$$

where q is an electron charge, t_{Si} is the Si film thickness, W and L are the device width and length, respectively. The extracted "effective lifetime" τ_{geff} will reflect the dominant generation rate in the device. In fact, here we used the same approach, as proposed recently[5,6] for extracting the "effective" recombination lifetime in FD SOI devices from the recombination current peak measured in slightly forward-biased SOI gated-diodes in the double-gate or in gate-all-around devices.

Another set of measurement conditions, which simplifies parameter extraction involves scanning the voltage at one of the gates whilst keeping the opposite gate in strong accumulation or inversion. In this case, only two components are present: the front surface component (for the front-gate control),. and the volume component which increases gradually with gate voltage (see Fig.1(c), V_{gb}= -20 V). Therefore, the rise edge of the experimental $I_R(V_{gd})$-curve will represent the surface generation current at the front interface and can be used for extracting the surface generation velocity at the front interface. In the same manner the back surface generation velocity can be obtained from $I_R(V_{gb})$-measurements. The current measured in the plateau range, where one interface is in strong inversion and the other is in strong accumulation, can be used for the evaluation of the generation lifetime in the Si film volume. In these conditions the generation current saturates as a function of gate bias. Besides, under the above conditions, the potential varies almost linearly across the film thickness. By assuming the surface potential in strong accumulation to be pinned at $\varphi_s(acc) \approx 0$ and in strong inversion at $\varphi_s(inv) \approx 2\varphi_F + |V_R|$, the width of the active generation layer can be defined as the distance where the potential changes by $|V_R|$. This approximation has been compared to the results of a two-carrier numerical simulation performed for room temperature and 300^0 C. Such a comparison reveals that a better approximation for the generation width, in particular, at high temperatures, is:

$$t_{geff} = t_{Si} \frac{|V_R|}{2\varphi_F + |V_R| + 6kT/q} \tag{2}$$

This reflects the fact that pinning of the surface potential in accumulation and inversion occurs at values differing from 0 and $2\varphi_F + |V_R|$ by a few kT/q. In this regime, the carrier generation lifetime in the volume of the Si film, τ_{gv}, is obtained from the measured plateau current, $I_{plateau}$ as follows:

$$\tau_{gv} = \frac{q \cdot t_{geff} \cdot n_i \cdot W \cdot L}{I_{plateau}} \tag{3}$$

The extracted lifetime value τ_{gv} would reflect the quality of the Si film only, since surface generation at both interfaces is effectively suppressed.

4.   EXPERIMENTAL RESULTS

The method has been applied to study generation processes in devices fabricated on two types of SOI substrates (UNIBOND, ZMR) with distinctly different properties.

Fig.2 shows the experimental room temperature front-gate measurements (a) and back-gate measurements (b) of the reverse current in the p^+-n$^-$- n$^+$ ZMR SOI gated-diode for various biases at the opposite gate. Part of the curves in Fig.2 has sharp rise and fall regions, which could be erroneously assigned to the surface generation. However, the height of these rise and fall regions strongly change with bias at the opposite gate. Furthermore, when the opposite interface is biased in strong accumulation or inversion, the left-side and right-side plateau current saturates and does not vanish. These features are indicative of the volume-dominated generation current behavior. Careful analysis of the observed curves and their comparison with simulations presented in Fig.1(c) show that these characteristics are entirely determined by generation in the volume of the Si film. This is supported by the fact that, in this case, the front-gate voltage and back-gate voltage curves are almost identical. The volume generation lifetime in the ZMR SOI film evaluated from Fig.2 is found to be 3.5×10^{-7} s.

Fig.3 shows similar measurements of a UNIBOND SOI device, clearly illustrating the surface-dominated generation behavior. The rise and fall edges of the curves in Fig.3a measured for large positive or large negative back gate biases are due to carrier generation at the front interface, whereas the difference between right-side and left-side of these curves is due to generation in the Si film volume, which is small compared to the surface contribution. The plateau regions, which are observed only when the back interface is depleted and disappear when the back interface is biased in accumulation or inversion, represent generation at the back interface. The central curve measured under conditions close to the double-gate regime gives the sum of all generation components in the device, in a given case, the sum of two surface components with almost no contribution of generation in the Si film volume. These conclusions are in excellent agreement with back-gate measurements shown in Fig.3 (b). The front and back surface generation velocities extracted from Fig.3 are $s_f = 5.1$ cm/s and $s_b = 1.5$ cm/s, respectively.

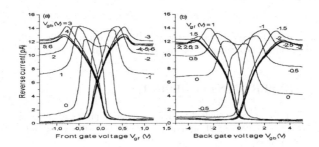

Figure 2. Experimental reverse current measurements in a ZMR p-i-n SOI gated-diode being the example of volume-dominated generation current behavior ($V_R = -1V$; $T = 250^0$ C; $t_{Si} = 300$nm, $t_{of} = 80$ nm, $t_{ob} = 300$ nm, $N_D = 2 \times 10^{15}$ cm^{-3}, W/L=500 μm /20 μm).

An important point is that in thin-film SOI devices the technique is applicable at high temperatures due to strong suppression of the diffusion current. The front-gate measurements in a UNIBOND device at temperatures of 200^0 C, 250^0 C and 300^0 C are shown in Fig.4. Though the background current increases with temperature due to increasing diffusion from the highly doped regions, the front and back gate-controlled components remain clearly pronounced, allowing separate evaluation of the front and back-surface components.

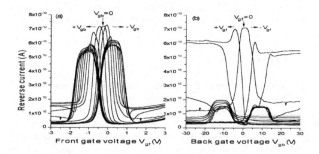

Figure 3. Experimental reverse current measurements in the UNIBOND n+-p-p+ SOI gated-diode illustrating surface-dominated generation current behavior: (a) front-gate measurements for various V_{gb} varied from -20V to + 20 V with step of 2 V; (b) back-gate measurements for various V_{gf} varied from –4 V to +3 V with step of 0.5 V (V_R= 0.5 V; T=150^0 C; t_{Si}=85 nm, t_{of}=30 nm, t_{ob}=380 nm, N_A=3x10^{16} cm^{-3}, W/L=1600/3).

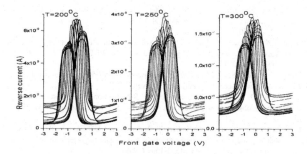

Figure 4. The reverse current measured as a function of V_{gf} at different temperatures in the UNIBOND SOI gated-diode for various V_{gb} varied from –20 V to +20V with the 2V step.

It is interesting to note, that in the UNIBOND devices under study the surface generation velocity at the back interface was found to be 3-4 times lower than that at the front interface. One of the probable explanations for the observed difference may be hydrogen passivation of the interface states at the buried oxide/silicon film interface. An alternative explanation is the reduced average capture cross section of surface states at the buried oxide/silicon film interface, for example, if near-interface oxide traps are involved. In order to decide between the above two hypotheses, independent measurements of interface state densities at both interfaces would be needed.

5.   CONCLUSIONS

The reverse gated-diode technique has been revised for application in thin-film SOI devices. The gate-controlled volume and surface generation components have been modeled and analyzed. Based on the performed analysis, simple methods for the extraction of generation parameters in thin-film SOI devices are proposed. The method has been illustrated using ZMR and UNIBOND SOI devices, being, respectively, the examples of the volume- and surface dominated generation current behavior. Furthermore, the technique has been used to characterize UNIBOND SOI devices at high temperatures (up to 300^0 C).

REFERENCES

1. D. E. Ioannou, S. Cristoloveanu, M. Mukherjee, and B. Mazhari, Characterization of carrier generation in enhancement-mode SOI MOSFET's, *IEEE Electron Device Lett.,* **11(9)**, 409-411 (1990).
2. N. Yasuda, K. Taniguchi, C. Hamaguchi, Y. Yamaguchi, and T. Nishimura, New carrier lifetime measurement method for fully depleted SOI MOSFET's, *IEEE Trans. Electron Devices* **39(5)**, 1197-1201 (1992).
3. S. Cristoloveanu, and S. Li, *Electrical Characterization of Silicon-On-Insulator Materials and Devices* (Kluwer Academic Publishers, Norwell, 1995).
4. A. Grove, and D. Fitzgerald, Surface effect on P-N-junction characteristics of surface space charge regions under nonequlibrium conditions, *Sol. State Electron.* **9(8)**, 783-806 (1966).
5. T. Ernst, S. Cristoloveanu, A. Vandooren, T. Rudenko, and J.-P Colinge, Recombination current modeling and carrier lifetime extraction in dual-gate fully-depleted SOI devices, *IEEE Trans. Electron Devices* **46(4)**, 1503-1509 (1999).
6. T. Rudenko, A. Rudenko., V. Kilchytska, S. Cristoloveanu, T. Ernst, J-P. Colinge, V. Dessard, D. Flandre, Determination of film and surface recombination in thin-film SOI devices using gated-diode technique, *Sol. State Electron.,* **48(3)**, 389-399 (2004).

BACK-GATE INDUCED NOISE OVERSHOOT IN PARTIALLY-DEPLETED SOI MOSFETS

N. Lukyanchikova[1], N. Garbar[1], A. Smolanka[1], E. Simoen[2] and C. Claeys[2, 3]

[1]Institute of Semiconductor Physics, Prospect Nauki 41, 03028 Kiev, Ukraine; [2]IMEC, Kapeldreef 75 B-3001 Leuven, Belgium; [3]also at E.E. Dept. KU Leuven, Belgium

Abstract: The back-gate induced noise overshoot revealed in partially-depleted SOI nMOSFETs and pMOSFETs is described. It is shown that the Lorentzian noise is responsible for that overshoot. The parameters of such Lorentzians are measured for different back-gate and front-gate voltages including the case where the Electron-Valence-Band (EVB) tunneling occurs. A model is proposed that attributes the back-gate induced Lorentzian to the Nyquist noise of the body-source conductance. It is assumed that the accumulating back-gate voltage induces the p^{++}-n^{--} junction near the back interface and its leakage current increases that conductance.

Key words: Lorentzian noise; floating-body effects; linear kink effect; noise model; Silicon-on-Insulator; partially-depleted; accumulating back-gate voltage.

1.   INTRODUCTION

It is known that noise overshoot phenomena are typical for floating-body SOI MOSFETs[1-4]. Those phenomena manifest themselves as a strong increase in the low-frequency noise, whereby the noise spectrum changes typically from $1/f$ noise to a Lorentzian type and are observed at sufficiently high drain voltages[1,2] or, in devices with an ultra-thin gate oxide, at sufficiently high front-gate voltages (so called LKE Lorentzians)[3,4]. In this paper, the noise overshoot of a third type is described. This effect has been revealed in partially-depleted (PD) and fully-depleted SOI MOSFETs and can be called a back-gate induced (BGI) noise overshoot because it is observed at sufficiently high accumulating back-gate voltages.

D. Flandre et al. (eds.), Science and Technology of Semiconductor-On-Insulator Structures and Devices Operating in a Harsh Environment, 255-260.

2.   EXPERIMENTAL

The characteristics of the BGI noise overshoot are evaluated on n- and p-channel PD MOSFETs that have been fabricated in a 0.1 μm SOI process on 200 mm diameter UNIBOND wafers, using a PELOX isolation scheme, a 2.5 nm Nitrided gate Oxide, a 150 nm polysilicon gate and 80 nm nitride spacers. The final film thickness was 100 nm. The buried oxide thickness was 400 nm. A high-dose HALO implantation ($3 \cdot 10^{13}$ cm^{-2}) was applied to control the short-channel effects. The device width was $W=10$ μm. The noise measurements were performed with the front-gate bias V_{GF} varied from weak to strong inversion, and the drain bias V_{DS} kept at 25 mV in absolute value. The back-gate bias V_{GB} was changed from $V_{GB}=0$ V to $|V_{GB}|=49.8$ V where the sign of V_{GB} (negative for nMOSFETs and positive for pMOSFETs) corresponds to accumulation at the back interface.

3.   RESULTS

Figure 1a demonstrates the BGI noise overshoot typical for the pMOSFETs. It is seen that the spectral density of the drain current noise S_I measured at a given frequency f and a given front-gate voltage passes though a maximum as far as $|V_{GB}|$ is increased and the maximum shifts to higher $|V_{GB}|$ with increasing f.

Figure 1b shows the noise spectra measured at $|V_{GF}|=0.69$ V for different $|V_{GB}|$. As is seen, the Lorentzian component enters the spectra at $|V_{GB}| \neq 0$ V. Such a component can be described by $S_I(f)=S_I(0)/[1+(2\pi f \tau)^2]$ where $S_I(0)$ is the plateau amplitude, $\tau=(2\pi f_0)^{-1}$ is the time constant and f_0 is the turnover frequency determined by $S_I(f_0)=[S_I(0)/2]$. It is seen from Fig. 1b that $S_I(0)$ decreases while f_0 increases with increasing $|V_{GB}|$. The latter means that τ decreases with increasing $|V_{BG}|$.

The typical dependences of $S_I(0)$ on τ measured at $|V_{GF}|=$const for pMOSFETs and nMOSFETs are shown in Fig. 2a from which it follows that $S_I(0) \propto \tau$ where τ decreases with increasing $|V_{GB}|$. This explains the observation of the BGI noise overshoot shown in Fig. 1a. It is also seen from Fig. 2a that the value of $[S_I(0)/\tau]$ for nMOSFET is higher than for pMOSFET of the same channel length.

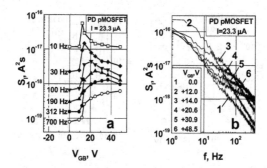

Figure 1. Current noise spectral density vs. back-gate voltage for different frequencies (a) and the noise spectra for different back-gate voltages (b) measured at V_{GF}=-0.69 V and V_{DS}=-25 mV in a 10 μm×0.08μm PD SOI pMOSFET.

Figure 2b shows the dependences of τ on $|V_{GB}|$. It is seen that in the case of nMOSFETs $\tau = \tau_n \propto exp\ (-\beta\ |V_{GB}|)$ where $\beta \approx 0.04$ V^{-1} in the whole range of $|V_{GB}|$ investigated. At the same time, the curve $\tau(V_{GB})$ for pMOSFETs is composed of two sections. The first one corresponds to very rapid decrease of $\tau = \tau_p$ with increasing $|V_{GB}|$ at $|V_{GB}| \leq 18$ V, while the second one corresponds to very slow decrease observed at higher $|V_{GB}|$. It is also seen from Fig. 2b that $\tau_n > \tau_p$ for the same value of $|V_{GB}|$.

The family of the BGI Lorentzians measured at different values of $|V_{GF}|$ and a sufficiently high value of $|V_{GB}|$ is shown in Fig. 3a. The following conclusions can be drawn from that figure: (i) $S_I(0)$ increases with increasing V_{GF} as I^2 in weak inversion and this increase saturates in strong inversion which suggests that $S_I(0) \sim (g_m)^2$ where g_m is the transconductance; (ii) the BGI Lorentzians are characterized by the same value of f_0 at different V_{GF} up to $|V_{GF}|$=1.2 V which means that τ is independent of V_{GF} up to rather high $|V_{GF}|$.

However, if $|V_{GB}|$ is not too high, the BGI Lorentzian is observed only at $|V_{GF}|<1$ V while with increasing $|V_{GF}|$ at $|V_{GF}| \geq 1$ V (that is under conditions of the EVB tunneling) the BGI Lorentzian is transformed into the LKE Lorentzian[3,4] characterized by the higher turnover frequencies and, hence, by the lower values of τ. The higher is the accumulating back-gate voltage, the higher is the value of $|V_{GF}|$ at which this transformation occurs.

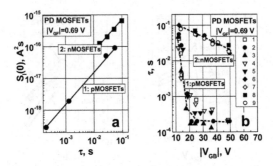

Figure 2. Plateau amplitude vs. time constant for BGI Lorentzians in p- and nMOSFETs of L=0.22 μm (a) and BGI Lorentzian time constant vs. absolute value of the accumulating back-gate voltage for pMOSFETs of L=0.60 μm (1), 0.22 μm (2), 0.13 μm (3) and 0.08 μm (4 and 5) and nMOSFETs of L=0.22 μm (6), 0.20 μm (7), 0.12 μm (8) and 0.08 μm (9) (b).

It has been found that the LKE Lorentzian time constant decreases with increasing $|V_{GB}|$ at a constant drain current (Fig. 3b) and, hence, at V^*=(V_{GF}-V_{th})=const where V_{th} is the threshold voltage. Therefore, that decrease takes place at a constant EVB tunneling current. In addition, a decrease of $[S_I(0)/\tau]_{LKE}$ with increasing $|V_{GB}|$ by a factor of 2 has been observed.

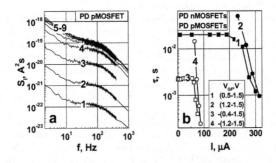

Figure 3. Noise spectra measured in a pMOSFET of 0.6 μm at V_{GB}=−25 V and V_{GF}=−0.38 V (1), −0.43 V (2), −0.50 V (3), −0.6 V (4), −0.69 V (5), −(0.80–1.2) V (6–9) (a) and Lorentzian time constants vs. drain current in the nMOSFET (1,2) and pMOSFET (3,4) of L=0.22 μm measured at V_{GB}=−49.8 V (1), +18.5 V (3) and 0 V (2,4); the last four points on the high-current portion of each curve correspond to $|V_{GF}|$=1.2, 1.3, 1.4 and 1.5 V, respectively (b).

It should be noted that the BGI and LKE Lorentzians show a number of common features. Among them are the higher values of τ and $[S_I(0)/\tau]$ for nMOSFETs than for pMOSFETs. Moreover, the ratio of the value of $[S_I(0)/\tau]_{LKE}$ measured by changing $|V_{GF}|$ at $|V_{GB}|=0$ V to the value of $[S_I(0)/\tau]_{BGI}$ measured in strong inversion by changing $|V_{GB}|$ at $|V_{GF}|$=const appears to be less than 2.5 times for one and the same device.

4. DISCUSSION

The fact that $S_I(0) \propto (g_m)^2$ for the BGI Lorentzians and the similarity with the behaviour of the LKE Lorentzians suggest that the fluctuations of the body-source potential[4] give rise to the BGI Lorentzians. The Nyquist noise of the body-source conductivity G_0 can be the reason for such fluctuations at $|V_{DS}|$=25 mV and in the absence of the EVB tunneling currents (that is practically under equilibrium conditions). At the same time, the equilibrium is disturbed at $|V_{GF}| \geq 1$ V where the forward current flows over the source junction[4]. As a result, the characteristics of the body-source voltage noise can change with increasing $|V_{GF}|$ to such an extent that the BGI Lorentzian is converted into the LKE one.

It can be shown that the following formulae are valid for such a model:

$$S_I(0) = 4kT\tau g_m^2 \beta^2 [(q/kT)I_T + G_0]/C_{eq}[(\gamma+\alpha)I_T + G_0] \qquad (1)$$

$$\tau = C_{eq}/[(\alpha+\gamma)I_T + G_0] \qquad (2)$$

where k is the Boltzmann constant, T is the temperature, $\beta = \partial V_{th}/\partial V_{BS}$ is the body factor, V_{BS} is the body-source potential, q is the electron charge, I_T is the EVB tunneling current, C_{eq} is the equivalent body-source capacity, $\gamma = (17$ V$^{-1})^4$, $\alpha = q/nkT$, n=1-2.

By taking I_T=0 A in Eqs. (1)–(2), one finds for the BGI Lorentzians observed at $|V_{GF}|$<1 V:

$$S_I(0) = 4kT\tau g_m^2 \beta^2 / C_{eq} \qquad (3)$$

$$\tau = C_{eq}/G_0 \qquad (4)$$

In accordance with Eq. (4), the only reason for the decrease of τ with increasing $|V_{GB}|$ is the increase of G_0. The model assumes that the accumulating back-gate voltage induces a $p^{++}\text{-}n^{--}$ junction near the back interface in parallel with the source junction and the excess tunneling current flowing over that $p^{++}\text{-}n^{--}$ junction increases the value of G_0.

The above model explains the other experimental results as follows. The dependences $S_I(0) \propto I^2 \tau$ in weak inversion and $S_I(0) \propto \tau$ in strong inversion are predicted by Eq. (3) where $S_I(0) \propto (g_m)^2 \tau$. It follows from Eq. (2) that BGI Lorentzians can be observed even at $|V_{GF}| \geq 1$ V if $G_0 > (\gamma + \alpha)I_T$. The decrease of $\tau = \tau_{LKE}$ with increasing $|V_{GB}|$ is due to the increase of G_0 in Eq. (2). The higher value of $[S_I(0)/\tau]$ for the LKE Lorentzian can be explained by the higher value of g_m in Eq. (1) due to the linear kink effect[4].

5.   CONCLUSIONS

A low-frequency noise overshoot induced by the accumulating back-gate voltage is found to be typical for the PD SOI MOSFETs. The effect is explained in the framework of a model considering the filtered Nyquist fluctuations of the body-source potential as a source of the excess Lorentzian noise responsible for the noise overshoot. The increase of the contribution of such a Lorentzian component into the noise spectra with increasing $|V_{GB}|$ is attributed to the increase of the body-source conductivity due to the additional tunnel current flowing through the $p^{++}\text{-}n^{--}$ junction which is induced by the accumulating back-gate voltage near the back interface.

REFERENCES

1. W. Jin, P. C. H. Chan, S. K. H. Fung and P. K. Ko, Shot-noise induced excess low-frequency noise in floating-body partially depleted SOI MOSFET's, *IEEE Trans. Electron Devices* **46**(6), 1180-1185 (1999).
2. Y.-C. Tseng, W. M. Huang, M. Mendicino, D. J. Monk, P. J. Welch and J. C. S. Woo, Comprehensive study on low-frequency noise characteristics in surface channel SOI CMOSFETs and device design optimization for RF ICs, *IEEE Trans. Electron Devices* **48**(7), 1428-1434 (2001).
3. F. Dieudonné, J. Jomaah, F. Balestra, Gate-induced floating body effect excess noise in partially depleted SOI MOSFETs, *IEEE Electron Device Letters* **23**, 737-739 (2002).
4. N. B. Lukyanchikova, M. V. Petrichuk, N. P. Garbar, A. Mercha, E. Simoen and C. Claeys, Electron valence-band tunneling-induced Lorentzian noise in deep submicron silicon-on-insulator metal-oxide-semiconductor field-effect transistors, *J. Appl. Phys* **94**(7), 4461-4469 (2003).

SiGe HETEROJUNCTION BIPOLAR TRANSISTORS ON INSULATING SUBSTRATES

S. Hall[1], O. Buiu[1], I.Z. Mitrovic[1], H.A.W. El Mubarek[2], P. Ashburn[2], M. Bain[3], H.S. Gamble[3], Y. Wang[4], P.L.F. Hemment[4] and J. Zhang[5]
[1]Department of Electrical Engineering & Electronics, The University of Liverpool, Brownlow Hill, Liverpool L69 3GJ, UK; [2]School of Electronics & Computer Science, University of Southampton, Southampton SO17 1BJ, UK; [3]The Northern Ireland Semiconductor Research Centre, The Queen's University of Belfast, Belfast BT7 1NN, UK; [4]School of Electronics & Physical Sciences, University of Surrey, Guildford GU2 7XH, UK; [5]Imperial College London, Department of Physics, Blackett Laboratory, Prince Consort Road, London SW7 2BW, UK

Abstract: This paper reviews progress in SiGe HBT technology as well as work on Si bipolar transistors on insulator (SOI) along with current work on SiGe HBTs on SOI. The state-of-the-art results on self-aligned selective epitaxially grown SiGe HBTs and SiGe:C HBTs clearly indicate the extendibility of these technologies into high-speed wired communication applications. Special emphasis is put on Silicon-on-Insulator HBT devices in vertical and lateral design. Research work on SOI SiGe HBT technology by a UK consortium has come up with a number of novel solutions, which are outlined. Moreover, issues regarding SOI operation in harsh environments are discussed.

Key words: SiGe Heterjunction Bipolar Transistors (SiGe HBT); Silicon-on-Insulator (SOI); BiCMOS; Lateral Bipolar Transistors (LBTs); radiation hardness.

1.   THE HIGHLIGHTS OF SiGe HBTs

SiGe HBT technology has been a leading contender among the Si-based approaches for the high frequency applications owing to its advantages over CMOS technology such as higher operation speed, higher transconductance, higher output impedance, lower 1/f noise, better device matching, and better power performance. The first advance – replacing the implanted base with epitaxial boron-doped SiGe – increased the speed performance initially to about 50 GHz and then to 100 GHz[1,2]. More recently, the addition of carbon has advanced speeds first to over 100 GHz[3] and then through optimisation to

D. Flandre et al. (eds.), Science and Technology of Semiconductor-On-Insulator Structures and Devices Operating in a Harsh Environment, 261-272.

200 GHz[4], 350 GHz[5] and to 375 GHz[6]. These rapid improvements in high-speed and high frequency performance are attributed to thinning the base width combined with an ingenious device structure. The improvements in structure reduce device parasitics, enabling higher maximum oscillation frequency (f_{max}) in addition to higher cutoff frequency (f_T). The progression in f_T for IBM[7] and Hitachi[8] SiGe HBTs generations is illustrated in Fig. 1.

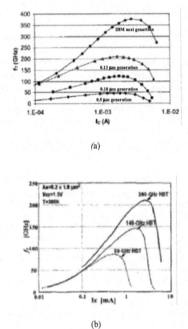

(a)

(b)

Figure 1. f_T vs. I_C as a function of a) IBM[7] and b) Hitachi[8] SiGe HBTs generations.

The speed improvement represented by the cutoff frequency (Fig. 1) demonstrates not only a steady increase of performance but also that the performance of SiGe HBTs is now equivalent or better than III–V compounds as indicated by the recent 375 GHz f_T SiGe HBT result[6]. Konig summarized f_T dependence on base width W_B, and found that f_T was saturated at around 100 GHz in 1998[9], as shown in Fig. 2. In recent years, however, coupled with the development of advanced device structures, a twofold higher f_T-W_B curve can be drawn. The measured f_T values increase with decreasing W_B down to about 10 nm in the self-aligned SEG HBTs, and further increase of f_T up to about 400 GHz by reducing W_B is predicted by analytical calculation[10].

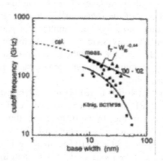

Figure 2. f_T vs. W_B (after Washio[10]).

Largely responsible for the silicon-base HBT advancement are material innovations, the first of which is 10-25% Ge incorporation into the silicon lattice. Another more recent material innovation is the additional incorporation of < 1% C into the SiGe epitaxy[11], which dramatically reduces the base layer boron diffusion through subsequent thermal processing. Fig. 3 shows a simplified band diagram, which illustrates the principal of operation of a graded SiGe profile. The graded Ge profile introduces a quasielectric field in the base (indicated by the sloping conduction band), which greatly minimises the base delay and speeds up the transport of carriers across the

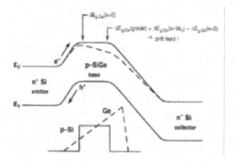

Figure 3. Band diagram of a graded base SiGe HBT.

base region. The collector-base depletion region delay is minimised by increasing the collector doping to decrease the collector-base depletion width and hence decrease the electron transit time, although high collector doping concentrations have the disadvantage of degrading the common emitter breakdown voltage BV_{CEO} of the transistor[11]. A recent study[12] investigates germanium incorporation into polysilicon emitters for gain control in SiGe HBTs. The inclusion of the poly SiGe layer in the emitter provides a trade-off of the very high HBT gain for f_T, without compromising BV_{CEO}. This is because the concept employs emitter engineering rather than base engineering, to tune the gain.

Using selective epitaxial growth (SEG) in the fabrication of HBTs has resulted in impressive RF performance[10]. Selective deposition opens a possibility to minimize parasitics by fabricating self-aligned structures with small overlap. The schematic cross-sections of the state-of-the-art self-aligned SEG SiGe HBTs[6] and SiGe:C HBTs[13] are shown in Fig. 4.

Gate delays of ring oscillators as well as static and dynamic frequency divider performance are a standard benchmark used to qualify and compare high-speed technologies[13]. Gate delays as low as 4.2 ps[4], 3.9 ps[14] and 3.6 ps[13] have been reported recently for SiGe HBTs. As found[10], the trends of ECL and CML gate delay are improving by half every five years, so it can be predicted that gate delay will reach about 3 ps in 2005. The fastest reported SiGe HBT dynamic dividers operate at up to 110 GHz[15] and 100 GHz[16], while static frequency dividers operate at 86.2 GHz[15] and 62 GHz[16]. Based on SiGe BiCMOS technologies all critical nonoptical elements of the high-speed 40-Gb/s system (multiplexer, demultiplexer, amplifiers, oscillators, modulator driver) have been fabricated[17].

SiGe BiCMOS technology has shown to be an extremely effective and flexible technology platform. The process is very mature. Its most obvious advantage over other platforms is the ability to integrate different functions on a single chip. As stated recently[7], from now on, two things could be focused on for developing SiGe BiCMOS technologies. One is increasing f_{max} of a SiGe HBT and the other is realizing full compatibility with standard CMOS platforms[18]. An advanced RF SiGe BiCMOS technology (f_T/f_{max}=73GHz/61GHz) based on a standard 0.18- μm CMOS process has been reported.[19] Innovations in process integration have led to a base-after-gate[20] process flow (for bulk CMOS based BiCMOS) that is widely followed in the industry. The challenges for SiGe BiCMOS process integration will be in adapting to advanced processes and structures required to address the scaling issues in CMOS. This may include changes in the substrate such as silicon-on-insulator (SOI) leading to lucrative SOI/BiCMOS technology.

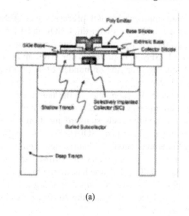

(a)

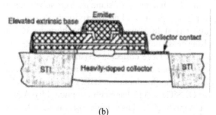

(b)

Figure 4. The state-of-the-art self-aligned SEG SiGe HBTs with elevated extrinsic base and: a) the highest f_T/f_{max}=375 GHz/210 GHz[6]; b) the shortest gate delay of 3.6 ps[13].

2. SOI HBT DEVICES & CONCEPTS

2.1 Vertical design

It is more and more frequently stated that SOI is capable to expand the predictable frontiers of the bulk-silicon technology[21]. It is even speculated that SOI transistors will be the unique survivors of the CMOS world. SOI substrates are becoming established for both bipolar and MOS integrated circuits[22]. The initial driving force was radiation hardness[23,24], but today it is device performance. Different methods of SOI fabrication have been summarized[25]. Two main technologies for producing SOI substrates are: i) Separation by Implantation of Oxygen (SIMOX) and ii) direct wafer bonding. The bonded approach to SOI processes like BESOI (Bond-and-Etch back SOI) used for above micron thick SOI and ELTRAN (Epitaxial Layer TRANsfer facilitated by porous silicon) especially suited for sub-micron SOI are used. BESOI is preferred for HBT application due to the better Si and SiO_2 quality. Variants on bonded SOI for advanced ICs have been reviewed by Gamble[22].

The potential of the SOI technology in the nanometre regime from the device level to the circuit level has been demonstrated[26], and, as well the merits of using bonded SOI versus standard junction isolation have been discussed[27]. The novel 0.2- μm SOI/HRS SEG SiGe HBT/CMOS technology has been demonstrated[28] for fabrication high performance HBTs, CMOS devices, poly-Si resistors, varactors, MIM capacitors and high-Q octagonal spiral inductors. The SOI was based on a high-resistivity substrate (SOI/HRS). Recently, a 60 GHz f_T super self-aligned selectively grown SiGe-base (SSSB)[29] HBT technology for the 20 Gb/s optical transmitter IC's has been reported by NEC[30].

Two of the main disadvantages of standard SOI substrates are that (i) the buried oxide layer has poor thermal conductivity and so, self-heating can be a problem[31-34] and (ii) at high frequencies the buried oxide is electrically transferred resulting in signal transmission losses and cross talk problems[35]. A further disadvantage is large collector resistance. As a viable compromise between high performance and low self-heating, a technological approach that employs a high-energy implantation (HEI) was developed[36].

A review on SOI SiGe HBTs has been conducted by Ashburn et al.[37]. A UK consortium comprising the Universities of Liverpool, Queen's Belfast, Southampton, Surrey, and the Imperial College (with industrial partners) is conducting research into SOI HBT technology and has come up with a number of novel solutions. A schematic diagram of the platform test transistor is shown in Fig. 5. In particular, bonded wafer technology has been developed to allow incorporation of buried silicide layers both above and below the buried oxide. The upper silicide layer produces a low collector resistance and the layer below the oxide has been shown to reduce drastically the coupling of noise from device to substrate. A 20 dB reduction in cross talk was realized with this technology[38]. The self-heating problem of HBTs has been addressed by the incorporation of thermal vias. Thermal via variants incorporating oxide, nitride and undoped poly-Si have been investigated[22].

Bonded SOI wafers for this work comprise 1μm buried oxide layer with the surface Si layer thinned to a nominal thickness of 1.5μm. Deep, poly-Si filled trenches provide isolation through to the buried oxide layer.

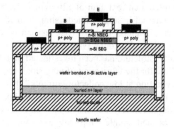

Figure 5. Platform SOI SiGe HBT device architecture (after UK consortium[39]).

The patterned SOI layer is used to provide the heavily doped buried collector as well as the crystalline seed layer for subsequent epitaxial layer growth of the silicon collector and SiGe base layers. The transistor layers are grown using SEG for the Si collector, followed in the same growth step by non-selective epitaxial growth (NSEG) for the SiGe base (nominally 12% Ge) and the n-Si emitter cap[40]. The advantages of this approach are that the basic transistor structure is grown in a single epitaxy step and the growth interface is kept away from the transistor active regions[39,41]. It is worth noting that fully optimized HBT BiCMOS would require thinner buried oxide to obviate short channel effects (SCE)[25] for the CMOS parts. The ability to realize a buried ground plane provides a major advantage in suppressing SCE associated with drain fringing fields for decananometer MOS transistors. The main additional advantage for the SOI HBT technology discussed here is the EMC screening offered by the buried silicide ground plane, which can be integrated with the buried oxide.

2.2 Lateral design

SOI technology is readily adapted for use in LBTs due to its capabilities to overcome the main difficulties namely, (i) definition of a lateral thin base, (ii) definition of a fully aligned base contact, and (iii) device isolation. The formation of a thin base can be achieved by spacer techniques, dopant diffusion, and novel angled ion implantation[42,43]. The spacer technique is the most commonly used[43,44]. Early LBT (npn) devices with very low current gains were fabricated on poor-quality SOI substrates[45,46]. However, higher current gains were achieved for LBT's using a 0.5-μm CMOS process on high-quality laser-recrystallized SOI[47]. Further, the lateral transistor structure on SOI with a base resistance of less than 20 Ω was developed[48]. Double-diffused, lateral npn bipolar transistors that were fabricated using SIMOX

SOI substrates exhibited a current gain of 120, with a peak f_T of 4.5 GHz[49]. The LBTs with an external base polysilicon strap formed vertically on the SOI layer[50,51] have the advantage of a low base resistance, resulting in f_{max} of 22 GHz and f_T of 14 GHz[52]. The other type of lateral BJT's[51,53] exhibited f_{max} of 31 GHz[53]. The latter structure has been investigated thoroughly[54].

LBTs on SOI allow for the fabrication of very small emitter areas leading to high frequency operation at low collector currents[51]. BiCMOS technology on TFSOI (Thin-Film-Silicon-On-Insulator) allows expansion of TFSOI technology into areas previously addressed by bulk BiCMOS applications, such as fast SRAM and single chip solutions for telecommunication products[52]. A novel lateral bipolar transistor fabricated on TFSOI and based on the spacer technique[43] achieved the highest f_{max} of 67 GHz among the SOI bipolar transistors.

As yet, SiGe has not been successfully incorporated into lateral bipolar junction transistors. However, simulations of lateral SiGe HBTs have been reported[42,56]. One of the problems of applying SiGe technology to lateral bipolar transistors is finding a suitable method to incorporate germanium into the base. The epitaxial lateral overgrowth (ELO)[50] and confined lateral selective epitaxial growth (CLSEG)[57] are two possible techniques. The CLSEG is a more difficult method to employ but it is more promising[58]. Vertical growth must be prevented by producing a growth front within some form of cavity in CLSEG. The schematic of lateral SiGe HBTs[59] proposed by Southampton is shown in Fig. 6.

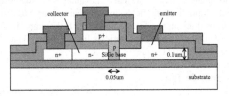

Figure 6. Schematic of proposed lateral SiGe HBT[59] (courtesy of Southampton University).

2.3   SOI & radiation hardness

In the context of operation in harsh environments, SOI has many well-known advantages that have been well documented[60]. A continuous research effort takes place in analysing the functioning and the performance of BiCMOS circuits for various applications, such as: 1) nuclear industry, space applications, high energy nuclear physics experiments[61,62]; 2) integrated microsensor for the automotive industry[63], environment monitoring[64,65]; 3) power devices/high temperature applications[32,66,67] for which various standards have been developed[68]. Degradation of static performances in bipolar transistors exposed to ionizing radiation is generally related to the effects of photon interaction with the passivating oxide. For low dose rate environments a significant enhancement in gain degradation was discovered[69]. This effect – known as the enhanced low dose rate sensitivity (ELDRS) – represents a serious problem for radiation hardness assurance tests. There are various factors affecting the response of the bipolar transistors to a radiation environment such as emitter perimeter to area ratio, the layout of the circuits, process choices. For lateral bipolar transistors on SOI substrates, the geometrical factor (i.e. emitter perimeter to area ratio) has been proved to be important for the case of X-ray and γ irradiations, while having a negligible effect in the case of neutrons environments[23]. The impact of the layout, total dose and dose rate has been demonstrated[61] to play an important role.

3. SUMMARY

In this paper the highlights of SiGe HBT technology as well as different SOI HBT devices and concepts are reviewed. In particular, a process for realising SiGe HBTs on SOI bonded wafer substrates is presented. Further to vertical SiGe HBTs on SOI, the discussion of fabricating a lateral SiGe HBT structure outlined a need for the development of suitable epitaxial technique (CLSG) for the incorporation of germanium into the base. Interestingly, recent simple extrapolation to smaller lithographies and SiGe HBT technologies with higher intrinsic f_T's reveals that a lateral SiGe HBT[70] has the potential to exhibit an f_{max} in the neighborhood of 500 GHz using 0.1 μm lithography at significantly lower bias currents compared to vertical SiGe HBT technology. Such devices would represent a new class of ultralow power high-frequency silicon-based transistors that could have a significant impact on present and future low power integrated RF/microwave communication and sensing systems.

ACKNOWLEDGEMENT

The authors thank the EPSRC for the financial support.

REFERENCES

1. G. Freeman et al., A 0.18 μm 90GHz f_T SiGe HBT BiCMOS, ASIC-compatible, copper interconnect technology for RF and microwave applications, *IEDM 1999 Tech. Dig.*, 569-572 (1999).
2. K. Washio et al., A 0.2-μm Self-Aligned SiGe HBT Featuring 107-GHz f_{max} and 6.7-ps ECL, *IEDM 1999 Tech. Dig.*, 557-560 (1999).
3. A. Joseph et al., A 0.18 μm BiCMOS Technology featuring 120/100 GHz (f_T/f_{max}) HBT and ASIC-compatible CMOS using copper interconnect, *BCTM 2001*, 143-146 (2001).
4. B. Jagannathan et al., A 4.2-ps ECL ring-oscillator in a 285-GHz f_{MAX} SiGe technology, *IEEE Electron Device Letters* 23(9), 541-543 (2002).
5. J.-S. Rieh et al., SiGe HBTs with Cut-off Frequency of 350 GHz, *IEDM 2002 Tech. Dig.*, 771-774 (2002).
6. J.-S. Rieh et al., Performance and design considerations for high speed SiGe HBTs of f_T/f_{max}=375GHz/210GHz, *Intern. Conf. On Indium Phosphide and Related Materials*, 2003, 12-16 May 2003, 374-377 (2003).
7. D.L. Harame et al., The revolution in SiGe: impact on device electronics, *Appl. Surf. Sci.*, 224(1-4), 9-17 (2004).
8. T. Hashimoto et al., Directions to improve SiGe BiCMOS technology featuring 200-GHz SiGe HBT and 80-nm gate CMOS, *IEDM'03 Tech. Dig.* 129-132 (2003).
9. U. Konig, SiGe and GaAs as competitive technologies for RF-applications, *BCTM 1998*, 87-92 (1998).
10. K. Washio, SiGe HBT and BiCMOS technologies, *IEDM'03 Tech. Dig.*, 113-116 (2003).
11. B. Martinet et al., 100GHz SiGe:C HBTs using non-selective base epitaxy, *Proc. ESSDERC 2001*, 97-100 (2001).
12. V.D. Kunz et al., Polycrystalline silicon-germanium emitters for gain control, with application to SiGe HBTs, *IEEE Trans. on Electron Devices* 50(6), 1480-1486 (2003).
13. H. Rucker et al., SiGe:BiCMOS technology with 3.6 ps gate delay, *IEDM'03 Tech. Dig.*, 121-124 (2003).
14. B. Jagannathan et al., 3.9 ps SiGe HBT ECL ring oscillator and transistor design for minimum gate delay, *IEEE Electron Device Letters* 24(5), 324-326 (2003).
15. H. Knapp et al., 86 GHz static and 110 GHz dynamic frequency dividers in SiGe bipolar technology, *2003 IEEE MTT-S Intern. Microwave Symposium Dig.* 2, 1067-1070 (2003).
16. A Rylyakov et al., 100 GHz dynamic frequency divider in SiGe bipolar technology, *Elec. Letters* 39(2), 217-218 (2003).
17. G. Freeman et al., 40-Gb/s circuits built from a 120-GHz f_T SiGe technology, *IEEE Journal of Solid-State Circ.* 37(9), 1106-1114 (2002).
18. D. Knoll et al., HBT before CMOS, a New Modular SiGe BiCMOS Integration Scheme, *IEDM 2001 Tech. Dig.*, 499-502 (2001).
19. F. Sato et al., A 0.18-μm RF SiGe BiCMOS technology with collector-epi-free double-poly self-aligned HBTs, *IEEE Trans. on Electron Dev.* 50(3), 669-675 (2003).
20. S.A. St. Onge, A 0.24 μm SiGe BiCMOS mixed-signal FR production technology featuring a 47 GHz f_T HBT and 0.18 μm LEFF CMOS, *BCTM 1999*, 117–120 (1999).
21. S. Cristoloveanu, Architecture of SOI transistors: what's next?, *Proc. of 2000 IEEE Int. SOI Conf.*, Oct. 2000, 1-2 (2000).
22. H.S. Gamble, in: *Silicon-On-Insulator Technology and Devices X*, edited by S. Cristoloveanu et al. (Electrochemical Society Proceedings Vol. 2001-3), pp. 1-12.
23. O. Flament et al., Radiation Effects on SOI analog devices parameters, *IEEE Trans. On Nucl. Sci.* 41(3), 565-571 (1994).

24. J. R. Schwank, Advantages and limitations of SOI technology in radiation environments, *Microelectronic Engineering* **36**, 335-342 (1997).

25. G.K. Celler and S. Cristoloveanu, Frontiers of SOI, *J. Appl. Phys.* **93**(9), 4965-4978 (2003).

26. J.-O. Plouchart, SOI nano-technology for high-performance system on –chip applications, *Proc. of 2003 IEEE Int. SOI Conf.*, Sept. 29-Oct. 2, 2003, 1-4 (2003).

27. S. Feindt, J. Lapham, J. Steigerwals, Complementary bipolar processes on bonded SOI, *Proceedings of 1997 IEEE International SOI Conference*, 4-6 (1997).

28. K. Washio et al., A 0.2-μm 180-GHz-f_{max} 6.7-ps-ECL SOI/HRS Self-Aligned SEG SiGe HBT/CMOS Technology for Microwave and High-Speed Digital Applications, *IEEE Trans. Elec. Dev.* **49**(2), 271-278 (2002).

29. F. Sato et al., Sub-20 ps ECL circuits with high-performance super self-aligned selectively grown SiGe base (SSSB) bipolar transistors, *IEEE Trans. El. Dev.* **42**(3), 483-488 (1995).

30. F. Sato et al., A 60-GHz f_T super self-aligned selectively grown SiGe-base (SSSB) bipolar transistor with trench isolation fabricated on SOI substrate and its application to 20-Gb/s optical transmitter IC's, *IEEE Trans. on El. Dev.* **46**(7), 1332-1338 (1999).

31. L.J. McDaid et al., Physical origin of negative differential resistance in SOI transistors, *IEE Electronic Letters* **25**(13), 827-828 (1989).

32. J. Olsson, Self-heating effects in SOI bipolar transistors, *Microelectr. Eng.* **56**, 339-352 (2001).

33. P. Palestri et al., Thermal resistance in $Si_{1-x}Ge_x$ HBTs on bulk-Si and SOI substrates, *BCTM 2001*, 98-101 (2001).

34. T. Vanhoucke et al., Revised method for extraction of the thermal resistance applied to bulk and SOI SiGe HBTs, *IEEE Elec. Dev. Lett.* **25**(3), 150-152, (2004).

35. S. Zhang et al., The effects of geometrical scaling on the frequency response and noise performance of SiGe HBTs, *IEEE Trans. on Electron Devices* **49**(3), 429-435 (2002).

36. M. Mastrapasqua, et al., Minimizing Thermal Resistance and Collector-to-Substrate Capacitance in SiGe BiCMOS on SOI, *IEEE TED* **23**(3), 145-147 (2002).

37. P. Ashburn et el. in: *Silicon-On-Insulator Technology and Devices X*, edited by S. Cristoloveanu et al. (Electrochemical Society Proceedings Vol. 2001-3), pp. 433-444.

38. J.S Hamel et al., Substrate crosstalk suppression capability of silicon-on-insulator substrates with buried ground planes (GPSOI), *Microwave and Guided Wave Lett.* **10**(4), 134-135 (2000).

39. S. Hall et al., SiGe HBTs on Bonded Wafer Substrates, *Microelectr. Eng.* **59**, 449-454 (2001).

40. H.A.W. El Mubarek et al., Non-selective growth of SiGe heterojunction bipolar transistor layers at 700°C with dual control of n- and p-type dopant profiles, *Journal of Materials Science: Materials in Electronics* **14**(5-7), 261-265 (2003).

41. J.F.W. Schiz et al., Leakage Current Mechanisms in SiGe HBTs Fabricated Using Selective and Nonselective Epitaxy", *IEEE TED*, **48**(11), 2492-2499 (2001).

42. R. Gomez et al., On the design and fabrication of novel lateral bipolar transistor in a deep-submicron technology, *Microelectronics Journal* **31**, 199-205 (2000).

43. H. Nii et al., A Novel Lateral Bipolar Transistor with 67 GHz f_{max} on Thin-Film SOI for RF Analog Applications, *IEEE Trans. on Electron Devices* **47**(7), 1536-1541 (2000).

44. B. Edholm et al., A self-aligned lateral bipolar transistor realized on SIMOX-material, *IEEE Transactions on Electron Devices* **40**(12), 2359-2360 (1993).

45. M. Rodder et al, SOI bipolar transistors, *IEEE Elec. Dev. Lett.* **4**, 193-195 (1983).

46. T.I. Kamins and D.R. Bradbury, Trench-isolated transistors in lateral CVD epitaxial Silicon-on-Insulator films, *IEEE Electron Device Letters* **5**(11), 449-451 (1984).

47. J.P. Collinge, Half-micrometer-base lateral bipolar transistors made in thin-silicon-on-insulator films, *Electronic Letters* **22**(17), 886-887 (1986).
48. J.C. Sturm et al., A lateral SOI bipolar transistor with a self-aligned base contact, *IEEE Electron Device Letters* **8**(3), 104-106 (1987).
49. S.A. Parke et al., A high-performance lateral bipolar transistor fabricated on SIMOX, *IEEE Electron Device Letters* **14**(1), 33-35 (1993).
50. G.G. Shahidi et al., A novel high-performance lateral bipolar on SOI, *IEDM 1991 Tech. Dig,.* 663-666 (1991).
51. R. Dekker et al., An ultra low power lateral bipolar polysilicon emitter technology on SOI, *IEDM 1993 Tech. Dig.*, 75-78 (1993).
52. J.A. Babcock et al., Low-frequency noise dependence of TFSOI BiCMOS for low power RF mixed-mode applications, *IEDM Tech. Dig. 1996*, 133-136 (1996).
53. T. Shino et al., A 31 GHz f_{max} lateral BJT on SOI using self-aligned external base formation technology, *IEDM Tech. Dig. 1998*, 953-956 (1998).
54. T. Shino et al., Analysis on high-frequency characteristics of SOI lateral BJTs with self-aligned external base for 2-GHz RF applications", *IEEE Trans.Elec. Dev.* **49**(3), 414-421 (2002).
55. W.-L.M.Huang et al., TFSOI complementary BiCMOS technology for low power applications, *IEEE Trans. Elec. Dev.*, **42**(3), 506-512 (1995).
56. Y.T. Tang, *Advance Characterisation and Modeling of SiGe HBT's*, PhD, Southampton University, United Kingdom, 2000.
57. P. J. Schubert and G. W. Neudeck, Confined lateral selective epitaxial growth of silicon for device fabrication, *IEEE Electron Device Letters* **11**, 181-183 (1990).
58. K. Osman, PhD, University of Southampton, United Kingdom, 2003.
59. J. S. Hamel and Y. T. Tang, Numerical simulation and comparison of, vertical and lateral SiGe HBT's for RF/microwave applications, *ESSDERC 2000*, 620-623 (2000).
60. J.P. Colinge, *Silicon-on-Insulator Technology: Materials to VLSI* (Kluwer, 2nd Ed. 1997).
61. M. Manghisoni et al., Gamma Ray response of SOI BJTs for, fast radiation tolerant front-end electronics", *Nucl. Instr. & Meth. In Phys. Res.* **A518**, 477-481 (2004).
62. M. Dentan et al, Industrial transfer and Stabilization of a CMOS-JFET-Bipolar Radiation Hard Analogue - digital SOI technology, *IEEE Trans. Nucl. Sci.* **46**(4), 822-828 (1999).
63. N. Iwamori et al., Mixed process IC on SOI wafer for automotive application, *J. of Soc. Automot. Eng. (JSAE)* Review **22**, 217-219 (2001).
64. T. Fujita and K. Maenaka, Integrated multi–environmental sensing system for the intelligent data carrier", *Sensors and Actuators* **A97-98**, 527–534 (2002).
65. P. Losantos et al., "Magnetic field sensor based on a thin film SOI transistor", *Sensors and Actuators* **A67**, 96-101 (1998).
66. S. Adriaensen and D. Flandre, Analysis of the thin film SOI lateral bipolar transistor and optimisation of its output characteristics for high temperature applications, *Solid State Electronics* **46**, 1339-1343 (2002).
67. J. Boch, et al., "Impact of Mechanical Stress on Total-Dose Effects in Bipolar ICs", *IEEE Trans. Nucl. Sci.* **50**(6), 2335-2340 (2003).
68. D.M. Fleetwood and H.A. Eisen, Total Dose Radiation Hardness Assurance, *IEEE Trans. Nucl. Sci.* **50**(3), 552-564 (2003).
69. E. W. Enlow et al., Response of advanced bipolar processes to ionizing radiation, *IEEE Trans. Nucl. Sci.* **38**, 1342-1351 (1991).
70. J.S. Hamel et al., Technological requirements for a lateral SiGe HBT technology including theoretical performance predictions relative to vertical SiGe HBTs, *IEEE Transactions on Electron Devices* **49**(3), 449-456 (2002).

SILICON-ON-INSULATOR SUBSTRATES WITH BURIED GROUND PLANES (GPSOI)

M.Bain*, S. Stefanos[†], P.Baine*, S.H. Loh*, M. Jin* , J.H.Montgomery*, B.M.Armstrong*, H.S.Gamble*, J.Hamel[†], D.W.McNeill*, M. Kraft[†] and H. Kemhadjian[†]

* School of Electrical & Electronic Engineering, Queen's University Belfast, Belfast BT9 5AH Northern Ireland, United Kingdom,[+] Department of Electrical and Computer Engineering, Universtiy of Waterloo, Canada, [†] School of Electronics & Computer Scienc,. University of Southampton, Southampton SO17 1BJ, United Kingdom.

Abstract: Advanced integrated circuits may employ SOI substrates and incorporate both analogue and digital systems on a single chip. These system-on-chip integrated circuits are susceptible to cross talk noise generated by the digital components. This paper addresses the issue and describes an SOI substrate produced by wafer bonding which incorporates a tungsten silicide ground plane layer. This ground plane layer suppresses the cross talk yielding a20 dB improvement in performance compared with alternative techniques. Double gate MOS capacitor structures have been manufactured on these GPSOI substrates and the overlying silicon layer has been shown to be of high quality, unaffected by the underlying silicide. The buried insulator layer incorporates undoped polysilicon which has been shown to act as a dielectric layer.

Key words: MOS, SOI, ground planes, tungsten silicide, double gate MOSC

1. INTRODUCTION

Highly integrated "system-on-chip" rf/microwave circuits generally require co-integration of high-speed digital and high frequency low noise analog circuits. As the level of integration and operating frequency continue to increase in future portable communication products, cross-talk through the common substrate between the noisy digital circuitry and the highly sensitive analog circuitry becomes more problematic[1]. Silicon-on-insulator (SOI) substrates can potentially offer a solution for co-integration of such components, where the under-lying substrate can be more resistive (e.g. 15 –

273

D. Flandre et al. (eds.), Science and Technology of Semiconductor-On-Insulator Structures and Devices Operating in a Harsh Environment, 273-278.

20 ohm-cm) than employed in bulk CMOS and the active silicon layer above the substrate can be highly conductive. It has been shown[2] that the buried oxide layer in combination with a higher resistivity substrate provides a degree of isolation between adjacent circuits which can be increased with increased separation. Unfortunately, at frequencies above a few hundred MHz, the buried oxide layer in SOI substrates with standard resistivities (e.g. 15 – 20 ohm-cm) becomes transparent to a.c. signals thereby failing to offer improved signal isolation compared to bulk substrates at frequencies above a few hundred MHz[2]. Improvements in cross-talk suppression at higher frequencies can be attained in SOI substrates through the use of diffused capacitive guard rings in the active silicon layer between adjacent circuits and/or the use of more highly resistive substrates (e.g. above 200 ohm-cm)[2]. This work employs a buried WSi_2 layer in the SOI structure to act as a ground plane leading to cross-talk suppression. The structure is also of importance for Ground Plane MOS transistors where Drain Induced Barrier Lowering (DIBL) can be reduced.

2.   EXPERIMENTAL

The basic structure of the GPSOI wafer is shown in Figure 1. The handle wafer was coated with a CVD layer of WSi_x of thickness 200 nm. WF_6 and SiH_4 chemistry was employed in the deposition process to provide a silicon rich layer with x = 2.6[3]. The silicide layer was annealed to provide a sheet resistance of 2 ohms per square. The silicide layer is polycrystalline, relatively rough and unsuitable for wafer bonding. As the tungsten silicide layer cannot easily be polished, it was therefore coated with an undoped LPCVD polycrystalline silicon layer of thickness ~200 nm. The polysilicon

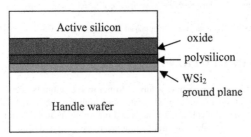

Figure 1. The GPSOI substrate

was then polished to provide a bondable surface. The handle wafer was bonded to the active wafer which was then ground and polished back to the required thickness. The back insulator therefore consists of 100 nm of silicon dioxide and 150 nm of undoped polycrystalline silicon. Substrates produced in this way had silicon layers of thickness 0.6 to 1.5 microns.

The cross-talk transmission structure consists of two signal pads where one becomes the transmitter and the other the receiver in microwave network analyser s21 measurements. The magnitude of the s21 transmission parameter using this method is considered to be a standard technique used by others [1] and provides a basis for meaningful comparison between different technologies from different research laboratories. The signal pads consist of diffused diodes manufactured using standard silicon processing techniques. Some designs include a Faraday cage enclosure consisting of an oxide isolated silicide lined trench feature. An SEM view of a test structure with the Faraday Cage is shown in Figure 2. The silicide lined trench made direct contact to the underlying ground plane layer. The Faraday Cage was held at ground potential either by the ground plane or by an additional top contact. A cross-section view including Faraday cage is shown in Figure 3.

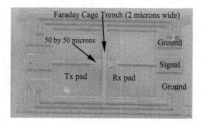

Figure 2. SEM of a crosstalk test structure

The test structures have been manufactured on both standard SOI substrates and GPSOI substrates. Crosstalk measurements between the diffused diode transmitter electrode and the receiver diode are shown in Figure 4 for the frequency range 0.5 – 50 GHz. The standard SOI results provide a reference for comparison of performance. It is clearly observed that the inclusion of the silicide ground plane alone makes significant reduction in the level of crosstalk. A reduction of 30-40dB is observed across the frequency range of measurement. This is because the coupling path through the capacitance of the underlying handle wafer has been eliminated as the energy is shunted to ground. The remaining low levels of crosstalk are by propagation through the thin active silicon layer and buried

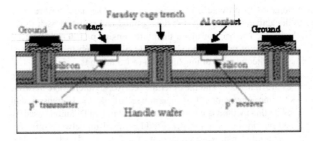

Figure 3. 3 Cross-section of a crosstalk test structure

insulator. The inclusion of the Faraday Cage structure suppresses transmission through the active silicon layer and a further reduction of about 30 dB is observed. Any remaining crosstalk is now through the air and most likely between the measurement probes. This technology provides about 20dB better performance than other crosstalk suppression schemes[4].

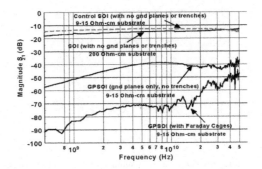

Figure 4. Crosstalk measurement results

The exploitation of this type of substrate in mixed signal integrated circuits requires some demonstration that the silicon quality is as good as that of bulk silicon or standard SOI. It is important to show that the presence of the underlying silicide layer does not lead to stress or defect generation in the buried insulator or the active silicon. In order to quantify the quality of the buried insulator and the active silicon layer, MOS capacitors have been manufactured on GPSOI substrates. The active silicon was patterned into square islands (of side 0.125 mm) and thermally oxidised to give an oxide thickness of 60 nm. The buried insulator was patterned to expose the underlying silicide ground plane. Aluminium was deposited and patterned to form circular electrodes 1mm in diameter. This yielded the double gate MOS capacitor structure shown in Figure 5. The top capacitor electrode area was therefore 7.85×10^{-3} cm^2, while the bottom electrode area was 1.56×10^{-2} cm^2. The active layer was p type silicon of thickness $\sim$ 600 nm. The buried insulator was made up of 100 nm of thermal oxide and 150 nm of undoped polysilicon. Typical high frequency C-V characteristics are shown in Figure 6. The ground plane and substrate were maintained at ground potential at all times and the upper electrode was driven by the swept signal. In figure 6(a) the bias voltage was swept from +10 V to 10 V. In this case the top MOS capacitor was swept from inversion to accumulation, while the back MOS capacitor was simultaneously swept from accumulation to deep depletion. The characteristics therefore do not show values of C_{max} but exhibit a peak at the cross over point of the two C-V characteristics. In Figure 6(b) the bias voltage was swept from -10 V to $+10$ V. In this case the top capacitor was swept from accumulation to deep depletion, while the

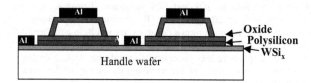

Figure 5. The double gate MOS capacitor structure

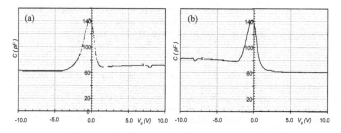

Figure 6. High frequency C-V profiles for double gate MOS capacitors

bottom capacitor was swept from inversion to accumulation. The threshold voltage of the top capacitor is 1.2 V while that of the bottom capacitor was 2 V. At these voltages the silicon layer is not fully depleted so the capacitor has a floating body. The minimum capacitance with one capacitor in inversion has been employed to extract the silicon layer doping. In this case the theoretical values of top and bottom insulator capacitance have been employed to enable extraction of C_{si}. This yields a value for N_a of 2×10^{15} cm^{-3} in good agreement with the active wafer doping level. The bottom insulator which consists of undoped polysilicon and thermal oxide clearly performs as a dielectric stack. The C-V characteristics indicate the manufacture of high quality MOS capacitor structures in this substrate. Further work will examine the pulsed C-V, C-t and Zerbst analysis to characterise the GPSOI substrate in greater detail.

3.   CONCLUSIONS

GPSOI substrates incorporating a buried tungsten silicide layer have been manufactured by wafer bonding.   Crosstalk test structures have been manufactured in both GPSOI and SOI substrates. Crosstalk has been suppressed by 30 – 40 dB by the GPSOI substrate.   The inclusion of a silicide lined trench feature to form a Faraday Cage results in a further 30 dB reduction in crosstalk yielding the lowest reported values to date. Double gate MOS capacitors have been successfully manufactured. High quality C-V characteristics have been achieved verifying the quality of the insulator and silicon layers

REFERENCES

1. K. To, P. Welch, S. Bharatan, H. Lehning, T. Huynh, R. Thomas, D. Monk, W. Huang, and V. Ilderem, "Comprehensive study of substrate noise isolation for mixed signal circuits," *IEEE Electron Device Meeting Digest*, pp519 – 522, December 2001
2. J. Raskin, A. Viviani, D. Flandre, J. Colinge, "Substrate crosstalk reduction using SOI technology," *IEEE Transactions on Electron Devices*, vol. 44, pp. 2252-2261, December 1997.
3. M. F. Bain, B. M. Armstrong & H. S. Gamble, "The deposition and characterization of CVD tungsten silicide for application in microelectronics", *Vacuum* 64(200) 227-232
4. S. Stefanou, J. Hamel, P.Baine, M. Bain, B.M. Armstrong, H. Gamble, M. Kraft and H.A. Kemhadjian "Ultralow Silicon Substrate Noise Crosstalk using Metal Faraday Cages in an SOI Technology" *IEEE Trans. On Electron Devices*, vol. 51, no. 3, pp 486491, Mar 2004.

HIGH-VOLTAGE HIGH-CURRENT DMOS TRANSISTOR COMPATIBLE WITH HIGH-TEMPERATURE THIN-FILM SOI CMOS APPLICATIONS

P. Godignon[1], M. Vellvehi[1], D. Flores[1], J. Millán[1], L. Moreno Hagelsieb[2] and D. Flandre[2]

[1]Centro Nacional de Microelectrónica, CNM-CSIC, Campus UAB, Barcelona, Spain[2] Microelectronics Lab. - DICE, Université catholique de Louvain, Belgium

Abstract: The goal of this work is to explore different ways for co-integrating a power DMOS device in the bulk Si-substrate which underlies the SOI buried oxide and thin Si overlayer, providing optimal performance and isolation of both kinds of devices. A first phase has consisted of the design and fabrication of a power DMOS, defining and optimising 3 mixed DMOS/SOI-CMOS process based on existing power VDMOS and FD SOI CMOS technologies. 150V power VDMOS have been fabricated. One of these process clearly gives better results than the other two proposed technologies. The compatibility of both power VDMOS devices and SOI circuitry is then demonstrated.

Key words: FD-SOI, VDMOS, Power Integrated Circuit

1. INTRODUCTION

A large number of applications may benefit from the development of electronics capable of operation in harsh environments featuring ambient temperatures up to several hundreds degrees Celsius.[1] Among the technologies considered as possible candidates for high-temperature electronics, Silicon-on-Insulator (SOI) is emerging as the most serious contender for operation at temperatures in the range 200-400°C.[2] An interesting example is the Actuator Control Electronics Module targeted at completely distributed control in future All Electrical Aircrafts and operation up to 200°C for the wings and empennage, 315°C for the engines.[3] ACE

D. Flandre et al. (eds.), Science and Technology of Semiconductor-On-Insulator Structures and Devices Operating in a Harsh Environment, 279-284.

modules will require low signal circuitry (high-speed data bus interface, large memories, microprocessor...) as well as power drivers.[4] A few circuits already aim in this direction, based for example on power n-channel LDMOS as in the integrated solenoid driver developed in a 1.2μm partially-depleted (PD) SOI CMOS technology.[5] However, although rated for high voltage, the former LDMOS circuit is not suitable for high currents since the thickness of the Si overlayer is only about 150 nm in this PD SOI process. In addition, when considering fairly complex circuits, the technology choice limits to fully-depleted (FD) SOI CMOS technology whose film thickness has to be reduced below 80 nm, thereby further limiting the direct integration of power devices. The goal of this work is therefore to explore ways for co-integrating a power DMOS device in the bulk Si-substrates which underlies the SOI buried oxide and thin Si overlayer, providing optimal performance and isolation of both kinds of devices. For this purpose, the first phase of development has consisted in the design and fabrication of a power DMOS, defining and optimising a mixed DMOS/SOI-CMOS process (Fig. 1) based on existing power VDMOS and FD SOI CMOS technologies.

2.   TECHNOLOGY MODELLING

The final goal of this work, illustrated in Fig.1, represents a number of technological difficulties to be studied and overcome. First of all, the availability of adequate SOI wafers incorporating the substrate N^+ and epitaxial N^- layers for power DMOS integration. Secondly, a complete etching of Si overlayer and buried oxide to reach the bulk material in the power device region. And finally the definition and adaptation of the power DMOS and SOI CMOS process flows to allow for sharing or ordering fabrication steps. Basically, the FD SOI CMOS process must not be changed

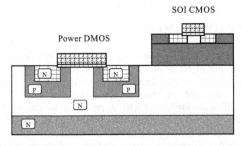

Figure 1. Schematic cross section of the Vertical power MOS together with the SOI circuitry

in order to keep the existing design and technology rules usable for the final circuit design. To integrate a VDMOS with this fixed technology, it is desirable to introduce new process steps before starting the CMOS fabrication rather than in the back-end process. The main process issues have been first identified. First of all, the deep junctions needed for VDMOS are not available within the SOI process and so must be performed in the initial process phase. Secondly, the gate oxide thickness of the power device must be at least twice that of the SOI CMOS circuitry, to withstand voltage peaks. A gate oxide grown in two steps will be required therefore. In addition, the thermal budget for the SOI process will significantly affect the previously implanted layers of the VDMOS process. Finally, for the back end process, the last Al metal layer must be thicker in the power VDMOS than the standard SOI metal layer to enable high current density capability.

This novel concept has been validated and optimised by two-dimensional technological simulations with TSUPREM-4. The electrical characteristics of the resultant 2D structure have then been obtained by simulation with MEDICI. The technological simulations have been dedicated to evaluation of the impact of SOI CMOS process thermal treatments on the deep junctions resulting from the power device process. The CMOS process contains more than 8 thermal treatments made at temperatures in the range of 850°C to 950°C with a total time higher than 6 hours. Consequently, we have reduced the deep P^+ and the P-body annealing times in order to fit the simulated profiles of the standard VDMOS with those obtained after novel VDMOS processing and the additional SOI CMOS thermal treatments. This has resulted in a 10% decrease for the annealing time of these P^+ and P wells with respect to the standard process. The simulated electrical characteristics (on-state I-V curves) of the new device are similar to the characteristics of

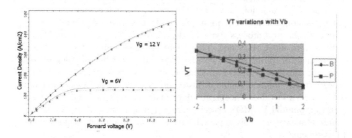

Figure 2. Simulated I-V on state characteristics of the standard VDMOS (cross) and the modified type 3 VDMOS

Figure 3. CMOS calculated threshold voltage V_T versus Vb for the two SOI substrate types (standard B and power device P).

the standard VDMOS. An example is given in Fig. 2 where I-V curves at a gate voltage of 15V are presented for both types of devices.

The impact on the SOI circuitry of using the epitaxial N^-/N^+ wafer required for the substrate of the power device has been also analysed by simulation. Currently, SOI CMOS devices are generally fabricated using a P-type Boron 1e15 cm^{-3} substrate with a buried oxide layer of 400nm. However, for the integration of SOI-CMOS with the power DMOS the CMOS devices will be built in a N-type Phosphorus 1e15 cm^{-3} 15µm thick epilayer over a substrate with very high Phosphorus doping (1e20 cm^{-3}). To facilitate the integration of the two processes, two structures were used; a standard device with p-type SOI film and substrate, and one with a p-film over N+/N- type SOI substrate. The drain currents for the two approaches were compared at different gate bias for increasing drain voltage. We have observed that the difference in current is insignificant; the current being slightly lower for the p-type substrate. On Fig. 3 we can observe the simulated values of threshold voltage V_T, versus applied substrate voltage V_b. A dependence on substrate type is evident with V_T being slightly higher when using the P-type substrate. The transconductance on I_d ratio versus I_d has been also calculated. We have observed that the maximum values corresponding to the sub-threshold slope and the current drive in strong inversion are almost the same in all cases.

3.   DEVICES CHARACTERISATION

A fabrication process has been carried out in order to validate the new device concept and the processes optimised by simulation. Both standard VDMOS and modified novel 150V VDMOS using 3 technological processes optimised by simulation have been fabricated. The variations between the processes concern the formation of the double diffused P-well and N^+ source. In the first process (type 1), the VDMOS N^+ source is completely formed from the CMOS one. This approach results in a non-self-aligned MOS transistor. In the second process (type 2), the double diffusion is done as in the standard VDMOS process, before CMOS formation. The third process (type 3) uses a double diffusion process but the source is formed in two phases: a low dose Phosphorus implantation double diffused with the P-well in the initial VDMOS process, completed with the CMOS N^+ Arsenic source implantation. This approach results in a self-aligned channel with a LDD structure.

When measuring the different fabricated wafers, we have observed a larger than usual dispersion of characteristics for processes type 1 and 2. Specifically, devices from process type 1 have a very high on-resistance. It seems that the channel is not correctly formed and we may have no overlapping of the polysilicon with the N⁺ source (since they are not self aligned). This process is then unlikely to produce a good yield and should be avoided as a possible solution for integration of VDMOS and SOI devices. Devices from process type 2 show the highest on-resistance with large dispersion also for this wafer. The VDMOS characteristics (Fig. 4a) from process 3 are similar to those of the reference VDMOS process, except the threshold voltage is 2.7V compared to 4.5V because of the thinner gate oxide for the former device. The results from this process are the most uniform in terms of device on-resistance. This last process clearly gives better results than the other two proposed technologies. The forward I-V curve shows a rectifying behaviour probably due to a high source contact resistance. This high contact resistance is likely to be due to dopant depletion at the surface which we have previously been observed when using arsenic for the source formation. This problem can be solved by modifying the thermal budget after the source formation.

In the reverse mode, the breakdown voltage is not affected by the type of process. The three processes provide similar results of breakdown voltage. The breakdown voltage is non destructive at 155V (Fig. 4(b)). The termination of the VDMOS is a multi-step field plate optimised for 180V. The resulting termination efficiency is 85%. The leakage currents at 90V are in the range of 30-80nA. The process does not influence the breakdown voltage and the on-resistance varies due to the non-uniformities of the wafers. In Table 1, a summary of the characterisation results is presented.

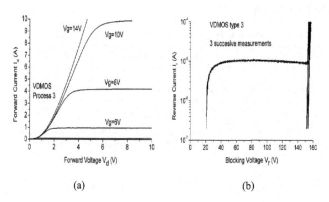

(a)   (b)

Figure 4. Forward (a) and reverse (b) I-V curves of VDMOS fabricated with type 3 process.

Table 1. Summary of electrical results from the different fabricated devices

Process	Breakdown voltage	Threshold voltage	On-resistance
Reference	153 V	4.5 V	514 mΩ - 620 mΩ
Type 1	149 V	-	no conduction
Type 2	154 V	2.6 V	500 mΩ - 880 mΩ
Type 3	154 V	2.6 V	510 mΩ - 720 mΩ

4.   CONCLUSIONS

Compatibility of both power VDMOS and FD SOI circuitry has been demonstrated. The substrate required to form the power device has no significant influence on the CMOS device operation. Three common processes have been defined using 2D simulations. The best solution seems to be process 3 whereby a phosphorus source is implanted during VDMOS process and an arsenic N^+ contact is implanted within the CMOS process.

ACKNOWLEDGEMENTS

This work was funded by the Spanish Ministry of Science and Technology (TIC1999-1222 Project) and the European Union for the Access to Research Infrastructure Action (HPRI-1999-00107 MICROSERV).

REFERENCES

1. Shorthouse G. and Lande S., The global market for high temperature electronics, *Proc. 3rd Int. High Temperature Electronics Conf.* (Albuquerque, USA), pp. I.3-8, (1996).
2. HITEN Network, Semiconductors for high temperature electronics, *HITEN News Int. (Ed. AEA Technology)*, **2**, n° 4, pp. 2-3, July 1997. (see also *http://www.hiten.com*)
3. K. Reinhardt, and M. Marciniak, Wide-bandgap power electronics for the more electric aircraft, *Proc. 3rd Int. High Temperature Electronics Conf. (*USA), pp. I.9-15, (1996).
4. C.M. Carlin and J.K. Ray, The requirements for high temperature electronics in & future high speed civil transport (HSCT), *Proc. 2nd Int. High Temperature Electronics Conf.* (Charlotte, USA), pp. I.19-26, (1994).
5. J.B. McKitterick et al, An SOI smart-power solenoid driver for 300°C operation, *Proc. 3rd Int. High Temperature Electronics Conf.* (Albuquerque, USA), pp. XV.17-22, (1996).

A NOVEL LOW LEAKAGE EEPROM CELL FOR APPLICATION IN AN EXTENDED TEMPERATURE RANGE (-40 °C UP TO 225 °C)

S. G. M. Richter[1], D.Kirsten[1], D. M. Nuernbergk[1] and S. B. Richter[2]

[1]*Institute for Microelectronic and Mechatronic Systems gGmbH, Ilmenau, Germany;* [2]*X-FAB Semiconductor Foundries AG*

Abstract: The increasing demand for high temperature circuits for applications in automotive, aerospace and oil/geothermal drilling industries over the last few years has created a demand for high temperature memory circuits. Silicon-on-insulator technologies, though well suited for the design of high temperature applications, prove to be problematic for the design of EEPROM circuits. SOI specific leakage currents lead to data loss at high temperatures. To successfully design high temperature EEPROM, new memory cell structures have to be developed.

Key words: EEPROM / non-volatile memory / high-temperature electronics / low-leakage high-voltage transistor / Silicon-on-insulator

1.   INTRODUCTION

The increasing demand for high temperature circuits for applications in automotive, aerospace and oil/geothermal drilling industries over the last few years has resulted in an increased demand for high temperature memory circuits. In particular non-volatile memories such as EEPROM are sought-after for reliable data collection over long periods of time and as reference data memories in sensor applications.

To render the development of memories for extended temperature ranges possible, both new process technologies and adapted memory cell concepts have to be employed. Silicon-on-insulator (SOI) technologies are very well suited for the fabrication of circuits for high temperature applications. The feasibility of EEPROM memories in SOI process technologies has already been proven.[1-4] However, increasing leakage currents at high temperatures

D. Flandre et al. (eds.), Science and Technology of Semiconductor-On-Insulator Structures and Devices Operating in a Harsh Environment, 285-290.

impose serious data retention problems, which can be solved with extra circuitry[2] at the cost of an increase in silicon area.

In this paper an alternative approach for a low leakage EEPROM cell with compact layout in a SOI technology is presented. The process technology used for the design is a 1.0 μm partially-depleted SOI technology provided by X-FAB AG.[5] This process features one polysilicon layer and three metal layers with a high temperature metallisation module (tungsten metallisation). It is optimised for the design of high temperature circuit applications, with raised threshold voltages to ensure functionality of transistors above 200 °C. An EEPROM module with the necessary tunnel oxide generation is under development and could be used for the memory cell design.

2.   EEPROM CELL DESIGN IN SOI

An EEPROM memory cell consists of a floating gate transistor (storage transistor) and a select transistor to allow access to specific cells in a memory cell array. Floating gate transistors for single-poly SOI and CMOS technologies have previously been introduced.[1][6] A design similar to the one described by Gogl,[1] using an implant layer as control gate instead of a second polysilicon layer, has been adapted for the technology previously mentioned. Figure 1 shows the schematic and operating conditions of a standard SOI EEPROM cell.

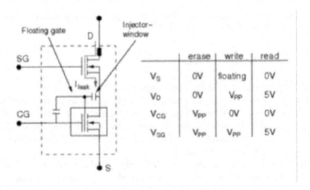

	erase	write	read
V_S	0V	floating	0V
V_D	0V	V_{PP}	5V
V_{CG}	V_{PP}	0V	0V
V_{SG}	V_{PP}	V_{PP}	5V

Figure 1. Schematic and operating conditions of an EEPROM cell

The select transistor must be a high-voltage device because it has to withstand the programming voltage of 16V-18V. To avoid parasitic action due to the floating body effect both storage and select transistor must include body ties in the form of film contacts, attaching the body to the source region.[7,8] For the select transistor this is necessary to ensure stable switching under all operating conditions. In the storage transistor the body tie cancels out the Kink-effect and any history dependency of the device thus allowing for reliable data readout. Although this method is very useful at lower temperatures, it creates new problems at temperatures greater than 200°C, where the increased temperature induced leakage currents lead to a loss of the stored information in the deselected cells.[2] The reason for this data loss is the unwanted programming of the cells caused by drain to body/source leakage. As can be seen in figure 4, a high-voltage transistor with film contacts is permanently open at high temperatures, regardless of the gate voltage. A solution to this problem was presented by Gogl,[1] and is shown in figure 2.

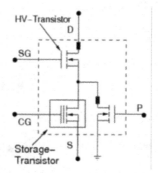

Figure 2. EEPROM cell for use at high temperature

The reach-through from drain to source at the select transistor is compensated by an extra transistor which connects the internal node of the memory cell to ground whenever the cell is deselected. The extra device suppresses unwanted programming very efficiently, but results in increased cell area, particularly as a high voltage transistor has to be used.

3.   LOW LEAKAGE HIGH-VOLTAGE TRANSISTOR

To overcome the problem of increased cell size, a new transistor design was developed. This new device is a bi-directional high-voltage transistor with separate body contact directly below the gate and extension areas on both source and drain (see figure 3). It allows reliable switching in all operating states. Furthermore, the separately connected body can be used to suppress data loss at high temperatures. Therefore, the body will be connected to ground during read and in idle mode, reverse biasing the body-to-source diode of the high voltage transistor. This will tie down the floating drain of deselected cells to a maximum of one diode voltage drop, preventing charge loss from the floating gate and thereby data loss reliably. Layouts and schematics including parasitic elements of a conventional HV-transistor and the new bi-directional device are shown in figure 3. The increase in device area is tolerable, considering that with this new design the applicability of the EEPROM cell at high temperatures could be improved without additional devices.

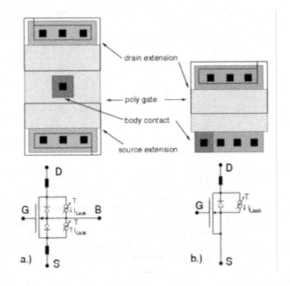

Figure 3. Layout and schematic of a.) bi-directional HV transistor, b.) standard HV transistor

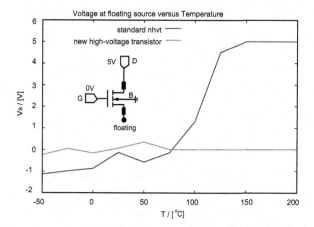

Figure 4. Measurements of drain to source reach-through in high-voltage transistors

Measurement results confirming the efficiency of the new device are presented in figure 4. The measurements on the transistor were made as if it was operating as the select transistor of a deselected memory cell. It can be seen that standard transistors pass the drain voltage regardless of the gate voltage at temperatures above 150°C, while the bi-directional transistor operates reliably at all temperatures.

4.   LOW LEAKAGE EEPROM CELL DESIGN

The previously described bi-directional high-voltage transistor allows the assembly of a two transistor EEPROM cell for high temperature applications (see figure 5). The new leakage paths, shown in figure 5, do not affect the cell at lower temperatures, but actively suppress data loss at 150°C and above. The temperature induced leakage current pulls the sensitive internal node of the cell towards ground, thus serving the same purpose as the additional transistor discussed before.

However, in this design, regarding breakdown, the body now becomes the weakest point of the transistor in all modes of operation.

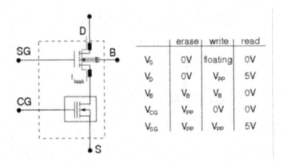

	erase	write	read
V_S	0V	floating	0V
V_D	0V	V_{pp}	5V
V_B	V_B	V_B	0V
V_{CG}	V_{pp}	0V	0V
V_{SG}	V_{pp}	V_{pp}	5V

Figure 5. Two transistor EEPROM cell for application at high temperatures

Therefore, the occurrence of high body-to-source and body-to-drain voltages must be prevented. However, this has to be achieved whilst maintaining the body potential below the source potential to secure transistor operation. A solution to this problem is to either connect the body to VDD=5V during erase and write operations, if the programming voltage does not exceed 17V, or by introducing a new voltage above VDD but considerably less than the programming voltage, thus preventing breakdown.

REFERENCES

1. D. Gogl, G. Burbach, H.-L. Fiedler, M. Verbeck, and C. Zimmermann, A single-poly EEPROM cell in SIMOX technology for high-temperature applications up to 250 °C, *IEEE Electron Device Letters*, vol.18, 541-543 (1997).
2. D. Gogl, H.-L. Fiedler, M. Spitz, and B. Parmentier, A 1-Kbit EEPROM in SIMOX technology for high-temperature applications up to 250 °C, *IEEE Journal of Solid-State Circuits*, vol.35, 1387-1395 (2000).
3. D. Gogl, Untersuchungen zur Realisierung hochtemperaturtauglicher EEPROM-Speicher in SIMOX Technologie, Dissertation (Gerhard-Mercator-Universität, Duisburg, 1997)
4. J.-W. Zahlmann-Nowitzki, Ein Beitrag zur Zuverlässigkeit von EEPROM-Zellen für den erweiterten Betriebstemperaturbereich, Dissertation (Christian-Albrechts-Universität, Kiel, 2001)
5. X-FAB, Erfurt (April 15, 2004); http://www.xfab.com
6. J.-I. Miyamoto, J.-I. Tsujimoto, N. Matsukawa, S. Morita, K. Shinada, H. Nozawa, and T. Iizuka, An experimental 5-V-only 256-Kbit CMOS EEPROM with a high performance single-polysilicon cell, *IEEE Journal of Solid-State Circuits*, vol. 21, 852-859 (1986).
7. J.B. McKitterick, The floating body in SOI, *Proc. 6th Int. Symp. Silicon-on-Insulator Technology and Devices*, Electrochem. Society, 278-289 (1994).
8. J.-P. Colinge, *Silicon-on-Insulator Technology: Materials to VLSI* (Kluwer, Boston, 1991).

DESIGN, FABRICATION AND CHARACTERIZATION OF SOI PIXEL DETECTORS OF IONIZING RADIATION

D.Tomaszewski[1], K.Domański[1], P.Grabiec[1], M.Grodner[1], B.Jaroszewicz[1], T.Klatka[2], A.Kociubiński[1], M.Kozieł[2], W.Kucewicz[2], K.Kucharski[1], S.Kuta[2], J.Marczewski[1], H.Niemiec[2], M.Sapor[2], M.Szeleźniak[2]

[1]*Institute of Electron Technology, Al.Lotnikow 32/46, 02-668 Warszawa, Poland;* [2]*AGH-University of Science and Technology, Al.Mickiewicza 30, 30-059 Krakow, Poland*

Abstract: Development of a novel monolithic active pixel image sensor based on SOI technology is presented. Active pixel test matrices have been recently manufactured and are under extensive examination. This paper describes the concept of the device and shows the most recent results.

Key words: Ionizing radiation; Pixel detectors; SOI; CMOS

1.   INTRODUCTION

In the paper novel approach to the integration of CMOS technology with the a technique to fabricate silicon pixel detectors of ionizing radiation is described. According to this approach, SOI wafers with a relatively thick (1.5 μm) low-resistivity Si-film and a very high resistivity (> 4 kΩcm) 300 μm thick substrate of superior quality are used. In the substrate area the sensor (a matrix of diodes, which are fully depleted under operating conditions) is fabricated. The top Si-film is used for manufacturing of read-out circuitry using the CMOS process. Both subcircuits are DC-coupled via openings in the 1 μm thick BOX (buried oxide) film.

D. Flandre et al. (eds.), Science and Technology of Semiconductor-On-Insulator Structures and Devices Operating in a Harsh Environment, 291-296.
© 2005 *Kluwer Academic Publishers. Printed in the Netherlands.*

2.   MOTIVATION

Protons or light ions (carbon) are used in radiotherapy for irradiation of deep-seated tumors following their contours with millimetric accuracy. For safety reasons the beam should be precisely controlled during the treatment. A system called SLIM (Secondary Emission for Low Interception Monitoring) has been recently developed for on-line control of the irradiation beam[1]. In this system an aluminum foil hit by the particles produces secondary electrons that are collected by an imaging/focusing system. An analog pixel detector is one of the most important components of this system.

There are several techniques for the manufacture of pixel detectors manufacturing. However they are not well-suited for the detection of low-energy electrons, for example CMOS MAPS devices[2] would require an expensive and sophisticated back thinning step, in order to allow particles to reach the sensitive area. On the other hand SOI sensors do not need any backside entrance window since they are fully depleted. Thus, the whole silicon volume between the BOX/substrate interface and the bottom ohmic contact may be efficiently used for detection of ionizing radiation. This facilitates detection of 20 keV electrons which penetrate to a depth of about 3 μm in Silicon. The fabrication of SOI sensors for the SLIM project has been one of the driving forces for reported studies[3,4].

3.   DETECTOR DESIGN

The detector has been arranged as an array of pixels, which are accompanied by read-out cells. The structure of a single cell is illustrated in Fig.1. The read-out electronics of the SOI detector has been designed to detect the charge generated by the ionizing radiation in the range up to 300 MIPs (for 1 MIP=22000 e). A simple 3-transistor circuit has been completed with a diode protecting the input and a transmission gate with MOS transistors of both p- and n-type. This solution, though advantageous, was impossible for CMOS MAPSs. A special read-out sequence has been developed. It is compatible with external correlated double sampling (CDS) processing.

Any pixel may be addressed by row-selection switches located in every cell of the array, and column-selection switches connected with every column of the array.

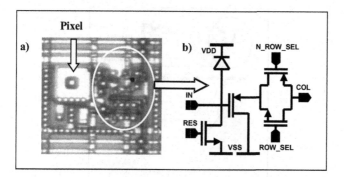

Figure 1. a) Microscopic image of the SOI sensor (a pixel surrounded by metallization can be seen as the biggest element); b) Microelectronic circuit representing read-out cell of the sensor.

4. FABRICATION CHALLENGES

The realization of the above described SOI sensor requires development and validation of a new non-standard technology, which allows creation of the connection between the electronics and the detector layers as well as fabrication of the devices on both sides of the buried oxide. For this reason a dedicated technological sequence that consists of more than 100 individual processes has been defined at the Institute of Electron Technology in Warsaw.

Mutual interferences between the two technological sequences required for the manufacturing of the pixel matrix and the front-end circuitry have been accounted for. The main limitations are as follows:

- the thermal budget of the whole sequence must be minimized; moreover the long-lasting high temperature process steps (e.g. p-well diffusion) must be performed before making openings in the Si-film and BOX layer above the pixel junction areas (see Fig. 2a),
- the process sequence determining the shape of a pixel cavity must be carefully designed (as a combination of the wet and dry etching steps) to assure continuity of the metallization lines sloping down to the pixel contacts (see Fig.2a); the shape of pixel cavity is significantly determined by a mechanism of selective anisotropic etching of silicon,
- backside contact must be shallow enough (because the detector may be used for detection of low energy beta radiation, as in the SLIM system).

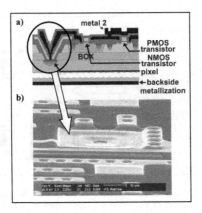

Figure 2. a) Schematic cross-section of the final version of the device (without a passivation layer); b) SEM photograph of the corresponding final pixel area (indicated by an arrow).

It is worthwhile to mention that, after the p-n pixel junctions are created, a poly-Si barrier is deposited as a protection from contamination during the next process steps. Moreover, since the leakage current was not expected to be a major problem, pixels are DC coupled to the front-end electronics, simplifying the process. Figs 1a and 2b show selected results of technological sequence.

5.   CHARACTERIZATION AND MEASUREMENTS

Test elements (MOSFETs, p-n junctions, resistors, capacitors) on a dedicated test structure have been measured using semi-automatic as well as manual measurement setups, depending on accuracy requirements. The parameters, that describe technology have been identified using a dedicated extraction software MOSTXX developed in the IET[5]. Fig.3 shows measured and modelled I-V characteristics of the n-channel 50μm×3μm MOSFET. A set of extracted parameters has been used in the detector simulation and design.

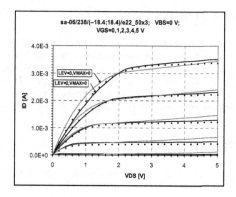

Figure 3. I-V char. of 50x3 nMOSFET; points - meas., solid lines – extracted SPICE models (LEVEL=2,3).

Preliminary measurements of the sensors have been completed. The backside of the sensor has been illuminated with a pulsed infrared laser (λ=850 nm). Each pulse of light is 5 μs wide, injecting a charge equivalent to 2 MIP. A series of four pulses has been generated during a 500 μs long integration time and the output voltage has been tracked. The result, after pedestal subtraction, is shown in Fig.4 and clearly confirms the sensitivity of the pixel matrix to direct charge carrier generation in the high resistivity substrate.

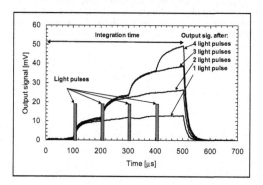

Figure 4. SOI sensor responses for infrared laser pulses (shown on the time scale); integration time is 500 μs.

6.   CONCLUSIONS

The CMOS read-out electronics of a novel pixel image sensor has been adapted to SOI technology and integrated with manufacturing of pixel junctions in a high resistivity substrate of a SOI wafer. Preliminary results obtained from measurements of test matrices are shown. These results demonstrate that this novel detector is an interesting alternative to other silicon monolithic devices like CMOS MAPS or DEPFETs.

ACKNOWLEDGEMENT

Development of the SOI pixel detector is supported by the European Union GROWTH Project G1RD-CT-2001-000561.

REFERENCES

1. L. Badano et al., "SLIM (Secondary Emission for Low Interception Monitoring) - An Innovative Non-destructive Beam Monitor for the Extraction Lines of a Hadrontherapy Centre", Proc. 6th European Workshop on Beam Diag. and Instr. for Particle Accelerators, DIPAC 2003, pp 77-79, Mainz (2003).
2. G. Deptuch et al., "Design and Testing of Monolithic Active Pixel Sensors for Charged Particle Tracking", IEEE Trans. on Nuclear Sc., vol. 49, pp. 601-610, April 2002.
3. M. Amati et al., "Hybrid Active Pixel Sensors and SOI Inspired Option", Nucl. Instr. and Meth. A 511/1-2, pp. 265-270, (2003)
4. J. Marczewski et al., "SOI Active Pixel Detectors of Ionizing Radiation - Technology & Design Development", IEEE Conf.- Nucl. Sc. Symp. 2003, Portland, USA, 19-25.10.2003
5. D.Tomaszewski et al., "A versatile tool for MOSFETs parameters extraction", 6th Symposium D&Y, Warsaw, June 22-25, 2003

POLYSILICON-ON-INSULATOR LAYERS AT CRYOGENIC TEMPERATURES AND HIGH MAGNETIC FIELDS

Anatoly Druzhinin[1], Inna Maryamova[1], Igor Kogut[1], Yuriy Pankov[1], Yuriy Khoverko[1], Tomasz Palewski[2]

[1]*Lviv Polytechnic National University, Scientific-Research Center "Crystal", 1 Kotlyarevski Str., Lviv, 79013, Ukraine;* [2]*International Laboratory of High Magnetic Fields and Low Temperatures, Wroclaw, Poland.*

Abstract: Moderately and heavily boron doped polysilicon-on-insulator layers before and after laser recrystallization were studied at cryogenic temperatures in high magnetic fields up to 14 T. Piezoresistance and magnetoresistance of poly-Si layers with different carrier concentration were investigated. It was shown that laser-recrystallized poly-Si layers could be used to develop piezoresistive sensors, operating at cryogenic temperatures and high magnetic fields.

Key words: polysilicon, laser recrystallization, cryogenic temperatures, magnetic fields, piezoresistance, magnetoresistance, piezoresistive sensors.

1. INTRODUCTION

The study of polysilicon layers conductivity at cryogenic temperatures in high magnetic fields allows to obtain information about the carrier transport mechanism in polysilicon at low temperatures, when significant freezing of carriers is expected. It is interesting to study the effect of strain on the conductivity of polysilicon layers at cryogenic temperatures to predict the possibility of such structures application in mechanical sensors operating in harsh conditions (cryogenic temperatures, high magnetic fields). In our previous work[1] were presented the results of studies of the temperature dependence of the resistance for such poly-Si layers in the temperature range 4.2–300 K. The next step of our investigation is the study of the strain effect on the resistance of polysilicon layers at low temperatures, including cryogenic temperatures, i.e. the study of piezoresistance in polysilicon at low

D. Flandre et al. (eds.), Science and Technology of Semiconductor-On-Insulator Structures and Devices Operating in a Harsh Environment, 297-302.

temperatures, and magnetoresistance studies of polysilicon layers, unstrained and strained, in high magnetic fields at cryogenic temperatures.

2.   EXPERIMENTAL

The measurements were carried out on polysilicon-on-insulator layers, non-recrystallized and laser-recrystallized[2], with different boron concentration: moderately doped samples with carrier concentration $p_{300K}=2.4\times10^{18}$ cm^{-3} before recrystallization (5nr set) and $p_{300K}=4.8\times10^{18}$ cm^{-3} after laser recrystallization (5r set) and heavily doped samples with $p_{300K}=3.9\times10^{19}$ cm^{-3} before recrystallization (6nr set) and $p_{300K}=1.7\times10^{20}$ cm^{-3} after recrystallization (6r set).

Temperature dependence of resistance for strained and unstrained (free) moderately doped laser-recrystallized (LR) polysilicon samples (5r set) in the temperature range 4.2–77 K are presented in Fig. 1a. One can see from Fig. 1a that even small compressive strain ($\varepsilon=-2.1\times10^{-4}$ rel. units at 4.2 K) leads to significant decrease of samples resistance. The temperature dependence of the gauge factor for such layers calculated from the experimental data is shown in the inset of Fig. 1a. Thus for LR poly-Si with carrier concentration $p_{300K}=4.8\times10^{18}$ cm^{-3} the gauge factor at helium temperature has the value GF$_{4.2K}$=325, while at the room temperature its value equals GF$_{300K}$=35.3 that is in good agreement with our previous measurements[3].

In heavily doped LR polysilicon layers (6r set) we also observed the classic piezoresistance at low temperatures (Fig. 1b), but their gauge factor is smaller comparing to moderately doped samples (5r set). The temperature dependence of gauge factor for such samples (6r set) is shown in the inset of Fig. 1b. In this case GF$_{4.2K}$=28 and GF$_{300K}$≈19 and there is a weak dependence of gauge factor vs. temperature.

We studied also poly-Si layers at high magnetic fields up to 14 T at cryogenic temperatures. Fig. 2 demonstrates the transverse magneto-resistance (MR) of non-recrystallized layers with carrier concentration $2,4\times10^{18}$ cm^{-3} (set 5nr) at different temperatures. In such samples, in which the concentration of the electrical active impurities (boron) corresponds to the insulating side of metal-insulator transition (MIT), we observed negative magnetoresistance (NMR) in relatively low magnetic fields. The appearance of NMR could be explained by the peculiarities of carrier transport due to the potential barriers at the grain boundaries in the polycrystalline material. At high magnetic fields MR of these samples becomes positive and achieves the maximal magnitude at 4.2 K (Fig. 2, curve 1).

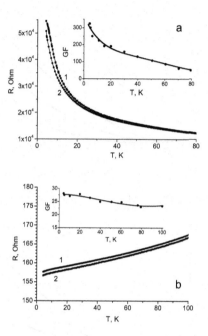

Figure 1. Temperature dependence of resistance for laser-recrystallized poly-Si layers with different p_{300K}: a – 4.8×10^{18} cm^{-3}, b – 1.7×10^{20} cm^{-3} for strained (1) and unstrained (2) samples; insets: gauge factor vs. temperature.

In LR poly-Si, which properties are approaching to the properties of monocrystalline silicon, only positive magnetoresistance (PMR) was observed. The transverse MR of LR moderately doped polysilicon layers (5r set) in the temperature range 4.2–20.2 K is illustrated by Fig. 3. Comparing Fig. 2 and Fig. 3, one could see that for moderately doped polysilicon MR of LR samples (5r set) is greater than for non-recrystallized polysilicon (5nr set).

To describe the MR of slightly doped samples with semiconductor-type conductivity the theory, which explain the PMR due to the deformation of wave functions of localized holes by the magnetic, field is used[4]. But one may notice that for semiconductors with the concentrations of major impurity near MIT there is no adequate description of MR effect.

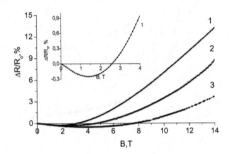

Figure 2. Transverse magnetoresistance of non-recrystallized poly-Si layers with $p_{300K}=2.4\times10^{18}$ cm^{-3} at different temperatures: 1 – 4.2 K, 2 – 6.8 K, 3 – 9.7 K.

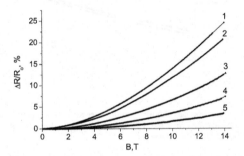

Figure 3. Transverse magnetoresistance of laser-recrystallized poly-Si layers with $p_{300K}=4.8\times10^{18}$ cm^{-3} at different temperatures: 1 – 4.2 K, 2 – 5.3 K, 3 – 6.8 K, 4 – 9.6 K, 5 –20.2 K.

The MR of strained LR samples from 5r set at cryogenic temperatures is shown in Fig. 4. From comparison Fig. 3 and Fig. 4 it is obvious that MR of strained polysilicon is greater than for unstrained polysilicon with $p_{300K}=4.8\times10^{18}$ cm^{-3}. In non-recrystallized poly-Si with $p_{300K}=2.4\times10^{18}$ cm^{-3} the effect of strain on the MR is smaller.

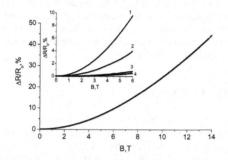

Figure 4. Transverse magnetoresistance for strained laser-recrystallized poly-Si layers with $p_{300K}=4.8\times10^{18}$ cm^{-3} at 4.2 K; inset: at different temperatures: 1 – 4.2 K, 2 – 6.7 K, 3 – 14 K, 4 – 20 K.

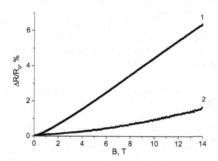

Figure 5. Transverse magnetoresistance of heavily doped poly-Si layers, non-recrystallized (1) and laser-recrystallized (2), at 4.2 K.

Heavily doped poly-Si layers have a small MR compared to the moderately doped layers. After laser recrystallization the magnetoresistance of heavily doped poly-Si (6r set) strongly decreases compared to the magnetoresistance of non-recrystallized samples (Fig. 5). In this case the change of poly-Si resistance is no more than 1 % in magnetic fields up to 14 T (Fig. 5, curve 2).

For heavily doped samples with metallic conductivity the appearance of PMR could be explained by weak localization of holes according to the theory of quantum correlations by Altshuler-Aronov[5]. In heavily doped polysilicon, both for non-recrystallized and LR samples, the effects of strain on the MR are negligibly small.

3.   CONCLUSIONS

From our experimental studies it could be concluded that:
- laser-recrystallized moderately doped poly-Si layers with carrier concentration p_{300K}=4.8×10^{18} cm^{-3} have the great piezoresistance in the temperature range 4.2–77 K and could be recommended for the piezoresistive sensor application in this temperature range;
- laser-recrystallized heavily doped poly-Si layers with p_{300K}>1×10^{20} cm^{-3} could be recommended for piezoresistive sensors operating in the wide temperatures range 4.2–300 K and high magnetic fields up to 14 T, because in this temperature range their gauge factor changes weakly and the resistance change for such strained layers is no more than 1 % in magnetic fields up to 14 T;
- moderately doped non-recrystallized poly-Si layers have NMR in weak magnetic fields at helium temperatures, and it becomes positive in high magnetic fields. In all other investigated layers was observed PMR. The maximal MR was observed 4.2 K.

REFERENCES

1. A. Druzhinin, E. Lavitska, I. Maryamova, Y. Khoverko, Laser recrystallized SOI layers for sensor applications at cryogenic temperatures, in *Progress in SOI Structures and Devices Operating at Extreme Conditions,* edited by F. Balestra et al. (Kluwer Acad. Publ., 2002), pp. 233–237.
2. A. Druzhinin, V. Kostur, I. Kogut et al., Microzone laser recrystallized polysilicon layers on insulator, in *Physical and Technical Problems of SOI Structures and Devices,* edited by J.P. Colinge et al. (Kluwer Acad. Publ., 1995), pp. 101–105.
3. A. Druzhinin, I. Maryamova, E. Lavitska, Y. Pankov, I. Kogut, Laser recrystallized polisilicon layers for sensor application: electrical and piezoresistive characteization, in *Perspectives, Science and Technologies for Novel Silicon on Insulator Devices,* edited by P.L.F. Hemment et al. (Kluwer Acad. Publ. 2000), pp. 127–135.
4. B.I. Shklovskyi, A.Z. Efros, *Electronic Properties of Doped Semiconductors* (Heidelberg, Berlin, 1984).
5. B.L. Altshuler, A.G. Aronov, A.I. Larkin, D.E. Khmelnitsky, On the anomalous magnetoresistance in semiconductors, *J. Exper. Theor. Fiziks (JETF)* **81**(2), 768–783 (1981) (in Russian).

PLANAR PHOTOMAGNETIC EFFECT SOI SENSORS FOR VARIOUS APPLICATIONS WITH LOW DETECTION LIMIT

V. N. Dobrovolsky[1] and V.K. Rossokhaty[2]
[1] Department of Radiophysics, Kiev University, 2 Glushkov Av., Bldg. 5, 03127 Kiev, Ukraine; [2] Department of Electrical and Computer Engineering, Concordia University, 1455 de Maisonneuve Blvd, W., Montreal, Quebec H3G 1M8, Canada

Abstract: Properties of the Planar PhotoMagnetic Effect (PPME) make it promising for the design of sensors with high sensitivity and low detection limit for various applications. We demonstrate the potential for application to magnetic field and chemical sensors. Theory is developed to show that PPME-based magnetic sensors can detect magnetic fields lower than current galvanomagnetic sensors, and the measured fields can approach values comparable to sensors using proton resonance and superconductivity. The applicability of PPME for chemical sensing is also demonstrated. The SOI technology is of especial interest for these sensors. However other semiconductor thin film technologies providing high bipolar photoconductivity and high electron and hole mobilities can also be used for the PPME sensor fabrication.

Key words: sensor, magnetic field, chemical sensing, low detection limit, SOI

1.   INTRODUCTION

The PhotoMagnetic Effect (PME) is a fundamental phenomenon of semiconductor physics. It was discovered in 1934 [1] and has been used in IR radiation detectors[2], and for determining recombination characteristics of semiconductors[3]. The effect is associated with the Lorentz force which diverges electrons and holes in their bipolar diffusion flow, to create an electric field E which is of the same order of magnitude or less then the diffusion field E_{dif}. Usually the field E and emf U of the PM effect are small. The emf is proportional to the sample length b, which is one of the limiting

D. Flandre et al. (eds.), Science and Technology of Semiconductor-On-Insulator Structures and Devices Operating in a Harsh Environment, 303-308.

factors for U (e.g., b must be within the region where the magnetic field is applied). This is not the case for the Planar PhotoMagnetic Effect (PPME)[4] as within an area of small linear sizes one can place a sample with large length b, and thus significantly increase the emf. Besides this property of PPME already mentioned[4], one can realize its other key property. Since $E \leq E_{dif}$, then[5], noise in the film is of the Johnson type, and the noise voltage across the sample terminals $U_{noise} \propto b^{1/2}$. However since $U \propto b$, then the signal-to noise ratio increases as $b^{1/2}$, $U/U_{noise} \propto b^{1/2}$, achieving large values in long samples. The SOI technology is of especial interest for these sensors. However other technologies providing high bipolar photoconductivity and high electron μ_n and hole μ_p mobilities in thin semiconductor films can also be used.

2.   FUNDAMENTAL RELATIONSHIPS

Consider the Si-film of the SOI structure with width $2a$, length b, and thickness c satisfying the relationships $b \gg 2a \gg c$ (Fig.1a). A light-tight shield covers one half of the film along its length. Light with photon energy $h\nu$ and absorption coefficient $\alpha \leq 1/c$ uniformly illuminates the film. Electron-hole pairs are uniformly generated with rate g, within the exposed part of the film and theses diffuse in the shaded part. The magnetic field B causes the electrons and holes to separate, so creating the PPME emf.

Subscripts "0" and "c" are referred to surfaces $z = 0$ and $z = c$, while "1" and "2" designate the illuminated and shaded parts of the silicon film, respectively. For weak magnetic field and $a \gg L_{dif\,1,2}$, theory analogous to that developed in [4] gives

$$U = \frac{cD\,e(\mu_n + \mu_p)Bg\,\tau_{ef\,1}}{R_0^{-1} + R_{ph}^{-1}}, \tag{1}$$

$$\frac{1}{R_{ph}} = \frac{c}{b}e(\mu_n + \mu_p)(a + L_{dif\,2} - L_{dif\,1})g\tau_{ef\,1}, \tag{2}$$

where e is the electron charge, R_0 is the equilibrium resistance and $1/R_{ph}$ is photoconductance of the film, $L_{dif\,1,2} = (D\tau_{ef\,1,2})^{1/2}$, D is the ambipolar diffusivity, $1/\tau_{ef\,1,2} = 1/\tau_b + (s_{01,2} + s_{c1,2})/c$, and τ_b is the carrier lifetime in the bulk of the film.

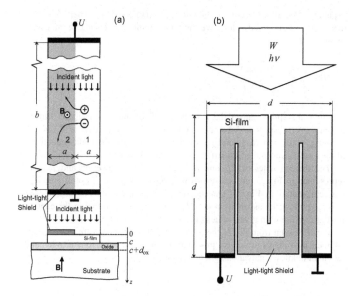

Figure 1. SOI structure of the PPME-based sensor (a), and snake-like shape of the sensor (b).

We will consider the film in a snake-like shape placed in the square with side $d = (2ab)^{1/2}$ (Fig. 1b). The electron-hole pair generation rate g under weak light absorption condition is $g = 2\alpha W/(d^2 h\nu)$, where W is the power of light flux incident upon the square[2].

3.   MAGNETIC FIELD SENSOR

Let the level of photoexcitation be high so that $R_{ph} \ll R_0$ and $D = 2kT\mu_n\mu_p/e(\mu_n+\mu_p)$, where k is the Boltzmann constant and T is the absolute temperature. We assume $\tau_{ef1} = \tau_{ef2} = \tau_{ef}$, $L_{dif1} = L_{dif2} = L_{dif}$, then from Eqs.(1) and (2) the emf and sensitivity of the sensor is as follows: $U = (b/a)DB$, $S = U/B = Dd^2/2a^2$, and

$$R_{ph} = \frac{d^4 h\nu}{4ce\left(\mu_n + \mu_p\right)a^2\alpha W\,\tau_{ef}}. \tag{3}$$

Equating the emf to the noise voltage $U_{noise} = 2(kT R_{ph} \Delta f)^{\frac{1}{2}}$, where Δf is the detection system bandwidth[5], we find the minimum detectable magnetic induction as

$$B_{min} = \frac{a}{L_{ef}} \sqrt{\frac{2hv\Delta f}{ca W \mu_n \mu_p}}. \tag{4}$$

Light absorption causes the temperature T, of the sensor silicon film to increase with respect to substrate temperature T_{sub}, so that $\Delta T = T - T_{sub} = 2\alpha W c d_{ox}/d^2 \chi_{ox}$, where d_{ox} and χ_{ox} are the thickness and thermal conductivity of the oxide layer in the SOI structure respectively (Fig.1a).

The dependencies of R_{ph} and B_{min} on light flux power W and T_{sub} are represented in Fig.2a, and corresponding dependencies of L_{dif} and b on T_{sub} are represented in Fig.2b. Dependency $b(T_{sub})$ arises due to dependency $L_{dif}(T_{sub})$ under the assumption $a = 2L_{dif}$. For values specified in the caption for Fig.2, estimation gives values of $S = 1.25$V/T, and of $\Delta T \le 1.4$ K.

It can be seen from Fig.2a that B_{min} of PPME sensors can approach values attained by sensors based on proton resonance and superconductivity, and less than in known galvanomagnetic sensors [9].

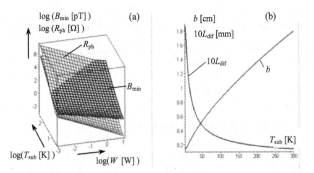

Figure 2. Dependency of minimum detectable magnetic field B_{min} and magnetosensor photoresistance R_{ph} on substrate temperature T_{sub} and incident optical power W calculated using (3) and (4): $a = 2L_{dif}$, $d = 1$ mm, $d_{ox} = 0.1\mu m$, $hv/c\alpha = 1$eV, $\tau_{ef} = 0.1\mu s$ [6] and $\Delta f = 1$Hz (a); and values of silicon film length b and diffusion length L_{dif} as functions of T_{sub} (b). Calculations of electron and hole mobilities take into account electron and hole scattering on phonons and ionized impurities[3] with concentration of 10^{13} cm^{-3}.

A value for B_{min} as low as about 1pT can be achieved with light illumination using the present-day semiconductor high power lasers[7,8] (optical power is units of Watts), at cryogenic temperatures

4.   CHEMICAL SENSORS

At a low level of photoexcitation when $R_{ph} \gg R_0$, the emf of PPME is dependent on the surface recombination rate s_{01}. Dependency of the latter on concentration C of molecules or ions absorbed or created at the surface can be used in chemical sensor design. Generally, the surface recombination rate varies due to change of both surface recombination center concentration and surface potential φ_s. For simplicity, to assess the potential of PPME for chemical sensor applications, we assume that only the second mechanism takes place.

Let φ_s be at the part of abrupt dependency $s(\varphi_s)$ [10] so we can use the approximate expression $s_{01} = s^0 \exp(\pm e|\Delta\varphi_s|/kT)$, where $\Delta\varphi_s$ is the change of the surface potential. Note that s_{01} can take two values depending on the sign of the exponential argument. According to experimental data [11-13], $|\Delta\varphi_s| = \eta \ln(C/C_0)$, where C_0 and η are constants. Using these relationships, after differentiating and substituting differentials with finite differences, we obtain an expression for the sensor sensitivity as

$$\left|\frac{\Delta U}{\Delta C/C}\right| = \frac{eBL_{dif1}^2 d^2 R_0 s_{01}\eta}{2kTac(a + L_{dif2} - L_{dif1})R_{ph}}. \tag{5}$$

A minimum value of the ratio $|\Delta C/C|$, which can be detected at the background noise level, can be found similarly to that for the magnetic field sensors as

$$\left|\frac{\Delta C}{C}\right|_{min} = \frac{4(kT)^{3/2}(\Delta f)^{1/2} ac(a + L_{dif2} - L_{dif1})R_{ph}}{eBL_{dif1}^2 d^2 s_{01}\eta R_0^{1/2}}. \tag{6}$$

The requirement for low level of photoexcitation and weak magnetic field limits values for R_{ph} and B. To satisfy these conditions, we assumed relationships $B = 1/(10\mu_n)$ and $R_0 = R_{ph}/10$ in Eq.(6). Then, for the case $L_{dif1} = L_{dif2} = L_{dif}$ and $a = 2L_{dif}$, and for typical values $T_{sub} = 300\,K$, $d = 1mm$, $c = 0.1\mu m$, $\tau_{ef1} = 0.1\,\mu s$, $s_{01} = 1000$ cm/s, $\alpha = 10^3\,cm^{-1}$, $h\nu = 1.2$ eV, $\Delta f = 1$ Hz, and $\eta = 25$ mV, using Eq.(2), we obtained the relationship $|\Delta C/C|_{min} = 0.5 \cdot 10^{-7} W^{-1/2}$ where the power of light flux W is measured in watts. It is apparent using this expression, that the detection limit for PPME-based chemical sensors can be between 1-2 orders of magnitude lower than that of known sensors [12,13].

5. CONCLUSION

The planar photomagnetic effect holds much promise for design of sensors for various applications where high sensitivity and low detection limit are required. We demonstrated this for magnetic field and chemical sensors. The sensors can be fabricated using SOI technology. They can also utilize methods to create bipolar flow of carriers other than the method considered in the work.

The necessity for a light source complicates PPME sensor design. However, on the other hand, it makes it possible to modulate optical excitation, and thus to apply the lock-in amplification technique for more effective signal detection. The magnetic field sensors can measure extremely low values of field induction. Due to this, they are promising for application in various areas despite some fabrication complexity.

REFERENCES

1. I.K. Kikoin and M.M. Noskov , Phys. Z. Sov. Un., **5**, 586,1934.
2. A. Rogalsky et al., *Infrared Photon Detectors*, SPIE (1995), p. 677.
3. R. Smith, *Semiconductors*, Cambridge University (1978), p. 523.
4. V.N. Dobrovolsky, *Microel. Eng.*, **36**, 133, 1997.
5. A. Rose, *Concepts in Photoconductivity and Allied Problems*, Wiley (1964), p. 188.
6. S. Cristoloveanu and S.S. Li, *Electrical Characterization of Silicon-On-Insulator Materials and Devices*, Kluwer Academic Publishers (1995), p. 379.
7. Model SDL-6380-A, JDS Uniphase Inc.; http://www.jdsu.com.
8. Model L8828-06, L7695-04, Hamamatsu Photonics, K.K.; http://usa.hamamatsu.com/.
9. *Semiconductor sensors*, Edited by S.M. Sze, Wiley (1994), p. 549.
10. D. Stevenson and R. Keys, *Physica*, **20**, 104, 1954.
11. B. W. Gopel, *Sensors and Actuators*, **B 52** , 125, 1998.
12. T. Yoshinobu, M. J. Schöning, R. Otto, et al., *Sensors and Actuators*, **B 95,** 352 (2003).
13. Yu. Ermolenko, T. Yoshinobu, Yu. Mourzina, et al., *Anal. Chim. Acta*, **459**, 1-9 (2002).

THEORETICAL LIMIT FOR THE SiO₂
THICKNESS IN SILICON MOS DEVICES

B. Majkusiak, J. Walczak
Institute of Microelectronics and Optoelectronis, Warsaw University of Technology, Koszykowa 75, 00-662 Warsaw, Poland

Abstract: Fundamental limitations for the minimum thickness of the SiO₂ layer as the gate insulator in silicon MOS devices, resulting from quantum-mechanical outflow of electron wave functions from the semiconductor region are considered.

Key words: Metal-Oxide-Semiconductor, Silicon-On-Insulator, silicon dioxide, tunneling

1.   INTRODUCTION

As shown in Fig. 1, the thickness of the SiO₂ layer as a gate dielectric in the most advanced integrated circuits is expected to be 1.2 nm in 2004 and 0.5 nm in the perspective of 15 years. The trend of thinning the gate dielectric layer with the reduction of the lateral dimensions of the devices, which is written in the scaling rules, results from its beneficial effect on MOS transistor performance. First of all, a thinner gate dielectric allows the number of carriers at the semiconductor surface to be maintained despite the reduced gate voltage. Second, it helps prevent short channel effects by reduction of the field coupling between the drain and the source-end of the channel.

Market reveals of product shipping are always behind the literature and conference publications, reporting achievements of university researchers and industry trials. In 1983 [1], a MOSFET with a 3.5 nm gate oxide was used as a measurement tool to separate the electron and hole tunnel currents. In 1985 [2], a 2.5 nm gate oxide was used in a MOSFET with an apt remark that reduction of the gate oxide thickness is a key factor for obtaining high transistor transconductance due to the direct effect on the higher carrier

D. Flandre et al. (eds.), Science and Technology of Semiconductor-On-Insulator Structures and Devices Operating in a Harsh Environment, 309-320.

density in the channel and indirect effect of a shorter applicable transistor channel. In turn in 1988 [3], operation of a MOSFET with a 2.3 nm gate oxide was reported with a suggestion that scaling of the gate lateral dimensions would help maintain undisturbed operation of transistors with even thinner oxides. These and other presentations have not induced wide comments since, as can be deduced from Fig. 1, the thickness of gate oxides in production ICs has been about one order of magnitude greater and the tunnel leakage current has not bother industry. The interest in the gate tunneling leakage current in MOS transistors erupted when the oxide thickness on production lines fell below 10 nm and was predicted to continue the trend beyond the Fowler-Norheim tunneling limit of $t_{ox} = \chi_c/qF_{br} \approx 3$nm. In 1994 [4], normal operation of an MOSFET with a 1.5 nm gate oxide was reported and in 2000 [5], 0.8 nm thick SiO_2 was used as the MOSFET gate dielectric. Now, as results from different presentations of technologies with a 1.2 nm gate oxide (e.g. [6]), the semiconductor industry is ready to fulfil the oxide thickness requirements of the roadmap predictions for the forthcoming years.

The issue of tunneling through ultrathin SiO_2 has been present in the literature for much more time than in the scale of Fig. 1, but mainly focusing on the physics of the MOS system with an ultrathin insulator known as the MOS tunnel diode and its possible applications. The diamonds in Fig. 1 reflect some exemplary experimental works on MOS tunnel diodes and tunnel devices (e.g. [7-11]). The thinnest experimental SiO_2 layer incorporated in a MOS device, that the authors have found in the literature, is 0.6 nm [11].

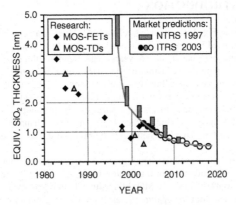

Figure 1. The NTRS/ITRS predictions of the gate oxide thickness in market ICs and examples of laboratory MOS transistors and MOS tunnel devices.

The gate SiO$_2$ thickness cannot be reduced infinitely and new materials with a high dielectric constant K_i are being searched to achieve the required equivalent oxide thickness $t_{eqv} = t_i K_{ox}/K_i$, where $K_{ox} = 3.9$ is the dielectric constant of SiO$_2$. Since the SiO$_2$ thickness in today's ICs is very close to its ultimate limit, many papers have been devoted to a discussion of the theoretical and practical limits for the minimum SiO$_2$ thickness and this is the aim of this paper.

In general, the minimum insulator thickness in a MOS device depends upon limitations of various character:

- physical structure: length of bonds and geometrical order,
- structural imperfections: pinholes, weak spots, surface roughness, thickness nonuniformity,
- reliability issues: hard and soft breakdown, defect generation [12], stress induced leakage current,
- manufacturability: oxide thickness reproducibility, boron penetration,
- device and circuit performance requirements: high gate leakage current[13], gate current induced threshold voltage fluctuations [14], global current and power dissipation[15], gate induced drain leakage[16], mobility degradation[17],
- quantum-mechanical constraints: wave function penetration into the insulator and the gate.

We are especially interested in the limits resulting from the quantum-mechanical physics of systems with an ultrathin barrier, which can be regarded as fundamental ones, independent of structural imperfections, electronic requirements and others. The considerations will be based on the effective mass approach, disregarding the SiO$_2$ amorphous and ultrathin structure. However, since the considerations are directly related to physical parameters of the ultrathin oxide, let us first discuss these issues.

2.   PHYSICAL AND ELECTRONIC STRUCTURE

Although SiO$_2$ as the gate dielectric in silicon technology is amorphous, the atomic structure of its basic block, i.e., the SiO$_4$ tetrahedron is well fixed - the average length of bonds between the Si and O atoms is equal to $d_{Si-O} = 0.161$ nm and the angle of the O-Si-O bonds is 109 degrees. Long range disorder results from the nearly free angle, i.e., varying from about 110 to 180 degrees, between the Si-O bonds belonging to the neighbouring tetrahedra, touching each other at their common corner oxygen atom. If the value of this angle is set up to 144 degrees as in α-quartz crystal, the average distance between the nearest neighbour silicon atoms is equal to $d_{Si-O-Si} \approx 0.31$ nm and this value can be regarded as the rough evaluation of the SiO$_2$

monolayer thickness. It has been found that the growth of an ultrathin SiO_2 in the thermal oxidation process proceeds in terraces [18] and the SiO_2/Si interface is abrupt in the atomic scale, changing periodically with progress of oxidation [19]. It is possible to build a model of an ideal SiO_2/Si interface with an abrupt transition and without excessive distortions of the bonds [20,21]. If acceptable distortion of the interface Si-O bonds is taken into account, the thickness of the thinnest possible oxide layer can be roughly estimated to 0.35-0.4 nm [21]. For thicker oxides, a variety of experiments indicate the existence of an intermediate SiO_x layer and the oxide nanoroughness dependent upon the silicon surface cleaning and oxide growth conditions.

The experimentally observed energy gap of ultrathin SiO_2 is basically the same as for the bulk oxide [22], i.e., about 8.95 eV although E_g apparently increases with decreasing oxide thickness from 3.2 to 1.0 nm [23]. Theoretical simulations [24] suggest that the energy band gap changes from the value for silicon to the bulk SiO_2 value over the distance between 0.1 nm and 0.4 nm from the structural interface, i.e., within the first SiO_2 monolayer. Also simulations in [22] point out that even for the thinnest oxides of about 0.4 nm, equal to two distorted Si-O bonds, the tunnel transmission of electrons can be described within a bulk band structure picture. This is consistent with analysis of Fowler-Nordheim current oscillations, according to which variation of the potential energy near the Si/SiO_2 interface is abrupt to within a few angstroms [25]. On the other hand, experimental data [26] points out that below 0.7 nm for ideal SiO_2 oxide an overlap of the electron wave functions of the silicon conduction band may effectively reduce the tunneling barrier. Also [23] suggests possible penetration of electronic states from the Si substrate into the SiO_2 up to 0.6 nm from the interface.

The electron effective mass $m_{ox} = 0.5\ m_0$ has been proposed [27] for the parabolic dispersion relation in SiO_2, as consistent with measurements of Fowler-Nordheim tunneling current from surfaces of different materials. Nevertheless, a variety of other experimental values have been proposed. Considerations in [28] predict an increase of the electron effective mass with reducing thickness in the ultrathin range. Such a behavior has been concluded also from experiments [29,8]. If this is true, the oxide thickness obtained from fitting the theoretical current to the measured one with the assumption of a constant bulk oxide effective mass would give a thickness overestimation for the thinnest oxides. Such an apparent increase of the electrical thickness was observed in [11] for oxides thinner than 1.2 nm and a difference between the electrical thickness obtained from *I-V* and *C-V* data and physical thickness obtained from non-electrical methods (ellipsometry, XPS, TEM) was equal to 0.32 nm for the oxide of 0.6 nm physical thickness.

3. QUANTUM-MECHANICAL CONSTRAINTS

The MOS structure constitutes a quantum-mechanical system - electrons in the semiconductor and in the gate are separated by a potential barrier set by the SiO$_2$ energy gap. Electrons are represented by wave functions fulfilling Schrödinger and Poison equations. This set of equations must be completed by boundary conditions for the wave functions and the potential energy. Frequently, the simplified boundary conditions are applied, for which the wave functions vanish at the semiconductor surfaces. Such a requirement is equivalent to the assumption of an infinitely high potential wall at the Si/SiO$_2$ interface (hard potential wall). Consequences of this assumption do not impose a limit for the minimum oxide thickness, but their impact on results of simulations of MOS/SOI devices with ultrathin oxides and ultrathin semiconductor layers should be considered. Due to the finite height and finite thickness of the potential barrier the electron wave functions outflow from the semiconductor to the gate. This process known as tunneling can impose performance limits upon the minimum oxide thickness if the tunnel leakage is too high. Further reduction of the oxide thickness will lead to the ultimate limit, when the washout of the probability distribution results in an indetermination of the electron localization in the structure.

3.1 Finite height of the oxide barrier

If the potential barrier at the Si/SiO$_2$ interface were infinite, as is frequently assumed to simplify computations, the wave functions would have to vanish at the silicon interface. Such a simplification means that the probability does not penetrate the oxide region. The barrier height for electrons $\chi_c \approx 3$ eV, seems to be high, especially in comparison with barriers typical for A^{III}B^V heterojunctions. However, the stationary energy levels in the semiconductor well under the assumption of hard potential walls, are significantly overestimated, as shown in Fig. 2 for simple theoretical considerations of a rectangular potential well of thickness t_s limited by abrupt potential barriers of 3 eV height. This overestimation is especially large for a lower effective mass energy ladder $m_s = m_t = 0.19m_0$ and very narrow potential wells. Thus, one might expect a significant influence of the wave function penetration of the oxide on the electron concentration and the drain current in the double gate SOI transistor with a silicon layer thinner than about 5 nm.

The assumption of a hard Si/SiO$_2$ potential wall is still used in many theoretical simulations and its consequences have not been recognized completely. Therefore, let us investigate the effect upon the magnitudes of

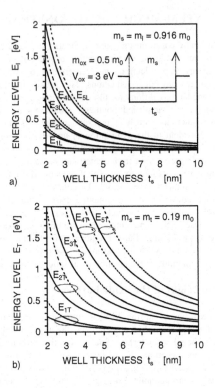

a)

b)

Figure 2. Five lowest stationary levels in the rectangular well with the finite potential walls 3eV (solid) and infinite walls (dotted) for two effective masses m_s in the well region.

an essential influence on the drain current in the double gate (DG) SOI transistor.

Figure 3 shows the dependence of the surface density of electrons confined in the silicon region of thickness $t_s = 5$ nm in the double gate SOI structure upon the oxide thickness and the gate voltage for two theoretical cases: for the infinite potential barrier and for the finite barrier at the Si/SiO₂ surfaces. The latter assumption results in an increase in the electron density due to the decrease of the energy levels and an increase of their occupation probability, but the effect is not as large as expected in the light of conclusions from Fig. 2, and vanishes for thick oxides.

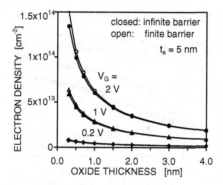

Figure 3. The electron surface density for two models of the Si/SiO₂ barrier.

These effects can be explained as follows. If the correct shape of the potential energy distribution within the oxide region is assumed in calculations of the allowed energy levels, they are lowered in comparison to the infinite barrier case. However, their higher occupation probability changes the potential distribution; the voltage drop across the oxide is increased and the band bending in the semiconductor is lowered. As a result, the semiconductor bands in the final self-consistent state go upwards, partially compensating the increase of the electron density. This compensation is weaker for thinner oxides due to the lower voltage drop across them.

However, changes of energy distances between the allowed energy levels should have a significant influence on the scattering probability of electrons, especially if the semiconductor is thinner than 5 nm and the oxide is ultrathin. Figure 4 compares the phonon limited electron mobility resulting from the relaxation time approximation for the two considered cases of the oxide barrier. Due to the smaller distances between energy levels for the finite barrier case the scattering probability increases and the electron mobility decreases, especially for low gate voltages.

3.2   Tunnel leakage current

If the quantum system is open, the electron wave functions outflow from the semiconductor region to the gate by tunneling through the oxide. The tunnel leakage current can been regarded as a limitation for the minimum oxide thickness in classical MOS devices due to the following effects:

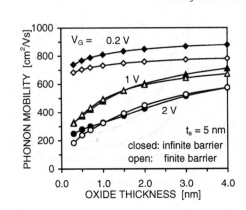

Figure 4. Phonon limited electron effective mobility in the DG SOI channel vs. the oxide thickness and the gate voltage for two models of the Si/SiO₂ barrier.

1. input current of the MOS transistor too high for some circuit applications,
2. total current budget and power consumption in the circuit chip too high,
3. gate current induced potential drop across the gate resistance weakens the transistor transconductance and changes the threshold voltage,
4. tunneling of electrons from the gate electrode to the drain-overlapped region disturbs the MOS transistor subthreshold characteristics,
5. outflow of electrons from the transistor channel cuts off the drain current.

   The first four constraints can be suppressed by appropriate circuit, system or technology solutions. The last constrain concerns the essential principle of the MOS transistor operation and it has been regarded as a serious limitation, until experimental work [30] proved that the effect was not so significant. Theoretical simulations in [31] were concluded with a prediction that the minimum gate oxide thickness would result from manufacturability issues rather than from the flow of too large gate tunnel current.

   Figure 5 shows the V_{GS} dependencies of the gate tunnel current and the drain current for a MOS transistor with an oxide thickness of 1.85 nm and a channel length of 20μm [32]. The gate current is dominated by tunneling of electrons from the channel to the gate (current I_{gc}) for large V_{GS} voltages or by tunneling of electrons from the gate electrode in the overlap region to the drain (current I_{gd}) for low gate and large drain potentials. The fit to the experimental data was obtained with the use of the model for the gate tunnel current [31] with the tunneling probability based on the WKB approximation[33]:

$$P = \frac{16\left(k_g/m_g\right)\left(\kappa_{ox,1}/m_{ox}\right)\left(\kappa_{ox,2}/m_{ox}\right)\left(k_s/m_s\right)}{\left[\left(k_g/m_g\right)^2+\left(\kappa_{ox,1}/m_{ox}\right)^2\right]\left[\left(\kappa_{ox,2}/m_{ox}\right)^2+\left(k_s/m_s\right)^2\right]}\exp\left[-2\int_{x_1}^{x_2}\kappa(E_x,x)dx\right] \quad (1)$$

and the Franz-type dispersion relation [34] for the imaginary wave vector κ

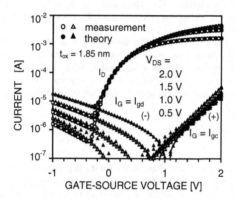

Figure 5. The gate and the drain currents upon the gate-source voltage for the experimental MOSFET: t_{ox} = 1.85 nm, L = 20μm[32].

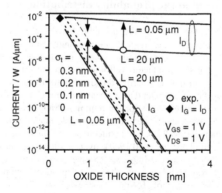

Figure 6. Electron channel-to-gate tunneling current and the drain current vs. the gate oxide.

The model is valid under the assumption that the electron tunnel current is much less than the channel current. A violation of this condition means that the tunnel outflow of electrons from the channel cuts off the drain current. Figure 6 extrapolates the experimental gate and drain currents with shortening of the channel length and thinning of the oxide up to the model failure point (diamonds). The ratio I_G/I_D at a given V_{DS} voltage is proportional to $1/L^2$. Thus, the intercept point for $L = 0.05$ μm is below $t_{ox} = 0.4$ nm.

However, it is difficult to predict the real behavior of the currents in this thickness range. Apart from possible changes of the potential barrier height and the effective mass, the tunnel current is very sensitive to the oxide thickness nonuniformity, increasing with an increase of the thickness standard deviation σ_t [35] according to the function stronger than exponential:

$$I_t = I_t(t_{av})\exp\left(-\frac{\beta^2\sigma_t^2}{2}\right), \qquad \beta(V_{ox}, t_{ox}) = \frac{2}{qV_{ox}} \int_{\chi_{cl}-E_{max}-V_{ox}}^{\chi_{cl}-E_{max}} \kappa(\phi)d\phi \qquad (2)$$

where β is the approximate slope of the $\ln I_t$-t_{ox} dependence, determined for the energy level E_{max} of the highest contribution to the tunnel current.

On the other hand, the oxide thickness nonuniformity is related to the roughness of the oxide surfaces, which in turn affect the electron mobility as the surface roughness [36] or remote roughness scattering mechanisms [37], resulting in the mobility degradation and the drain current decrease [17].

3.3   Washout of the probability distribution

With reducing thickness of the potential barrier, the electron wave functions penetrate the gate region more strongly and the probability represented by them is significantly washed out in the whole system. Simultaneously, the semiconductor well is penetrated by wave functions of the gate electrode. In the zero barrier thickness limit, the system is totally open and all states in the semiconductor are allowed for electrons.

In order to determine a limit for the minimum SiO$_2$ thickness resulting from the probability washout, we consider a simple model of the DG SOI. The central rectangular well $t_s = 5$ nm (silicon: $m_e = 0.19m_0$) is separated from two gate wells (polySi: $m_e = 0.19m_0$) of $t_g = 10$nm thickness by two rectangular potential barriers (SiO$_2$: $m_e = 0.5m_0$) $\chi_c = 3$eV. Fig. 7a shows the allowed energy levels in the system. There is a dependence upon the barrier thickness for t_{ox} less than about 0.4 nm. As shown in Fig. 7b, the probability representing electrons on the levels E_s originated from the central well decreases with reducing t_{ox} below 0.5 nm, while simultaneously the probability representing electrons on the levels E_g adequate for the gate wells penetrates the central region.

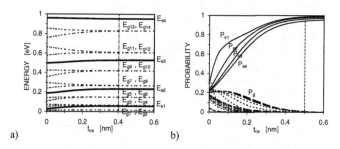

a)   b)

Figure 7. Semiconductor and gate energy levels dependence on barrier thickness (a), and probability distributed in the semiconductor and the gate states vs. the barrier thickness t_{ox} (b).

4.   CONCLUSIONS

The hard potential wall approximation results in an underestimation of the electron surface density due to the shift of the energy levels up and in an overestimation of the electron mobility due to the increase of the distances between the levels. The tunnel outflow of the channel electrons to the gate should not limit the thickness of SiO$_2$ layer in MOSFETs, but thickness nonuniformity can worsen the drain to gate current ratio. The ultimate theoretical limit for the minimum SiO$_2$ thickness in MOS devices is about 0.5 nm, which is fixed by the probability washout in the quantum-mechanical system.

ACKNOWLEDGMENTS

This work is supported by KBN, Poland, Grant No. 4 T11B 035 23 and by SINANO NoE 6FP Contract No. 506844.

REFERENCES

1. C. Chang, M.-S. Liang, C. Hu, R. W. Brodersen, *IEDM Tech. Dig.* 1983, 194.
2. S.Horiguchi, T.Kobayashi, M.Miyake, M.Oda, K.Kiuchi, *IEDM Tech. Dig.* 1985, 761.
3. K. Nagai and Y. Hayashi, *IEEE Trans. Electron Devices* **35**, 1145 (1988).

4. H. S. Momose, M. Ono, T. Yoshitomi, T. Ohguro, S. Nakamura, M. Saito, H. Iwai, *IEDM Tech. Dig.* 1994, 593.
5. R. Chau, J. Kavalieros, B. Roberds, R. Schenker, d. Lionberger, D. Barlage, B. Doyle, R. Arghavani, A. Murthy, G. Dewey, *IEDM Tech. Dig.* 2000, 45.
6. G. Timp, *et al.*, *IEDM Tech. Dig.* 1998, 1041.
7. J. Vuillod, G. Pananakakis, *Revue Phys. Appl.* **20**, 37 (1985).
8. B.Majkusiak, A.Jakubowski, A.Świt, *ESSDERC 1987*, Bologna, Italy, 683.
9. G.Timp, et. al., Progress toward 10 nm CMOS devices, *IEDM Tech. Dig.* 1998, 615.
10. C.Leroux, G.Ghibaudo, G.Reimbold, R.Clerc, S.Mathieu, *Microel. Eng.* **59**, 277 (2001).
11. C. L. Leroux, P. Mur, N. Rochat, D Rouchon, R. Truche, G. Reimold, G. Ghibaudo, *INFOS 2003*, Barcelona, Spain, p. GS6.
12. J. H. Stathis, *IBM J Res. & Dev.* **46**, 265 (2002).
13. T. Ghani, K. Mistry, P. Packan, S. Thompson, M. Stettler, S. Tyagi, M. Bohr, *Symp. on VLSI Technology*, 2000, 174.
14. M. Hirose, M. Koh, W. Mizubayashi, H. Murakami, K. Shibahara, S.Miyazaki, *Semicond. Sci. Technol.* **15**, 485 (2000).
15. W. K. Henson, N. Yang, E. M. Vogel, J. J. Wortman, K De Meyer, A. Naem, *IEEE Trans. Electron Devices* **47**, 1393 (2000).
16. T. Y. Chan, J. Chen, P. K. Ko, C. Hu, *IEDM Tech. Dig.* 1987, 718.
17. K.Chen et al., *Electron Device Lett.* **17**, 202 (1996).
18. U.Neuwald, H.E.Hessel, A.Feltz, U.Memmert, R.J.Behm, *Appl.Phys.Lett.***60**, 1307 (1992).
19. K.Ohishi, T.Hattori, *Jpn. J. Appl. Phys.* **33**, L675 (1994).
20. S.T.Pantelides, M.Long, in The Physics of SiO_2 and Its Interfaces, Ed. S.T.Pantelides, Pergamon, 1978, 339.
21. M. Städele, B. R. Tuttle, K. Hess, *J. Appl. Phys.* **89**, 348 (2001).
22. S.Miyazaki, H.Nishimura, M.Fukuda, L.Ley, J.Ristein, *Appl.Surf.Sci.* **113/114**, 585 (1997).
23. K. Takahashi, M.B. Seman, K. Hirose, T. Hattori, *Appl. Phys. Lett.* **190**, 56 (2002).
24. C.Kaneta,T.Yamasaki,T.Uchiyama,T.Uda, K.Terakura, *Microelectron.Eng.* **48**, 117 (1999).
25. G. Lewicki, J. Maserjian, *J. Appl. Phys.* **46**, 3032 (1975).
26. D.A.Muller, T.Sorsch, S.Moccio, F.H.Baumann, K.Evans-Lutterodt, G.Timp, *Nature* **399**, 758 (1999).
27. Z.A.Weinberg, *J. Appl. Phys.* **53**, 5055 (1982).
28. M.Städele, F. Sacconi, A. Di Carlo, P. Lugli, *J. Appl. Phys.* **93**, 2681 (2003).
29. S.Horiguchi, H. Yoshino, *J. Appl. Phys.* **58**, 1597 (1985).
30. B.Majkusiak, *IEEE Trans. Electron Dev.* **37**, 1087 (1990).
31. B.Majkusiak, *MIXDES 2001*, Zakopane, Poland, 2001, 69.
32. B.Majkusiak, A.Jakubowski, *J. Appl. Phys.* **58**, 3141 (1985).
33. W.Franz, in Handbook der Physik, S. Flugge, Ed. Berlin: Springer, 1956, **17**, 155.
34. B.Majkusiak, A.Strojwas, *J. Appl. Phys.* **74**, 5638 (1993).
35. T.Yamanaka, S.J.Fang, H.-C.Lin, J.P.Snyder, C.R.Helms, *IEEE Electron Dev. Lett.* **17**, 178 (1996).
36. J.Walczak, B.Majkusiak, *Microelectron. Eng.* **59**, 417 (2001).

COMPACT MODEL OF THE NANOSCALE GATE-ALL-AROUND MOSFET

David Jiménez[1], Benjamí Iñíguez[2], Juan José Sáenz[3], Jordi Suñé[1], Lluis Francesc Marsal[2], and Josep Pallarès[2]

[1]*Departament d'Enginyeria Electrònica, ETSE, Universitat Autònoma de Barcelona, 08193-Cerdanyola, Barcelona, Spain;* [2]*Departament d'Enginyeria Electrònica, Elèctrica i Automàtica, ETSE, Universitat Rovira i Virgili, 43007-Tarragona, Spain;* [3]*Departamento de Física de la Materia Condensada, Facultad de Ciencias, Universidad Autónoma de Madrid, 28049-Cantoblanco, Madrid, Spain.*

Abstract: We present a compact physics-based model for the nanoscale gate-all-around (GAA) MOSFET working in the ballistic limit. The current through the device is obtained by means of the Landauer approach, the barrier height being the key parameter in the model. The exact solution of the Poisson's equation is obtained in order to deal with all the operation regions tracing properly the transitions between them.

Key words: MOSFET, modeling, quantum wire

1.   INTRODUCTION

One of the major issues with the scaling down of the classical MOSFET is the control of short-channel-effects (SCE). A variety of non-classical MOSFETs have emerged to alleviate this problem, so extending the scalability of the CMOS technology as far as possible. These devices are based on the double-gate, the triple-gate, the pi-gate, or the gate-all-around structure, the latter offering the best control of SCE[1]. The small vertical electric field and the use of undoped silicon channels in these structures reduce the surface and Coulombic scattering,[2] permitting electronic transport close to the ballistic regime. In this article we present a compact model, based on the Landauer transmission theory,[3-6] for the undoped cylindrical nanoscale GAA-MOSFET (see Fig. 1). The model is suitable for design and

D. Flandre et al. (eds.), Science and Technology of Semiconductor-On-Insulator Structures and Devices Operating in a Harsh Environment, 321-326.

performance prediction of these devices. The proposed model is valid for "well-tempered" GAA-MOSFETs; i.e., for transistors with small SCE.

2.   MODELING THE CURRENT

The cylindrical nanoscale GAA-MOSFET can be seen as a quantum wire where the electrons are confined within a cylindrical potential well. Assuming semi-classical ballistic transport, the electrons with energies greater than $E_n{}^v(x_{max})$ have a unit transmission probability to cross the barrier, where $E_n{}^v$ is the energy of the bottom of the nth-subband, and x_{max} identifies the position of the maximum energy along the transport direction. The transmission probability is zero otherwise. There are two sets of silicon valleys (labelled by v) where electrons can lie; the first is four-fold degenerate ($g_1=4$) and the second two-fold degenerate ($g_2=2$). By means of the Landauer approach, the current can be expressed as

$$I = \frac{qkT}{\pi\hbar}\sum_v\sum_n g_v \ln\left\{\frac{1+e^{\frac{E_{FS}-E_n^v(x_{max})}{kT}}}{1+e^{\frac{E_{FD}-E_n^v(x_{max})}{kT}}}\right\} \tag{1}$$

where E_{FS} and $E_{FD}=E_{FS}-qV_{DS}$ are the source and drain Fermi levels respectively. The current in Eq. (1) is expressed in terms of the barrier height $E_{FS}-E_n{}^v(x_{max})$, which is related to the applied gate (V_{GS}) and drain voltages (V_{DS}).

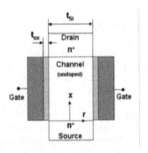

Figure 1. Cross section of the GAA-MOSFET

3. CALCULATION OF THE BARRIER HEIGHT

In this section we consider how the barrier height E_{FS}-$E_n{}^v(x_{max})$ depends on V_{GS} and V_{DS}. The potential distribution for the undoped cylindrical GAA-MOSFET along the radial direction is obtained by solving the one-dimensional Poisson's equation with the mobile charge term

$$\frac{d^2\psi}{dr^2} + \frac{1}{r}\frac{d\psi}{dr} = \frac{kT}{q}\delta e^{\frac{q\psi}{kT}}$$

(2)

subject to the boundary conditions

$$\frac{d\psi}{dr}(r=0) = 0, \qquad \psi(r = t_{Si}/2) = \psi_s$$

(3)

where $\psi(r)$ is the potential, ψ_s is the surface potential, $\delta = q^2 n_i/kT\varepsilon_{Si}$, and n_i is the silicon intrinsic concentration for a quantum wire.[4] The electrostatic analysis is only done for the radial direction, but it serves our purpose if we apply the results at x_{max}. At this special point, called the virtual cathode, the mobile charge is essentially controlled by the electric field along the r-direction provided that the SCE are not dominant. The exact solution of Eq. (2) is given by $\psi(r)=A-2kT\ln(Br^2+1)/q$ where A and B are constants satisfying $A=kT\ln(-8B/\delta)/q$.[7-8] This solution traces properly the transition between different operation regions. The constant B is related to ψ_s through the second boundary condition in Eq. (3). The mobile charge sheet density at the virtual cathode is controlled by the gate electrode. It can be written as $Q=C_{ox}(V_{GS}-\Delta\varphi-\psi_s)$, where $C_{ox}=2\varepsilon_{ox}/(t_{Si}\ln(1+2t_{ox}/t_{Si}))$ and $\Delta\varphi$ is the work-function difference between the gate electrode and intrinsic silicon. The surface potential ψ_s depends on the applied V_{GS}, and must satisfy the following relation:

$$C_{ox}(V_{GS} - \Delta\varphi - \psi_s) = Q = \varepsilon_{Si}\frac{d\psi}{dr}\bigg|_{r=t_{Si}/2} =$$

$$= \varepsilon_{Si}\frac{kT}{q}\sqrt{\frac{32}{t_{Si}^2}\left\{1 + \frac{1}{16}\delta t_{Si}^2 e^{\frac{q\psi_s}{kT}} - \sqrt{1 + \frac{1}{8}\delta t_{Si}^2 e^{\frac{q\psi_s}{kT}}}\right\}}$$

(4)

which states that the charge on the gate electrode is equal, but of opposite sign, to the charge in silicon. The charge term appearing on the right hand side of Eq. (4) is easily derived from the above solution of the Poisson's equation. This charge is provided by the source and drain reservoirs, by adjusting adequately the barrier height at the virtual cathode

$$Q = \frac{q\sqrt{2kT}}{2\pi^2 t_{Si}\hbar} \sum_v \sum_n g_v \sqrt{m_d^v} \left\{ \Im_{-1/2}\left(\frac{E_{FS} - E_n^v(x_{max})}{kT}\right) + \right.$$

$$\left. + \Im_{-1/2}\left(\frac{E_{FD} - E_n^v(x_{max})}{kT}\right) \right\} \tag{5}$$

where $\Im_{-1/2}$ denotes the Fermi integral of order $-1/2$, and m_d^v is the density of states effective mass ($m_d^1 = 0.19m_0$, $m_d^2 = 0.91m_0$). The derivation of Eq. (5) is obtained assuming a one-dimensional density-of-states, corresponding to a quantum wire. Note that Eq. (5) relates V_{GS} and V_{DS} with the barrier height. In the cylindrical GAA structure the electrons are confined by a cylindrical potential well. The separations between successive subbands are known[9] and only their position with respect to the Fermi levels must be determined.[4]

4.   COMPARISON TO EXPERIMENTAL RESULTS

In order to test our approach we have compared the results given by the compact model with an experiment reported in the literature.[10] The GAA-MOSFET of the experiment has a diameter (t_{Si}) of 69 nm, the insulator is silicon oxide with a thickness of 35 nm. The work-function of the gate material has been fixed to 4.33 eV in the compact model to match the observed experimental threshold voltage (0.48 V). To observe one-dimensional subband effects for this size, the operation temperature was 70 mK. In Fig. 2b we compare the measured transconductance (dotted line) with the simulated results by the compact model (solid line). The position of the five main peaks is well predicted by the compact model. A detailed analysis reveals six subbands below E_{FS} in the analyzed voltage range (Fig. 2a). When the gate voltage is large enough the subband bottom energy reaches E_{FD}. Every time one subband crosses E_{FS} (E_{FD}) a positive (negative) peak on the transconductance is recorded. The transconductance step is given by the derivative of the subband bottom energy with respect to the gate voltage multiplied by $4G_0$ or $2G_0$, depending whether the subband belongs to the first or second set of valleys, respectively.

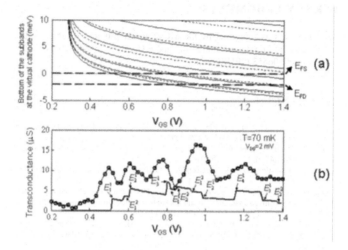

Figure 2. (a) Bottom of the subbands at the top of the barrier (virtual cathode) versus the gate voltage. The thin solid and dashed lines represent the bottom of the subbands from the first and second set of valleys, respectively. The Fermi level at the source (E_{FS}) is taken as a reference. In (b) comparison between experimental (dotted line) and simulated (solid line) transconductance of a cylindrical quantum wire GAA-MOSFET of 69 nm width.

5. CONCLUSIONS

In this work we have presented a compact physics-based model for the undoped cylindrical nanoscale GAA-MOSFET in the ballistic limit, derived from the Landauer transmission theory. The proposed model works in all operation regions, below and above threshold, for low and high temperatures, incorporating effects of multi-subband conduction, and taking into account the band structure of silicon. The quantum wire model was used to predict the transconductance structure of a real device, and showed reasonably good agreement with experimental results.

ACKNOWLEDGMENTS

This work was supported by the MCyT under project TIC2003-08213-C02-01 and the European Commission under Contract 506844 ("SINANO") and Contract 506653 ("EUROSOI").

REFERENCES

1. J.-T. Park and J.-P. Colinge, Multiple-gate SOI MOSFETs: device design guidelines, *IEEE Trans. Electron Dev.* **49**(12), 2222-2229 (2002).
2. H.-S. P. Wong, Beyond the conventional transistor, *IBM Journal of Research and Development* **46**(2/3), 133-168 (2002).
3. K. Natori, Ballistic metal-oxide-semiconductor field effect transistor, *J. Appl. Phys.* **76**(8), 4879-4890 (1994).
4. D. Jiménez, J. J. Sáenz, B. Iñíguez, J. Suñé, L. F. Marsal, and J. Pallarès, Unified compact model for the ballistic quantum wire and quantum well metal-oxide-semiconductor field-effect transistor, *J. Appl. Phys.* **94**(2), 1061-1068 (2003).
5. D. Jiménez, J. J. Sáenz, B. Iñíguez, J. Suñé, L. F. Marsal, and J. Pallarès, Compact modeling of nanoscale MOSFETs in the ballistic limit, in *Proc. ESSDERC Conf. 2003*, 187-190.
6. D. Jiménez, J. J. Sáenz, B. Iñíguez, J. Suñé, L. F. Marsal, and J. Pallarès, Modeling of nanoscale gate-all-around MOSFETs, *IEEE Electron Dev. Lett.* **25**(5), 314-316 (2004).
7. Y. Chen and J. Luo, A comparative study of double-gate and surrounding-gate MOSFETs in strong inversion and accumulation using an analytical model, in *Technical Proceedings of the International Conference on Modeling and Simulation of Microsystems* 2001, 546-549.
8. P. L. Chambré, On the solution of the Poisson-Boltzmann equation with application to the theory of thermal explosions, *J. Chem. Phys.* **20**(11), 1795-1797 (1952).
9. J. H. Davies, *The Physics of Low-Dimensional Semiconductors* (Cambridge University Press: Cambrigde UK, 1998).
10. M. Je, S. Han, I. Kim, and H. Shin, A silicon quantum wire transistor with one-dimensional subband effects, *Solid-State Electronics* **44**, 2207-2212 (2000).

SELF-ASSEMBLED SEMICONDUCTOR NANOWIRES ON SILICON AND INSULATING SUBSTRATES: EXPERIMENTAL BEHAVIOR

T. I. Kamins, S. Sharma, and M. Saif Islam
Quantum Science Research, Hewlett-Packard Laboratories, Palo Alto CA 94304 USA

Abstract: Metal-catalyzed semiconductor nanowires offer the possibility of combining nano-electronic structures with conventional electronics. They are formed by exposing a catalyst nanoparticle to a semiconductor precursor gas under conditions where the gas does not normally react. A column of the semiconductor (*ie,* the nanowire) is formed with diameter similar to that of the catalyzing nanoparticle. The growth of the nanowire depends on the size of the catalyzing nanoparticle, as well as the growth conditions. This approach offers the possibility of fabricating nanoscale electronic and sensing devices without costly and slow fine-scale lithography. When the connections to both ends of the nanowire are made during growth, advantages are obtained by combining "bottom up" fabrication of nanostructures with "top down" formation of the connecting electrodes using only conventional optical lithography. Self-assembled nanowires and "nano-bridges" constructed by the methods described here may provide some of the vital building blocks needed to enable the emerging technologies of nano-electronics and nano-sensors.

Key words: Nanowires; self-assembly; catalyst; sensor.

For more than three decades, the functionality of integrated circuits has increased exponentially, as noted by Gordon Moore.   This increased functionality has been achieved primarily by reducing the size of electronic devices and the interconnections between them. However, as feature sizes decrease, short-channel effects and inter-device interactions become increasingly troublesome. The cost of the fabrication facilities also increases exponentially ("Moore's second law"), limiting the number of companies that can economically participate in the continued evolution of conventional integrated-circuit technology.   New approaches are needed to allow the increasing functionality predicted by Moore's law to continue within sustainable economic limits.

D. Flandre et al. (eds.), Science and Technology of Semiconductor-On-Insulator Structures and Devices Operating in a Harsh Environment, 327-332.
© 2005 *Kluwer Academic Publishers. Printed in the Netherlands.*

To augment the functionality of conventional electronics and possibly to replace some of the electronic components, non-conventional systems relying on self-assembly are being developed.   One-dimensional, self-assembled "nanowires"[1,2] may provide the bridge connecting conventional electronics, microsystems, and nanoelectronics. They may serve as the electrodes of nanoelectronic elements such as molecular switches. The small diameter of the nanowires makes modulation by surface charges especially effective. Therefore, a nanowire can be used as the channel of an MOS transistor by forming a concentric dielectric and gate electrode around the nanowire. If the gate is omitted, the surface of the nanowire (or dielectric surrounding the nanowire) is very sensitive to charge in its neighborhood, making it an efficient sensor. The charge from the species being sensed modulates the conductivity of the nanowire.   The surface can be functionalized with a material that selectively interacts with the target species. Computational techniques can also be used to extract information about a particular species from an array of sensors with varying sensitivities to different species, even if each sensor is not totally selective to one species. For the needed isolation of these structures, the devices can be fabricated on SOI substrates.   In addition to Column IV nanowires with their varied applications, nanowires of compound semiconductors offer the possibility of optical emission.[3]

The diameter of one-dimensional, metal-catalyzed, self-assembled "nanowires" is determined by the nucleation and growth conditions, rather than by fine lithography.   To form the nanowires, small regions ("nanoparticles") of a catalytically active metal are first prepared and then exposed to a gaseous precursor of (typically) silicon under conditions where the normal growth rate of silicon is very low. The metal nanoparticle locally accelerates the reaction so that silicon atoms are deposited on the nanoparticle. The silicon atoms diffuse through or around the nanoparticle and precipitate at the interface between the nanoparticle and the substrate. As the silicon precipitates, it pushes the metal nanoparticle away from the substrate, forming a column of silicon (*ie,* a nanowire) with a diameter similar to that of the catalyzing nanoparticle and the desired high surface-to-volume ratio. The catalyst can be in the liquid phase during growth, as in the widely known vapor-liquid-solid (VLS) growth technique,[1] or it can remain in the solid state.[4] The nanowires can be further processed using well known and controlled integrated-circuit processing.

The nanowires can be high-quality, thin semiconductor structures with diameters from several nanometers up to a fraction of a micrometer and lengths up to hundreds of micrometers. For suitably chosen catalysts and growth conditions, the metal catalyst generally remains at the tip of the nanowire, and little is incorporated into the nanowire.   To control the nanowire diameter and distribution, therefore, the size and distribution of the catalyzing nanoparticles must be controlled. Ideally, the catalyst material should be compatible with Si IC technology, at least at the lower temperatures involved in nanowire formation.

Nanowires can be grown on silicon or on amorphous, insulating substrates, such as SiO_2. When grown on a crystalline substrate, the growth direction of the nanowire can be influenced by the orientation of the substrate.[5] When grown on amorphous SiO_2, of course, no directionality from the underlying substrate is obtained. When sensors are integrated onto the same substrate as the computing elements, good electrical and thermal isolation between the different elements is critical. Silicon-on-insulator (SOI) technology provides this isolation, as well as allowing directional growth.

Once the nanowires are formed, connection to them must be made. In some techniques of using the nanowires, they are detached from the substrate and must then be connected to electrodes — a complex and expensive process. Forming the connections automatically during nanowire growth provides an efficient, inexpensive and straightforward method of making connections. From previous work,[6] it is known that good connection is made during the growth process between the substrate and the nanowire growing from that surface. However, making good electrical connection to the opposite end of the nanowire has been difficult. We have developed a method of forming the connections between the two ends of a nanowire and the electrodes during nanowire growth.[7]

Nanowires often grow in <111> crystal directions.[8] On the commonly used, (100)-oriented Si wafers, they grow preferentially at angles corresponding to the <111> directions; on (111)-oriented Si surfaces, they grow preferentially perpendicular to the surface. If the (111) surface is vertical (*ie,* perpendicular to the surface of the wafer), the nanowires grow laterally (*ie,* parallel to the surface of the wafer).

To demonstrate the connection technique, we formed a trench bounded by vertical surfaces in a (110)-oriented silicon wafer.[7] One well-known technique of forming vertical (111) planes is anisotropic, wet chemical etching of a (110)-oriented Si wafer. This orientation of Si has two sets of (111) planes perpendicular to the surface. Because the features being defined by etching correspond to the electrodes connected to the micro-scale electronics, they can be patterned by conventional lithography. With (110)-oriented silicon, the vertical etching rate is much greater than the lateral etching rate (by as much as 100:1), leaving vertical, (111)-oriented silicon surfaces bounding the trench. In this work, the trenches are approximately 8-μm deep and 2–15-μm wide.

After forming the vertical (111) surfaces, the nucleating metal catalyst — titanium or gold on the order of 1 nm thick — is deposited by angled electron-beam evaporation onto the vertical surfaces of the etched grooves. Ti is especially attractive because of its compatibility with silicon technology.[4] Because of the geometry of the structure and the deposition angle, no catalyst is deposited on the bottoms of trenches narrower than 8 μm. After inserting the sample into the chemical vapor deposition reactor, it is annealed in hydrogen to form Au-Si alloy nanoparticles and to reduce the native oxide on the Ti and form $TiSi_2$. A mixture of SiH_4 and HCl is then introduced into the hydrogen ambient to grow the nanowires at ~640°C.

Because the nanowires grow perpendicular to the vertical (111) surfaces, they grow laterally across the trench toward the opposing (111)-oriented sidewall. When a nanowire reaches the opposite sidewall, it bonds to the surface of that sidewall, forming a robust mechanical connection. Thus, the nanowire "bridges" across the trench.

Lateral growth and mechanical bridging connection has been demonstrated both with Ti and with Au catalysts. Figure 1 illustrates Au-nucleated Si nanowires growing laterally from one (111)-oriented trench face toward the opposing face. 8-μm long nanowires extend only partially across the 15-μm wide trench illustrated. The longest nanowires are 10 μm long, while >90% have a length of ~8±0.7 μm. Typical nanowires have a diameter of 180±20 nm. For gaps 8-μm wide or less, many Au-nucleated nanowires extend completely across the trench, as shown for an 8-μm wide trench in Fig. 2. Most Au-nucleated nanowires are straight and ~70% of them intersect the opposing sidewall at an angle of 90°±0.5°.

For application of these nanowires in sensors, the connection between the sidewall and the impinging nanowire must be mechanically strong. In the case of Au-nucleated Si nanowires, the catalyzing nanoparticle is in the liquid state during nanowire growth. After impinging, the catalyst and, therefore, the accelerated Si growth spread along the sidewall a controllable distance, firmly connecting the nanowire to the sidewall. In the case of Ti-nucleated nanowires, the catalyzing nanoparticle is in the solid state during growth, so growth along the sidewall is less, but it is sufficient to firmly attach the nanowire to the sidewall. The strength of the connection is indicated by the nanowires often breaking along their length, rather than at the connection point, when they are stressed to failure.

To use these nanowires in a sensor, the two sides of the trench can be formed into electrodes. To achieve the needed electrical isolation, an SOI substrate can be used. The current flowing through the nanowire can then be measured. If the surface of the nanowire is properly treated, the gas being sensed will modulate the conductance of the nanowire and the current flowing between the two electrodes. By using a large number of such "nano-

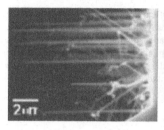 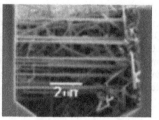

Figure 1. Cross section scanning electron micrograph of lateral epitaxial nanowire growth from (111) sidewall surface into a 15μm wide trench. (Scale bar: 2 μm.)

Figure 2 Lateral epitaxial growth of nanowires across an 8 μm wide trench and connection to opposing sidewall. (Scale bar: 2 μm.)

bridges" in parallel, the desired high surface area can be obtained in a small volume, forming a very sensitive gas sensor.

For either sensor or field-effect-transistor applications, the conductance of the nanowires must be modulated effectively. When used as a depletion-mode FET, the entire cross section of the nanowire must be depleted. When used as a sensor, the sensitivity can be increased if the conducting region is narrow. Therefore, the maximum thickness of the depletion region can set a limit on the range of useful combinations of doping and diameter. From a one-dimensional analysis in radial coordinates, the maximum thickness of the depletion region is given by the common formula for x_{dmax} of a plane structure multiplied by a factor of $\sqrt{3}$. Because the amount of charge that must be depleted to extend the depletion region further decreases as the depletion region extends into the nanowire, the maximum depletion region thickness is greater for a nanowire than for a plane surface. For a nanowire diameter of 20 nm, the maximum dopant concentration that allows complete depletion is mid-10^{19} cm^{-3}. For somewhat thinner nanowires, the maximum dopant concentration will be limited by the solid solubility of the dopant in Si. For narrow nanowires, quantum effects will also modify the bulk conductivity.

CONCLUSION

The metal-catalyzed semiconductor nanowires described here offer the possibility of combining nanoelectronic structures with conventional electronics. When the connections are made during growth, advantages are obtained by combining "bottom up" fabrication of nanostructures with "top down" formation of the connecting electrodes using only conventional optical lithography. This approach offers the possibility of fabricating nanoscale electronic devices without costly and slow fine-scale lithography. SOI substrates can provide the necessary isolation between the electrodes and between the nanowire array and computing electronics integrated on the same chip. Nanowires constructed by this method may provide some of the vital building blocks needed to enable the emerging technologies of nano-electronics and nano-sensors.

ACKNOWLEDGEMENT

The authors thank Tan Ha and Xuema Li for expert experimental assistance.

REFERENCES

1. J. Westwater, D. P. Gosain, S. Tomiya, S. Usui, and H. Ruda, J. Vac. Sci. Technol. B **15**, 554 (1997).
2. A. M. Morales and C. M. Lieber, Science **279**, 208 (1998).
3. M. T. Bjork, B. J. Ohlsson, T. Sass, A. I. Persson, C. Thelander, M. H. Magnusson, K. Deppert, L. R. Wallenberg, and L. Samuelson, Nano Letters **2**, 87 (2002).
4. T. I. Kamins, R. Stanley Williams, D. P. Basile, T. Hesjedal, and J. S. Harris, J. Appl. Phys. **89**, 1008 (2001).
5. T. I. Kamins and R. Stanley Williams, mstnews 3/03, 12 (2003).
6. Q. Tang, X. Liu, T. I. Kamins, G. S. Solomon, and J. S. Harris, Jr., Fall Mat. Res. Soc. Mtg, Boston, MA, Dec. 2002, paper F6.9.
7. M. Saif Islam, S. Sharma, T. I. Kamins, and R. Stanley Williams, Nanotechnology **15**, L5 (2004).
8. T. I. Kamins, X. Li, R. Stanley Williams, and X. Liu, Nano Letters **4**, 503 (2004).

FABRICATION OF SOI NANO DEVICES

Xiaohui Tang, Nicolas Reckinger, and Vincent Bayot
Microelectronics Laboratory, Université Catholique de Louvain, Louvain-la-Neuve, Belgium

Abstract: Some current fabrication technologies of SOI nano devices are reviewed in this paper. By means of arsenic-assisted etching and oxidation effects, we have fabricated several SOI nano devices: single-electron transistor, nano floating gate memory device and cell, Ω-gate elevated source/drain MOSFET. The application of this technique for fabricating a Schottky barrier MOSFET is also presented.

Key words: SOI nano device; single-electron transistor; nano floating-gate memory; elevated source/drain MOSFET; silicidation.

1.   INTRODUCTION

SOI wafers have come into the nano fabrication area from their traditional military and space applications. Almost all the reported silicon nano devices are fabricated on SOI wafers due to the intrinsic two-dimensional confinement and extended scalability in addition to high performance and reliability. The aim of this paper is to review current fabrication technologies of SOI nano devices and to demonstrate a new technology based on arsenic-assisted etching and oxidation effects for realizing these devices. In section 2, arsenic-assisted etching and oxidation effects will be first presented. Section 3 will briefly describe the single-electron transistor fabrication. Section 4 will focus on the fabrication of a floating gate memory device and cell. Section 5 will deal with elevated source/drain MOSFET fabrication. Finally, source/drain architecture as a feature of a Schottky barrier MOSFET will be discussed in section 6.

D. Flandre et al. (eds.), Science and Technology of Semiconductor-On-Insulator Structures and Devices Operating in a Harsh Environment, 333-344.

2.   ARSENIC-ASSISTED ETCHING AND OXIDATION EFFECTS

It is found that silicon etching[1] and oxidation rates[2] are the fastest in regions of high arsenic concentration. Figure 1 shows a SEM cross-section of an As doped silicon wire after reactive ion etching. It is found that the cross-section of the silicon wire is in the form of a trench that coincides with the depth of arsenic peak concentration. The trench formation can be explained as follows. The lateral etch rate differs in differently doped silicon layers and progressively increases with increasing arsenic concentration. This effect is hereafter called an "arsenic-assisted etching effect". The same effect can also be observed during oxidation processes. We call this an "arsenic-assisted oxidation effect". Both effects are used to achieve our nano floating gate memory, elevated source/drain MOSFET and Schottky barrier MOSFET.

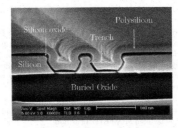

Figure 1. SEM cross-section of a silicon wire showing a trench in the region of the arsenic peak concentration. The wire is doped by arsenic implantation with a dose of 10^{15} cm^{-2}.

3.   SINGLE-ELECTRON TRANSISTOR FABRICATION

One of the most fundamental elements in the nanoelectronics era is the single-electron transistor, which is composed of a conductive island connected to source and drain by tunnel barriers and coupled to a capacitive gate. Silicon single-electron transistors operating at room temperature require the island smaller than 10 nm [3]. To reach this goal, various methods have been used by many experimentalists, including anisotropic wet etching[4-6], pattern-dependent oxidation[7-9], STM nano-oxidation process[10], side gate on thin SOI wafer[11,12] and point contact MOSFET[13]. At UCL, single-electron transistors were realized on a thin SOI wafer by intentionally

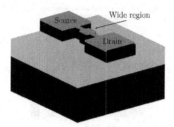

Figure 2. Layout of a single-electron transistor showing "large" source and drain connected by a narrow-wide-narrow quantum wire.

converting a quantum wire into an island connected to a source and a drain through two constrictions[14]. The layout is shown in figure 2.

"Large" source and drain are connected by a narrow-wide-narrow quantum wire. After oxidation, the wide region becomes a central island, while the narrow regions are changed into two constrictions in which the minimum electron energy level is higher than that in the central island, thereby forming two tunnel barriers enabling the Coulomb blockade effect.

Figure 3 plots conductance as a function of both drain and gate voltages in this single-electron transistor at 4.2 K. The drain current is zero in dark diamond shaped regions in the center of the figure, which confirms Coulomb blockade effect caused by electron injection into the central island through tunnel barriers.

4. NANO FLOATING GATE MEMORY DEVICE AND CELL FABRICATION

Nano-memory devices are considered to be a driver of the semiconductor industry during the next decade due to its advantages of high density, persistent programming and fast electrical erasability. In our work, this nano memory device consists of a narrow channel MOSFET with a nano floating gate embedded in the gate oxide[15]. The ultimate floating gate memory device is the single-electron memory, in which only one electron is stored in the floating gate for one bit of information. In order to meet the room temperature operating conditions, the floating gate is required to be less than 10 nm. Two main approaches have been proposed to fabricate this device since the first single-electron memory device was reported in 1993[16]: multiple and single nanoparticle (semiconductor or metallic dots) devices as shown in figure 4. In the former[17-23], the floating gate is created by randomly

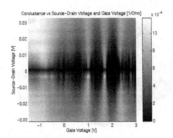

Figure 3. Conductance in the V_D/V_G plane for a fabricated single-electron transistor at 4.2 K.

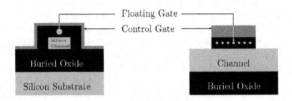

Figure 4. Schematic views of memory devices with multiple nanoparticle (right) and single nanoparticle (left) floating gate.

deposited nanoparticles. Each particle stores only one or a few electrons, depending on its size. The multiple nanoparticle device is more immune to spurious discharging as electrons are distributed among isolated particles. However, the randomness of nanoparticle deposition induces device characteristic fluctuations. The single nanoparticle device includes only one nanoparticle. The technological challenge is to fabricate this nanoparticle and to align it to the channel in a controlled manner. This usually requires very high resolution lithography tools[24-27]. Based on arsenic-assisted etching and oxidation effects, we fabricated a single-dot memory device[28], in which the floating gate is self-aligned with the channel. Moreover, the floating gate and the tunnel oxide are formed in a single oxidation step.

The layout of the processed memory device is the same as the one shown in figure 2. An SOI wafer is doped by arsenic at an energy of 110 keV with a dose of 10^{15} cm^{-2}. The arsenic-assisted etching results in a trench formation at the sidewalls of the silicon wire (see figure 1). After wet oxidation, a tunnel oxide is formed in the trench region shown in figure 5.

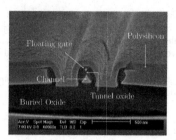

Figure 5. SEM cross-section of a fabricated nano memory device with a single-dot floating gate which is self-aligned to the channel. The floating gate has an equivalent diameter of 16 nm.

The tunnel oxide separates the top silicon dot from the bottom triangular wire. The dot and the triangular wire are used as the floating gate and the channel of the memory device, respectively. The floating gate is restricted to the central region since the top silicon dot is completely oxidized in the two constrictions. In the source and drain regions, the width of the silicon mesa is large enough to avoid the formation of the tunnel oxide at the peak concentration of arsenic. Thus, the floating gate is separated from the channel in a simple oxidation step. This results in a self-aligned structure. The floating gate has an equivalent diameter of 16 nm and the channel has a height and a base width of 80 nm. This process was found highly reproducible and could be used to realize memory circuits. Figure 6(a) illustrates a SEM image of the cross-section for a few cells in the central region, fabricated by means of this technology. It is found that the size of floating gate and channel is very uniform across each array. Indeed, for the feasibility of a memory circuit, uniformity across the array is a more relevant parameter than the memory device size itself. Figures 6(b) and 6(c) show SEM images of the cross-section of constriction region and electrode region,

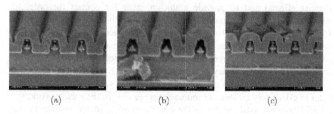

(a) (b) (c)

Figure 6. SEM cross-sections for a few cells in the central (a), constriction (b) and electrode (c) regions.

respectively. As mentioned above, only a triangular strip of silicon is formed in the constriction region (due to top silicon dot consumption), while the floating gate and the channel are not separated by oxide in the electrode region. Figure 7 gives typical characteristics of a fabricated memory device. A clear hysteresis in the drain current, which is only observed in the presence of a memory dot, confirms that the memory operation is caused by the injection of carriers into the floating gate.

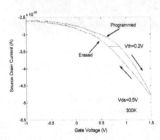

Figure 7. Transfer characteristics of a fabricated memory device before and after charges being stored onto the floating gate.

5.   ELEVATED SOURCE/DRAIN MOSFET FABRICATION

With the gate length reaches the nano range, MOSFET characteristics show undesirable "short channel effects". Various transistor architectures have been proposed to suppress these effects. A promising architecture is the Elevated Source/Drain MOSFET (ESD-MOSFET)[29] as shown in figure 8, in which source/drain regions are elevated relatively to the channel to allow the formation of shallow source/drain junctions (thereby attenuating short channel effects) and to provide a thicker layer for contact (in order to minimize parasitic series resistance). The methods to elevate the source and drain can be divided into two major categories: direct and relative elevation methods. In the former, the source/drain regions are elevated by selectively growing a thin epitaxial layer of monocrystalline silicon[30-32] or by selectively depositing a polysilicon layer[33]; while, in the latter, the channel region is recessed down to a predefined depth by etching or sacrificial oxidation[34] in order to effectively elevate the source/drain regions. In the present work, the channel was recessed thanks to the combination of an appropriate layout with arsenic-assisted etching and oxidation effects. The fabrication process of the ESD-MOSFET is similar to that of the memory device, but there is a

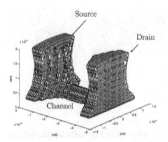

Figure 8. Simulated structure of an elevated source/drain MOSFET.

slight difference in the layout: "large" source and drain regions are connected by a narrow quantum wire. The wire width can be chosen in such a way that, under appropriate lateral oxidation conditions, the silicon above and around the trench is totally consumed, while the bottom region of the silicon becomes triangular (see figure 9(a)). Figure 9(b) shows a SEM image of a cross-section of the source/drain region after oxidation. The width of source/drain region is large enough to avoid the entire silicon consumption around and over the trench. It is very clear that the channel is recessed naturally by means of a single oxidation step. We want to lay stress on the fact that, although the fabrication process of the ESD-MOSFET requires a high implanted arsenic dose, the mean channel doping concentration is reduced by a factor of 5 at the end of process. Furthermore, the channel height can be adjusted by changing the arsenic doping energy to reach the design requirement. To increase the current drive, a multi-fingered structure is used in this work. Figure 10 shows SEM cross-section of three fingers. The silicon fingers are covered by gate oxide and polysilicon gate. It can be

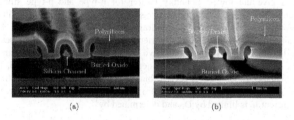

(a)   (b)

Figure 9. SEM cross-sections of the channel (a) and the electrode (b) of a fabricated ESD-MOSFET. The channel is about 100 nm below the source/drain, which results from the total consumption of the silicon above the arsenic peak concentration during oxidation.

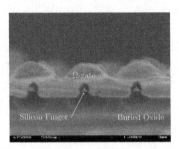

Figure 10. SEM cross-section of a multi-fingered structure with an Ω-gate.

seen that the finger is triangular and the gate is Ω-shaped. As mentioned by[35], Ω-gate MOSFET has the following advantages:
1. Suppressing short channel effects more effectively than double and triple gate MOSFETs;
2. Offering a much easier fabrication method than surrounding gate MOSFET (also called gate-all-around MOSFET).

6.   SOURCE/DRAIN ARCHITECTURE

When the device dimensions are reduced by a scale factor λ, the source/drain contact resistance is increased by λ^2. For nano devices, this resistance is the dominant one in the total series resistance and must accordingly be kept as small as possible to maintain a high current drive[36]. Recently, a Schottky barrier MOSFET has been proposed again as an alternative to the traditional MOSFET because of its low source/drain contact resistance[37]. Er and Pt silicides give the lowest Schottky barrier for n-type and p-type silicon, respectively. We have investigated Er silicide to replace the conventional heavily doped silicon in the source/drain electrodes of nano devices. The equivalent circuit of the Schottky barrier MOSFET is shown in the inset of figure 11. The current flows through two back-to-back Schottky diodes and a silicon series resistance, R_s. When a negative drain voltage is applied, the drain Schottky diode (D_d) is reverse biased and the source Schottky diode (D_S) is forward biased. If the voltage drop across D_S and R_s is negligible, the drain current I_D is limited by D_D and determined by[38]:

$$I_D = AA^*T^2 \exp(-q\Phi_B/(kT))$$

where Φ_B is the drain Schottky barrier height.

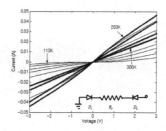

Figure 11. Current-voltage characteristics of a PtEr-stack silicide system on silicon from 110 K to 300 K. The inset is an equivalent circuit of a Schottky barrier MOSFET consisting in two back-to-back Schottky diodes connected by a silicon series resistance.

A higher current drive requires lower Schottky barriers. It is worth noting that Er potentially exhibits severe oxidation problems, which induce strong contact resistance and kill the performance gain expected from the low intrinsic resistivity of ErSi$_2$. A Pt layer (this choice allows CMOS applications) was deposited on top of the Er layer to protect the latter from oxidation. Figure 11 shows I-V curves taken at different temperatures for a PtEr-stack silicide system on a n-type silicon substrate with a concentration of 1.4×10^{16} cm^{-3}. It is observed that the PtEr-stack system has an ohmic behavior at room temperature. This implies very small Schottky barrier. With the help of an Arrhenius plot, where $\log(I_D/T^2)$ is plotted vs. $\log(1/T^2)$, the Schottky barrier height in PtEr silicide system is calculated to be less than 0.1 eV[39], which is much lower than the lowest reported value of 0.27 eV for the simple Er silicide system[40]. This shows that the PtEr silicide system exhibits a very low contact resistance for n-type silicon while Pt protects the silicide from oxidation and remains essentially unaffected in the formation of Er silicide (see figure 12).

7. CONCLUSION

We have reported a new technique relying on arsenic-assisted etching and oxidation effects for the fabrication of SOI nano devices. With this technique, the floating gate of the memory device is self aligned with the channel; the channel of the ESD-MOSFET is recessed naturally. This technique is compatible with SOI/MOS technology. We also investigate PtEr stack silicide system as an alternative material for the n-type source/drain electrodes.

(a)   (b)

Figure 12. TEM micrographs of the showing a cross-section of the PtEr-stack silicide before (a) and after (b) RTA. Pt is used as a protecting layer and does not influence Er silicide.

ACKNOWLEDGMENTS

The authors would like to acknowledge Pr. D. Flandre and Pr. A. Nazarov for their invitation. They also want to thank Pr. J.-P. Colinge, Pr. J. Katcki, Dr. E. Dubois, Dr. X. Baie, Mr. L. Moreno-Hagelsieb, Mr. A. Crahay, Mr. C. Renaux, Mr. P. Loumaye, Mr. David Spôte, Mr. N. Mahieu, Mr. B. Katschmarskyj and Mr. M. Zitout for their kind help in the realization of the present work.

REFERENCES

1. S. M. Rossnagel, Jerome J. Cuomo, and W. D. Westwood, *Handbook of Plasma Processing Technology - Fundamentals, Etching, Deposition, and Surface Interactions* (William Andrew Publishing/Noyes, 1990), p. 202.
2. C. P. Ho and J. D. Plummer. Si/SiO$_2$ Interface Oxidation Kinetics: A Physical Model for the Influence of High Substrate Doping Levels. *Journal of the Electrochemical Society* **126**(9), 1516–1522 (1979).
3. Y. Takahashi, H. Namatsu, K. Kurihara, K. Iwadate, M. Nagase, and K. Murase. Size Dependence of the Characteristics of Si Single-Electron Transistors on SIMOX Substrates. *IEEE Transactions on Electron Devices* **43**(8), 1213–1217 (1996).
4. H. Ishikuro, T. Fujii, T. Saraya, G. Hashiguchi, T. Hiramoto, and T. Ikoma. Coulomb blockade oscillations at room temperature in a Si quantum wire metal-oxide semiconductor field-effect transistor fabricated by anisotropic etching on a silicon-on-insulator substrate. *Applied Physics Letters* **68**(25), 3585–3587 (1996).
5. I. Gondermann, T. Röwer, B. Hadam, T. Köster, J. Stein, B. Spangenberg, H. Roskos, and H. Kurz. A triangle-shaped nanoscale metal-oxide-semiconductor device. *Journal of Vacuum Science and Technology B* **14**(6), pp. 4042–4045, November/December 1996.
6. H. Namatsu, M. Nagase, K. Kurihara, S. Horiguchi, and T. Makino. Fabrication of One-Dimensional Silicon Nanowire Structures with a Self-Aligned Point Contact. *Japanese Journal of Applied Physics* **35**(9B), 1148–1150 (1996).

7. H. Fukuda, J. L. Hoyt, M. A. McCord, and R. F. W. Pease. Fabrication of silicon nanopillars containing polycrystalline silicon/insulator multilayer structures. *Applied Physics Letters* **70**(3), 333–335 (1996).

8. Y. Takahashi, M. Nagase, H. Namatsu, K. Kurihara, K. Iwadate, Y. Nakajima, S. Horiguchi, K. Murase, and M. Tabe. Fabrication technique for Si single-electron transistor operating at room temperature. *Electronics Letters* **31**(2), 136–137 (1995).

9. L. Zhuang, L. Guo, and S. Y. Chou. Room Temperature Silicon Single-Electron Quantum-Dot Transistor Switch. *IEDM Technical Digest*, 167–169 (1997).

10. K. Matsumoto, M. Ishii, K. Segawa, Y. Oka, B. J. Vartanian, and J. S. Harris. Room temperature operation of a single-electron transistor made by the scanning tunneling microscope nanooxidation process for the TiO_x/Ti system. *Applied Physics Letters* **68**(1), 34–36 (1996).

11. A. Ohata and A. Toriumi. Coulomb Blockade Effects in Edge Quantum Wire SOI-MOSFETs. *IEICE Transactions on Electronics* **E79-C**(11), 1586–1589 (1996).

12. R. Augke, W. Eberhardt, C. Single, F. E. Prins, D. A. Wharam, and D. P. Kern. Doped silicon single electron transistor with single island characteristics. *Applied Physics Letters* **76**(15), 2065–2067 (2000).

13. H. Ishikuro and T. Hiramoto. Quantum Mechanical Effects in the Silicon Quantum Dot in a Single-Electron Transistor. *Applied Physics Letters* **71**(25), 3691–3693 (1997).

14. Xiaohui Tang, X. Baie, J.-P. Colinge, P. Loumaye, C. Renaux, and V. Bayot. Influence of device geometry on SOI single-hole transistor characteristics. *Microelectronics Reliability* **41**, 1841–1846 (2001).

15. Xiaohui Tang, X. Baie, J.-P. Colinge, C. Gustin, and V. Bayot. Two-Dimensional Self-Consistent Simulation of a Triangular P-Channel SOI Nano-Flash Memory Device. *IEEE Transactions on Electron Devices* **49**(8), 1420–1426 (2002).

16. K. Nakazato, R. J. Blaikie, J. R. A. Cleaver, and H. Ahmed. Single-electron memory. *Electronics Letters* **29**(4), 384–385 (1993).

17. J. De Blauwe, et al. A novel, aerosol-nanocrystal floating-gate device for non-volatile memory applications. *IEDM Technical Digest*, 683–686 (2000).

18. K. Han, I. Kim, and H. Shin. Characteristics of P-Channel Si Nano-Crystal Memory. *IEDM Technical Digest*, 309–312 (2000).

19. E. Kapetanakis, P. Normand, D. Tsoukalas, G. Kamoulakos, D. Kouvatsos, J. Stoemenos, S. Zhang, J. van den Berg, and D. G. Armour. MOS Memory Using Silicon Nanocrystals Formed by Very-Low Energy Ion Implantation. *ESSDERC Proc.*, 476–479 (2000).

20. K. Yano, T. Ishii, T. Hashimoto, T. Kobayashi, F. Murai, and K. Seki. Room-Temperature Single-Electron Memory. *IEEE Trans. on Electron Devices* **41**(9), 1628–1638 (1994).

21. Z. Liu, C. Lee, V. Narayanan, G. Pei, and E. C. Kan. Metal Nanocrystal Memories-Part I: Device Design and Fabrication. *IEEE Trans. on Elec. Dev.* **49**(9), 1606–1613 (2002).

22. Z. Liu, C. Lee, V. Narayanan, G. Pei, and E. C. Kan. Metal Nanocrystal Memories-Part II: Electrical Characteristics. *IEEE Trans. on Electron Devices* **49**(9), 1614–1622 (2002).

23. J. J. Lee, X. Wang, W. Bai, N. Lu, and D.-L. Kwong. Theoretical and Experimental Investigation of Si Nanocrystal Memory Device With HfO_2 High-k Tunneling Dielectric. *IEEE Transactions on Electron Devices* **50**(10), 2067–2072 (2003).

24. C. W. Gwyn, R. Stulen, D. Sweeney, and D. Attwood. Extreme ulraviolet lithography. *Journal of Vacuum Science and Technology B* **16**(6), 3142–3149 (1998).

25. J. P. Silverman. Challenges and progress in x-ray lithography. *Journal of Vacuum Science and Technology B* **16**(6), 3137–3141 (1998).

26. L. R. Harriott. Scattering with angular limitation projection electron beam lithography for suboptical lithography. *J. of Vacuum Science and Technology B* **15**(6), 2130–2135 (1997).

27. J. Melngailis, A. A. Mondelli, I. L. Berry III, and R. Mohondro. A review of ion projection lithography. *Journal of Vacuum Science and Technology B* **16**(3), 927–957 (1998).
28. Xiaohui Tang, X. Baie, J.-P. Colinge, A. Crahay, B. Katschmarsyj, V. Scheuren, D. Spôte, N. Reckinger, F. Van de Wiele, and V. Bayot. Self-aligned silicon-on-insulator nano flash memory device. *Solid-State Electronics* **44**, 2259–2264 (2000).
29. S. S. Wong, D. R. Bradbury, D. C. Chen, and K. Y. Chiu. Elevated Source/Drain MOSFET. *IEDM Technical Digest*, 634–637 (1984).
30. J. J. Sun, J.-Y. Tsan, and C. M. Osburn. Elevated n$^+$/p Junctions by Implant into CoSi$_2$ Formed on Selective Epitaxy for Deep Submicron MOSFET's. *IEEE Transactions on Electron Devices* **45**(9), 1946–1952 (1998).
31. S. Yamakawa, K. Sugihara, T. Furukawa, Y. Nishioka, T. Nakahata, Y. Abe, S. Maruno, and Y. Tokuda. Drivability Improvement on Deep-Submicron MOSFET's by Elevation of Source/Drain Regions. *IEEE Electron Device Letters* **20**(7), 366–368 (1999).
32. E. Augendre, R. Rooyackers, M. Caymax, E. P. Vandamme, A. De Keersgieter, C. Perello, M. Van Dievel, S. Pochet, and G. Badenes. Elevated Source/Drain by Sacrificial Selective Epitaxy for High Performance Deep Submicron CMOS: Process Window versus Complexity. *IEEE Transactions on Electron Devices* **47**(7), 1484–1491 (2000).
33. Zhikuan Zhang, Shengdong Zhang end Chuguang Feng, and Mansun Chan. An elevated source/drain-on-insulator structure to maximize the intrinsic performance of extremely scaled MOSFETs. *Solid-State Electronics* **47**, 1829–1833 (2003).
34. Mansun Chan, Fariborz Assaderaghi, Stephen A. Parke, Chenming Hu, and Ping K. Ko. Recessed-Channel Structure for Fabricating Ultrathin SOI MOSFET with Low Series Resistance. *IEEE Electron Device Letters* **15**(1), 22–24 (1994).
35. J.-P. Colinge. Multiple-gate SOI MOSFETs. *Solid-State Electronics* **48**, 897–905 (2004).
36. C.M. Osburn and K.R. Bellur. Low parasitic resistance contacts for scaled ULSI devices. *Thin Solid Films* **332**, 428–436 (1998).
37. E. Dubois and G. Larrieu. Low Schottky barrier source/drain for advanced MOS architecture: device design and material considerations. *Solid-State Electronics* **46**, 997–1004 (2002).
38. S. M. Sze. *Physics of Semiconductor Devices, 2nd Edition* (John Wiley & Sons, 1981).
39. Xiaohui Tang, J. Katcki, E. Dubois, N. Reckinger, J. Ratajczak, G. Larrieu, P. Loumaye, O. Nisole, and V. Bayot. Very low Schottky barrier to n-type silicon with PtEr-stack silicide. *Solid-State Electronics* **47**, 2105–2111 (2003).
40. Xu Zhenjia, in: *Properties of Metal Silicides*, edited by Karen Maex and Marc Van Rossum (INSPEC, Number 14 in EMIS Datareviews, 1995).

KEYWORD INDEX

AUTHOR INDEX